THE SOUTH AEGEAN ACTIVE VOLCANIC ARC

PRESENT KNOWLEDGE AND FUTURE PERSPECTIVES

THE SOUTH AEGEAN ACTIVE VOLCANIC ARC

PRESENT KNOWLEDGE AND FUTURE PERSPECTIVES

Edited by
Michael Fytikas
Aristotle University
Thessaloniki, Greece

and

Georges E. Vougioukalakis
Institute for Geology and Mineral Exploration
Athens, Greece

Milos Conferences

2005

ELSEVIER

Amsterdam – Boston – Heidelberg – London – New York – Oxford
Paris – San Diego – San Francisco – Singapore – Sydney – Tokyo

ELSEVIER B.V.
Radarweg 29
P.O. Box 211, 1000 AE Amsterdam
The Netherlands

ELSEVIER Inc.
525 B Street, Suite 1900
San Diego, CA 92101-4495
USA

ELSEVIER Ltd
The Boulevard, Langford Lane
Kidlington, Oxford OX5 1GB
UK

ELSEVIER Ltd
84 Theobalds Road
London WC1X 8RR
UK

First edition 2005

Library of Congress Cataloging in Publication Data
A catalog record is available from the Library of Congress.

British Library Cataloguing in Publication Data
A catalogue record is available from the British Library.

ISBN: 0-444-52046-5

∞ The paper used in this publication meets the requirements of ANSI/NISO Z39.48-1992 (Permanence of Paper).
Printed in The Netherlands.

THE SAAVA 2003 CONFERENCE

INTERNATIONAL PROGRAM COMMITTEE

Prof. Michael FYTIKAS, Aristotle University of Thessaloniki (GR) – Chair

Prof. Jon DAVIDSON, University of California (US)
Prof. Lorella FRANCALANCI, University of Florence, (IT)
Prof. F. INNOCENTI, Universita' di Pisa (IT)
Prof. Joerg KELLER, University of Freiburg (DE)
Prof. X. Le PICHON, Universite' P. Et M. Curie (FR)
Prof. Dimosthenis MOUNTRAKIS, Aristotle University of Thessaloniki (GR)
Prof. Vassilios PAPAZACHOS, Aristotle University of Thessaloniki (GR)
Prof. Georgia PE-PIPER, University of Saint Marry's (CA)
Assoc. Prof. Nikolaos SKARPELIS, University of Athens (GR)
Prof. Steve SPARKS, University of Bristol (UK)
Dr. Georges VOUGIOUKALAKIS, Institute of Geology & Mineral Exploration (GR)

LOCAL ORGANISING COMMITTEE

Emmanuel VOULGARIS "Milos Conference Center – George Eliopoulos"- Chair

Michael FYTIKAS, Aristotle University of Thessaloniki
Athanasios KEFALAS, S&B Industrial Minerals S.A.
Kostas KONSTANTINIDIS, Heliotopos Conferences

ORGANIZERS

The "Milos Conference Center – George Eliopoulos" and Heliotopos Professional Congress Organizers jointly organized this Conference within the framework of "Milos Conferences".

The Conference was under the auspices of the Aristotle University of Thessaloniki, the International Association of Volcanology and Chemistry of the Earth's Interior (IAVCEI), the Geological Society of Greece and the Institute of Geology and Mineral Exploration of Greece.

VENUE

The Conference was held at the "Milos Conference Center – George Eliopoulos" on Milos island, Greece.

CONFERENCE SECRETARIAT

Heliotopos Professional Congress Organizers
28, Ypsilantou str.
GR-172 36, Dafni, Athens, Greece
Tel.: +302109730697, Fax: +302109767208
E-mail: conf@heliotopos.net
http://heliotopos.conferences.gr

PREFACE

Our knowledge about of South Aegean Active Volcanic Arc (SAAVA) has grown substantially during the past 15 years. Therefore the time was ripe for reviewing past achievements and setting new goals for the future.

This book is a collection of selected papers from the Conference "The South Aegean Active Volcanic Arc: Present Knowledge and Future Perspectives" held in September 2003 in the framework of "Milos Conferences" at the "Milos Conference Center - George Eliopoulos" in Milos island, Greece. The aim of the conference was to serve as a forum for the presentation and constructive discussion of the state of the art and emerging issues on the South Aegean Volcanic Arc.

The Aegean area is one of the most rapidly deforming parts of the Alpine - Himalayan mountain belt delineated by the extremely high number of the seismic events occurring in this area. Deformation seems to be dominated by the effects of the westward motion of the Anatolian block, the south-westward motion of the southern Aegean and the vertical-rotational movements of big lithosphere portions.

Geophysical data record a thinned continental crust in all the Aegean area and an anomalous heat flow, implying a complex geodynamic - geotectonic situation.

Calc-alkaline volcanic activity was manifested at Plio-Quaternary, along a restricted belt which extends in an arc form, from Susaki (near the Korinth isthmus) to the west to Nisyros island to the east, the so-called South Aegean Active Volcanic Arc (SAAVA). This arc hosts the active (Methana and Santorini) and potentially active (Milos and Nisyros) Hellenic volcanic centres. The correspondence of the active volcanic arc location over a Benioff zone depth of about 130 km is evidence for a role of subduction in the genesis of the arc.

The volcanism is voluminous, with individual stratovolcanoes having volumes above sea level of the order of 10-40 km^3. The main volcanic centres in the central part of the arc appear to be associated with big tectonic lineaments and active faults trending NE.

The products of this volcanism form a typical calc-alkaline association which displays a continuous evolution from basalts to rhyolites. Their chemical characteristics are closely comparable with those of the volcanics of island arcs sited on thin continental margins.

The huge magmatic and hydrothermal activity created economic deposits and formations in most of the Aegean volcanic fields.

The 22 papers that are included in this volume review most of the above characteristics of the SAAVA, presenting and discussing also new data and ideas.

The three first papers (***Mountrakis***; ***Ranguelov***; ***Kassaras et al.***) discuss the tectonic- geodynamic setting and the present upper mantle structure of the Aegean area. The forth paper, by ***Papazachos et al.***, present an interesting interpretation of the data on spatial distribution of intermediate focal depth earthquakes, fault plane solutions and deep velocity structure, to further investigate active tectonics related to the deep structure of the southern Aegean volcanic arc. A low seismicity part (110 – 140km) of the subducted east Mediterranean lithospheric slab, that is sited under the volcanic arc, geochemical data and tomographic results lead them to suggest that the primary magma reservoir of the Hellenic volcanic arc is in the mantle wedge between the subducted Mediterranean and the overriding Aegean slabs, at depths 60 – 90km. The genesis of earthquakes at the shallow part of the subducted Mediterranean slab is attributed to dehydration embrittlement of basalt, the low seismicity at intermediate depths is due to increase of temperature and confining pressure and the increase again of seismicity in the lower active part of the slab to a second dehydration embrittlement of hydrous eclogite.

The next two papers deal with the general volcanological, petrological and tectonic characteristics of the SAAVA. ***Francalanci et al.*** present an extensive review of the volcanological, chemical and isotope data, using a large amount of new field and laboratory data. They conclude that volcanic center location was controlled by large tectonic lineaments, most of them still active, trending E-W to NW-SE for the western part and mainly NE-SW for the central and eastern parts of the arc. Volcanic fields developed along ellipse shaped areas with the longest axis oriented perpendicular to the subduction front. Volcanic activity is clearly controlled by both the active tectonic lineaments and the subduction kinematics dynamics, nevertheless, it is difficult to apply a simple evolutionary model, as the whole area is a patchwork of fragmented crustal blocks which move one respect to the other in a very complex relationship. Partial melting of a MORB-like astenospheric mantle, metasomatised by prevailing subducted sediments, is thought to produce the entire spectrum of parental magmas of SAAVA. Slab-derived fluids are generally reduced. Total amount of subducted sediments involved in the magma genesis decreases from West to East. ***Pe-Piper and Piper***, reviewing also a large amount of magmatological and tectonic data, conclude that two principal volcanic associations, together with a third minor association, occur in the South Aegean active volcanic arc, differing in magma type, age, spatial distribution, relationship to faulting, and petrogenesis. Variation in magmatism is related to changes in tectonics during the evolution of the arc, as a result of collision of African continental crust with the Aegean-Anatolian microplate, that set up changing patterns of strike-slip faulting in the arc. The ultimate cause of magmatism in the South Aegean active volcanic arc is concluded to be the subduction-related release of hydrous volatiles, but there are important differences between the petrogenesis of the older western and younger eastern parts of the arc.

The next paper, by ***Schuiling***, considering the Milos case, carries out some interesting calculations on the gas flow rates in volcanic areas and the provenance of these gasses.

Santo is presenting in the eighth paper a detailed study of the Nea Kameni (Santorini) plagioclase phenocrysts, with useful results for both the magmatic evolution and the volcanic hazard assessment of the most active SAAVA center.

The next eight papers deal with the volcanic hazard assessment and the monitoring efforts of the active SAAVA centers. The editors (***Vougioukalakis and Fytikas***) present an extensive review on the volcanic hazard assessment and the monitoring state of the SAAVA centers. Dimitriadis et al. and Stiros et al. present the results of the seismic and geodetic monitoring of the Santorini volcano. ***Dimitriadis et al.*** located two clusters of epicenters in the broader area of the Santorini Volcano. The first cluster is located in the Santorini caldera and is associated with the volcanic processes in the Kameni Islands. The second (larger) cluster is located NE of the Santorini Island group, at the Kolumbo submarine volcano, and is connected with the volcanic processes at this center. These clusters can be appropriately associated with the two main tectonic faults in the area under study. The first one (N60°E direction) corresponds to the continuation of the Amorgos fault in the area, while the secondary tectonic line (EW direction) is probably related with the southern edge of a submarine graben, which is located between the islands of Amorgos and Santorini. Between June 1994 and May 2003, ***Stiros et al.*** *(a)* registered a small-scale (up to 10cm), gradual inflation of the northern part of the caldera (between Nea Kameni and Thirasia), possibly associated with magma ascent along a dyke. ***Papadopoulos and Orfanogiannaki*** carried out a long-term prediction of the next eruption in Santorini volcanic field from conditional probability estimates.

Regarding Nisyros volcano, ***Stiros et al.*** *(b),* through a detailed geomorphological and biological study of the coasts of Nisyros and of the nearly Kos Island, combined with radiocarbon analysis of collected samples, provide evidence of land uplift along the northern and western coast of Nisyros Island, at a minimum, though increasing rate of 1.7 mm/yr during the last 2,000-3,000 years. This uplift correlates with late Quaternary uplift deduced from coastal and volcanological data, and reflects a new tendency for topography build-up due to caldera inflation, and consequently to potential for a new volcanic unrest. ***Galanopoulos and Kolettis*** present a DC resistivity study in order to investigate the formation of a 600 m long and 15-20 m deep fissure on the Nisyros caldera floor. The derived electrical models are compatible with the local geology and support the idea that the formation of the fracture is directly related with subsidence phenomena within the top 100 m depths. ***Di Filippo and Toro*** registered between 2001-2003, through monitoring with a microgravity network, variations that largely exceeded the measuring errors, suggesting that such variations are to be ascribed to mass variations of the volcanic complex. ***Teschner et al.*** present the results of a real time monitoring of gas-geochemical parameters in Nisyros fumaroles, during 2003.

The next two articles deal with hydrothermal deposits and processes in the SAAVA. ***Glasby et al.*** present a thorough study on the Vani manganese deposit (Milos island), a fossil stratabound hydrothermal deposit formed by the penetration of hydrothermal fluids through a lithified pyroclastic tuff. Two types of deposit have been recognized: "high-temperature" hydrothermal Mn deposits formed initially when the hydrothermal fluids penetrated faults and fissures within the volcaniclastic sandstone and bedded hydrothermal Mn deposits formed subsequently as the cooling hydrothermal fluids migrated along the bedding planes of the volcaniclastic sandstone. Both are late-stage, low-temperature deposits. ***Papoulis et al.*** present a study of the hydrothermal alteration of Kos rhyolitic rocks at using various vibrational spectroscopic techniques, covering a broad wavenumber range (Raman, FT-Raman and FTIR) along with XRD and SEM analytical methods.

The last four papers do not concern the SAAVA products and processes, but they have an interest and significance that link them to the SAAVA processes. ***Hannappel and Reischmann*** present data on the newly discovered rhyolitic dykes on Paros Island. ***Vamvoukakis et al.*** investigated the Lesvos non-pristine volcanic structures acting as probable hosts to epithermal gold mineralization, using combined satellite imagery and field data. ***Agostini et al.*** present an interesting paper on the Late Miocene high-Mg basaltic andesites to dacites found in Western Anatolia, interpreting the thermal anomaly affecting a depleted mantle source, invoked for the genesis of these products, that was produced by the ascent of deep sub-slab mantle, which replaced the underthrust lithosphere, already thinned and stretched by extensional process. ***Skianis et al.***, finally, present a mathematical model for the morphological evolution of a volcano on an island, which could act as a stimulus for the collection of field data for the evaluation of this theoretical model.

Closing this preface, we think that the papers presented in this volume are depicting a clear and global picture of our present knowledge on the South Aegean Active Volcanic Arc. This will help all scientists working in the area to devise (plan) their future research on the open questions that abound in the Aegean region, probably more than in any other active region in the world.

The Editors

CONTENTS

TECHNICAL PAPERS

The South Aegean Active Volcanic Arc
M. Fytikas and G.E. Vougioukalakis

Tertiary and Quaternary tectonics in Aegean area

D. Mountrakis

Department of Geology, Aristotle University, GR-54124, Thessaloniki, Greece, e-mail: dmountra@geo.auth.gr

ABSTRACT

The Hellenic orogen is a composite one, consisting of three orogenic belts. 1st the Cimmerian internal belt was created in pre-Late Jurassic times as the result of the northward drift of Cimmerian continental fragments from Gondwana towards Eurasia. 2nd the Alpine orogenic belt was created in Cretaceous-Tertiary times after the Neotethyan subduction beneath the unique Cimmerian-Eurasian plate and the collision of the Apulia to the great plate. 3rd the Mesogean orogenic belt along External Hellenic arc, due to the Mesogean-African underplate beneath the unique Alpine-Cimmerican-Eurasian plate in Miocene-Pliocene times and the exhumation of the Cretan-Southern Peloponesus tectonic windows. There is a clear deformation overprint of the Alpine deformation the Cimmerian belt and of the Mesogean one the Alpine belt. During Alpine and Mesogean orogenic processes in Tertiary a SW-ard migration of successive compressional and extensional tectonic events took place, producing nappe stacking and extensional exhumation of underplate rocks, successively in Rhodope (Eocene-Oligocene), Olympos-Ossa-Cyclades belt (Early Miocene) and Crete-Southern Peloponesus (Middle Miocene). Four deformation events based on numerous structural analysis and microscopic studies have been recognized in the broader Aegean region. The D1 extensional event in Late Eocene-Early Oligocene contemporaneous with the greenschist metamorphism, the D2 compressional event in Late Oligocene-Early Miocene contemporaneous with the 25 Ma HP/LT metamorphism in Crete and Southern Peloponesus, which produced the imbrication and the nappe stacking, the D3 extensional event in Early to Middle Miocene which produced low-angle shear zones, thinning of the crust, uplift and exhumation of the HP/LT metamorphic rocks as core-complexes and tectonic windows. The D4 tectonic event is the final deformation in Pliocene to recent times and it is the normal continuation of the extensional process in the broader Aegean region with brittle conditions, producing normal and strike-slip faults. In recent times the direction of the extension is generally N-S producing normal faults trending E-W and reactivated previous faults of other trends as strike-slip faults.

Keywords: Geodynamic evolution, imbrication, exhumation, active faults

1. INTRODUCTION

The geotectonic evolution of the broader Aegean area remains under a continues and intensive investigation since a long time. Recent research (Buick, 1991; Gautier et al., 1993; Gautier and Brun, 1994; Doutsos et al., 1994; Fassoulas et al., 1994; Kilias et al., 1994; Vandenberg and Lister, 1996; Kilias and Mountrakis, 1998) mainly insist to the geometry and kinematics of the late orogenic extension and collapse in the Hellenic orogen during Tertiary and Quaternary, since these extensional tectonics are the dominant particularly for the Southern Aegean area.

Herewith a brief geodynamic model of Greece and surrounding Aegean areas is presented, insisting particularly to the Tertiary and Quaternary tectonics and taking into account all recent relative investigations and views.

2. GEODYNAMIC EVOLUTION OF GREECE AND BROADER AEGEAN AREA

The Hellenic orogen is a composite one consisting of three orogenic belts (Fig. 1): 1st, The Cimmerian orogenic belt, including Rhodopian Serbomacedonian, Circum

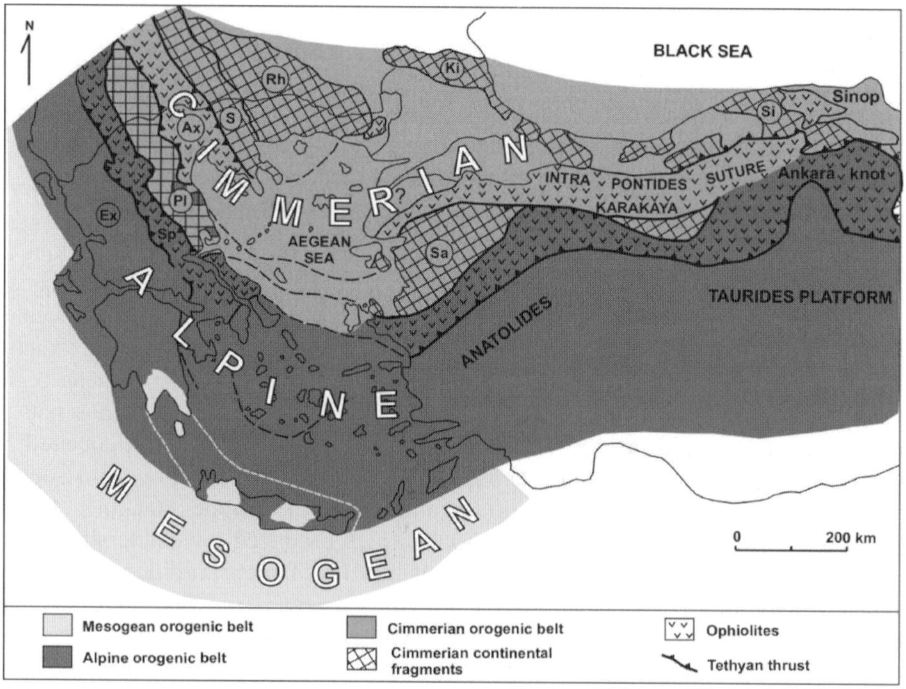

Fig. 1. Structural sketch-map showing Hellenides zones and the three belts of the Hellenic orogen and their extension to Turkey. Cimmerian continental fragmantes: Pelagonian (Pl), Serbomacedonian (S), Rhodope (Rh), Kirkarly (Ki), Sinop (Si), Sakarya (Sa). Ophiolitic sutures: Axios (Vardar) zone (Ax), Subpelagonian zone (Sp), Karakaya, Ankara, External Hellenides (Ex), Anatolides, Taurides.

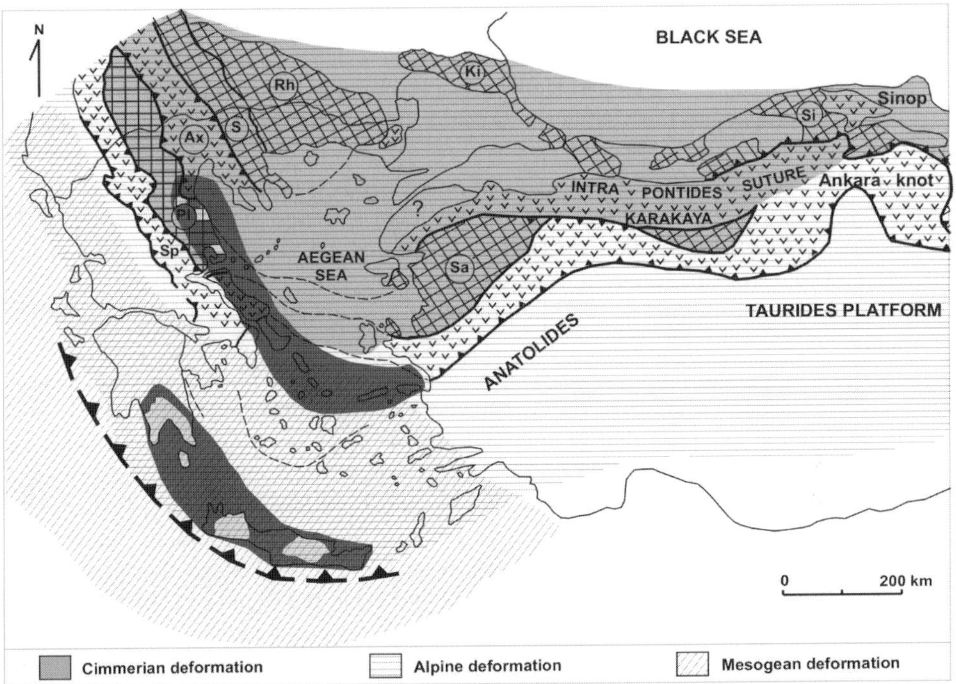

Fig. 2. Structural sketch-map showing the area covered by the orogenic process of each orogenic belt and the deformation overprint of each one to the other. Dark shadow areas correspond to both HP/LT metamorphic belts in Greece (see text).

Rhodope, Axios and Pelagonian zones, is the internal orogenic belt which has been created in pre-Late Jurassic times as a result of the northward drift of Cimmerian continental fragments from Gondwana towards Eurasia (Mountrakis, 1986; Robertson et al., 1996).

2nd, The Alpine orogenic belt, including External Hellenides and Pindos-Subpelagonian ophiolites and oceanic (Neo-Tethyan) sediments. The Alpine orogenic belt has been created in two periods: the first in Cretaceous and the second in Tertiary times, after the subduction of the Neotethyan oceanic crust beneath the unique Cimmerian – Eurasian plate and the collision of the Apulian micoplate to the great plate.

3rd, The Mesogean orogenic belt along the External Hellenic orogenic arc, due to the Mesogean-African underplate beneath the unique Alpine-Cimmerian-Eurasian plate in Miocene-Pliocene times and the exhumation of the Cretan-Southern Peloponesus tectonic windows.

Detailed structural analyses showed that there is clear deformation overprint of each one to the other orogenic process (Fig. 2). So that the Alpine deformation affected the Cimmerian belt (Mountrakis, 1982, 1986; Kilias et al., 1999), and the Mesogean tectonics certainly affected the Alpine belt as well (Fassoulas et al., 1994; Kilias et al., 1994).

Tertiary collisional tectonics were associated with the formation of two main HP/LT metamorphic belts with blue schists and sometimes eclogites, shaping two concentric ares: the internal and external ones (Fig. 2), reflecting the successive subduction processes. The internal of 45 Ma was formed in the Eocene and represents the suture zone between Internal and External Hellenides geological zones and the external belt of 25 Ma in Late Oligocene-Early Miocene within External Hellenides inbetween Gavrovo and Ionian zones (Seidel et al., 1982).

The Alpine orogenic process consists of two periods: the first in Cretaceous times with the subduction of Neo-Tethys, and the second in Eocene-Oligocene times the final continental collision of Apulian microplate to the unique Cimmerian-Eurasian plate. The Mesogean orogenic process took place in Miocene-Pliocene times, after the collision of Apulian microplate and the migration of the subduction southwards to the Mesogean ridge.

The schematic-palinspatic cross-sections of Fig. 3 show the plate motion reconstruction and the successive stages of the Alpine and Mesogean orogenic processes.

The collisional tectonics that is imbrication and nappe stacking formed the Nappe Pile of the Hellenic orogen in Tertiary (Fig. 4). A very important late-orogenic extension follows the nappe stacking and the lithospheric thickening during Tertiary and caused to the nappe pile a strong orogenic collapse and crustal thinning, which was associated with the exhumation of the deeper tectonic units and so that some very impressive metamorphic core-complexes were exposed in the island of Crete and South Peloponesus of the lower plate units or a series of tectonic windows of the External Hellenides carbonates in Cyclades islands and continental Greece (Avigad and Carfunkel, 1991; Fassoulas et al., 1994; Lister and Forster, 1996; Vendenberg and Lister, 1996; Ring et al., 1999).

Summarizing we can say that during Alpine and Mesogean orogenic processes a SW-ward migration of successive compressional and extensional tectonic events took place as a result of the successive subductions. Hense crustal thickening produced by compressional tectonics in each area was followed by an extensional exhumation of underplate rocks as tectonic windows.

The timing of the Tertiary late-orogenic extension is different in the Hellenides zones started in the more internal parts of the orogen and progressively reached the more external ones: first in Rhodope zone in Eocene-Oligocene times (Kilias and Mountrakis, 1990, 1998; Kilias et al., 1999), more externally along the HP/LT belt (Olympos-Ossa-Cyclades) in Early Miocene (Buick, 1991; Gautier et al., 1993; Ring et al., 1999) and finaly in Crete along the external HP/LT belt in Middle Miocene (Fassoulas et al., 1994; Kilias et al., 1994, 1999).

3. TERTIARY TECTONIC EVENTS

Taking into account numerous field data and microscopic studies on microstructures in many Aegean islands and Crete we suggest the following three deformation events which have been well recognized in Southern Aegean area, corresponding to the successive stages of the above described evolution model.

- A D_1 tectonic event in Late Eocene-Early Oligocene times produced to So

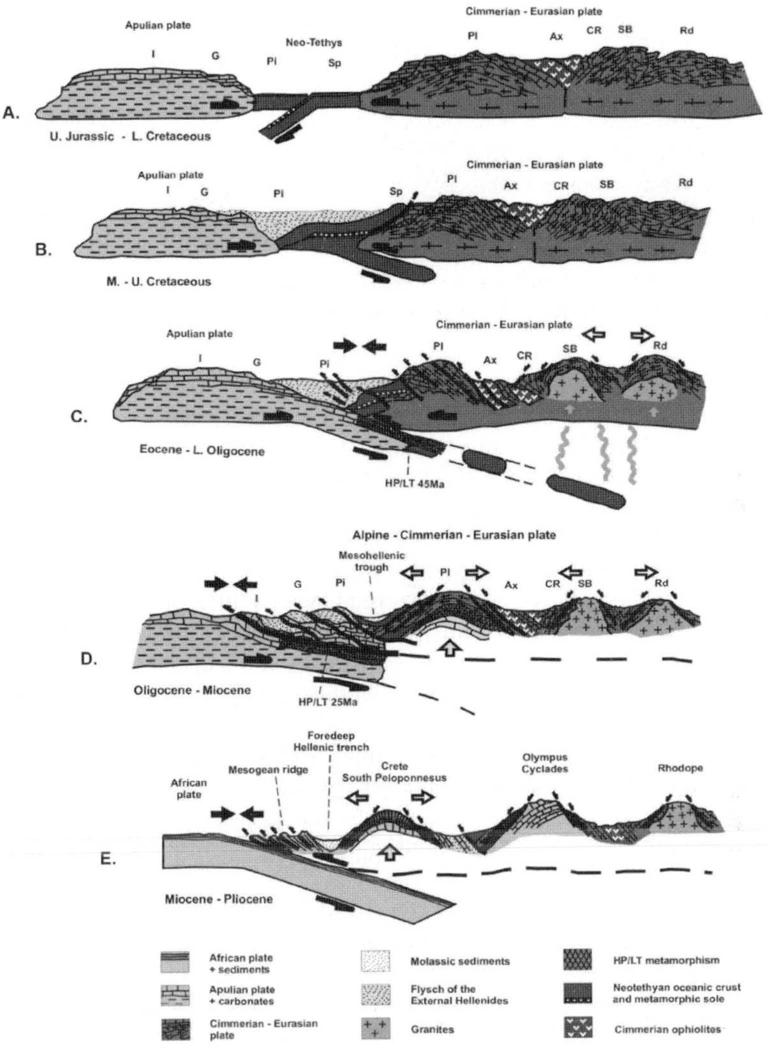

Fig. 3. Cross-sections showing the plate motion geodynamic reconstruction of the Hellenides during the two periods of the Alpine orogenic process (A, B, C, D), and the Mesogean one (E). During the first period the Neo-Tethyan intraoceanic subduction producing the metamorphic sole (A), the subduction of the oceanic crust beneath the unique Cimmerian-Eurasian plate and the obduction of the ophiolites onto Cimmerian (Pelagonian) continental margin (B), took place. In the second Alpine period the subduction process produced compressional tectonics, stacking of the nappes and crustal thickening along the plate convergence as well as the intraplate extensional tectonics, thinning of the crust, uplift and exhumation of the lower plate rocks and the plouton emplacement in Rhodope and Serbomacedonian zones. The successive migration of the compresional and extensional tectonics during Eocene-Early Oligocene times (C) and Oligocene-Miocene time (D) is also shown. The (E) cross-section shows the Mesogean orogenic process with the migration of the compressional deformation to the Mesogean ridge, the extensional tectonics in the Hellenides causing the uplift and exhumation of the tectonic windows in Crete, South Peloponese and Olympos area. Hellenides zones. Rd: Rhodope, SB: Serbomacedonian, Cr: Circum Rhodope, Ax: Axios, Pl: Pelagonian, Sp: Supelagonian, Pi: Pindos, G: Gavrovo, I: Ionian.

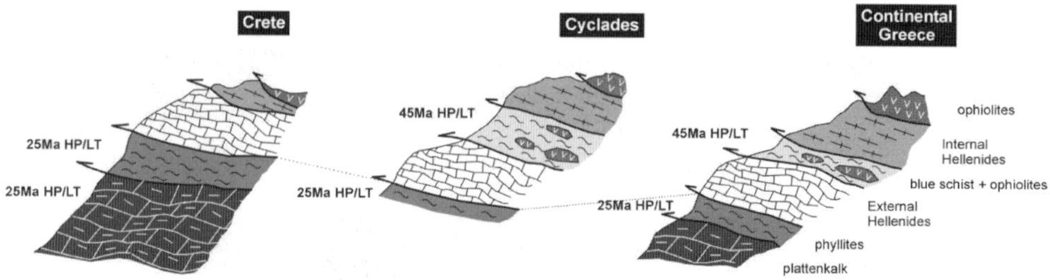

Fig. 4. The Nappe Pile of the Hellenic Orogen in three areas: Northern Continental Greece, Cyclades islands and island of Crete. Basic similarities and some significant differences between them are observed going from North to the South.

schistosity associated with a L_1 stretching lineation plunging SE. The contemporaneous with the So schistosity growth of greenschist minerals show that the greenschist metamorphism associated the D_1 event and affected the upper plate rocks of External Hellenides in Late Eocene-Early Oligocene times, after the HP/LT metamorphism of 45 Ma.

- A D_2 compressive event affected both upper and lower plate rocks, that is all units in Late Oligocene-Early Miocene times. Thrust faults produced a strong imbrication, nappe stacking and the final Nappe Pile in Cyclades islands and Crete. A relict S_2 schistosity and isoclinal intrafoliated folds are also observable structures of this D_2 event. This S_2 schistosity coincides with the growth of the blue amphiboles of 25 Ma suggesting that this HP/LT metamorphism took place at the same time with that D_2 event in Crete and Southern Cyclades.
- The most important tectonic event is the D_3 extensional one which affected all units during Early Miocene to present in Cyclades area and during Middle Miocene to present in Crete. The extensional process took place under ductile conditions in the lower plate units and semi-ductile to britle conditions in the upper plate units.

During the D_3 event a S_3 penetrative schistosity forms the dominant mylonitic fabric and due to the intense transposition along the S_3 planes the previous S_2 schistosity (formed contemporaneous with the HP/LT metamorphism) reorientated parallel to the S_3. And the isoclinal folds which had been formed in D_2 event, now reorientated parallel to a new stretching lineation L_3.

The growth of chlorite-epidote-stilpnomelane crystals in the S_3 schistosity, shows that the greenschist metamorphism, retrograde to the previous HP/LT 25 Ma one, took place still in Early to Middle Miocene together with the D_3 extensional event.

However the most impressive tectonic structures of the D_3 event are the discrete low-angle shear zones affecting in semi-britle conditions particularly the upper units. These shear zones sometimes overprint the mylonitic fabric as well.

In all cases kinematic indicators such as S-C structures, asymmetric mica fish, boudings, rotated clasts etc., demonstrate a top to N, NNE sense of shear in Cyclades area but with significant component of co-axial deformation particularily in Southern Cyclades, while in Crete a clear co-axial deformation in a double-vergent extensional regime is very well observable (Fassoulas et al., 1994; Kilias et al., 1994, 1999).

During the D_3 extensional regime some granitoid bodies intruded the alpine rocks in Cyclades islands and some of these bodies have also been affected by the extensional tectonic.

The crustal extension during this D_3 event caused thinning of the nappe pile, uplift and exhumation of the HP/LT metamorphic rocks of the lower plate as core-complex tectonic windows.

4. PLIO-QUATERNARY AND ACTIVE TECTONICS

The D_4 tectonic event, is the final deformation in Pliocene to recent times and it is the normal continuation of the extensional process in the Aegean area with britle conditions, producing normal high to mid-angle faults and strike-slip faults mainly trending NE-SW due to the great influence of the Anatolian westwards movement. These faults certainly help the volcanic activity in Plio-Quaternary times, but they also affected the volcanics with their reactivation in more recent times.

The D_4 extension in the broader Aegean area had during Pliocene a NE-SW direction and produced normal high angle faults trending NW-SE, while in Quaternary and recent times the direction of the extension is generally N-S producing normal faults trending E-W and reactivated the previous faults of other trends as strike-slip faults (Fig. 5).

Despite its N-S general direction, the active extension in Aegean and continental Greece shows some very interesting particularities as that is deduced from the focal mechanisms of the earthquakes (Papazachos and Papazachou, 2002; Papazachos and Kiratzi, 1996). The direction of the extension is completely different from the N-S along a narrow belt along the external Hellenic arc just inside of the compresional zone, following the direction of the compression which is almost E-W along continental Greece (Fig. 6) and continuous with the same E-W direction along Crete island.

Furthermore in Northern Greece the direction of the extension slightly differs from the general N-S direction in places, particularly in northwestern Macedonia where the extension becomes almost NW-SE (Fig. 6) as it is deduced from both focal mechanisms of large and small earthquakes and numerous geological field measurements along the active faults (Mountrakis et al., 2003). The influence of the westward movement of Anatolia along the North Anatolian strike-slip fault which reaches to North Aegean Sea and probably to continental Greece, is a possible explanation for these differentiations of the extensional direction, although further more field and instrumental evidence are need to make it clear.

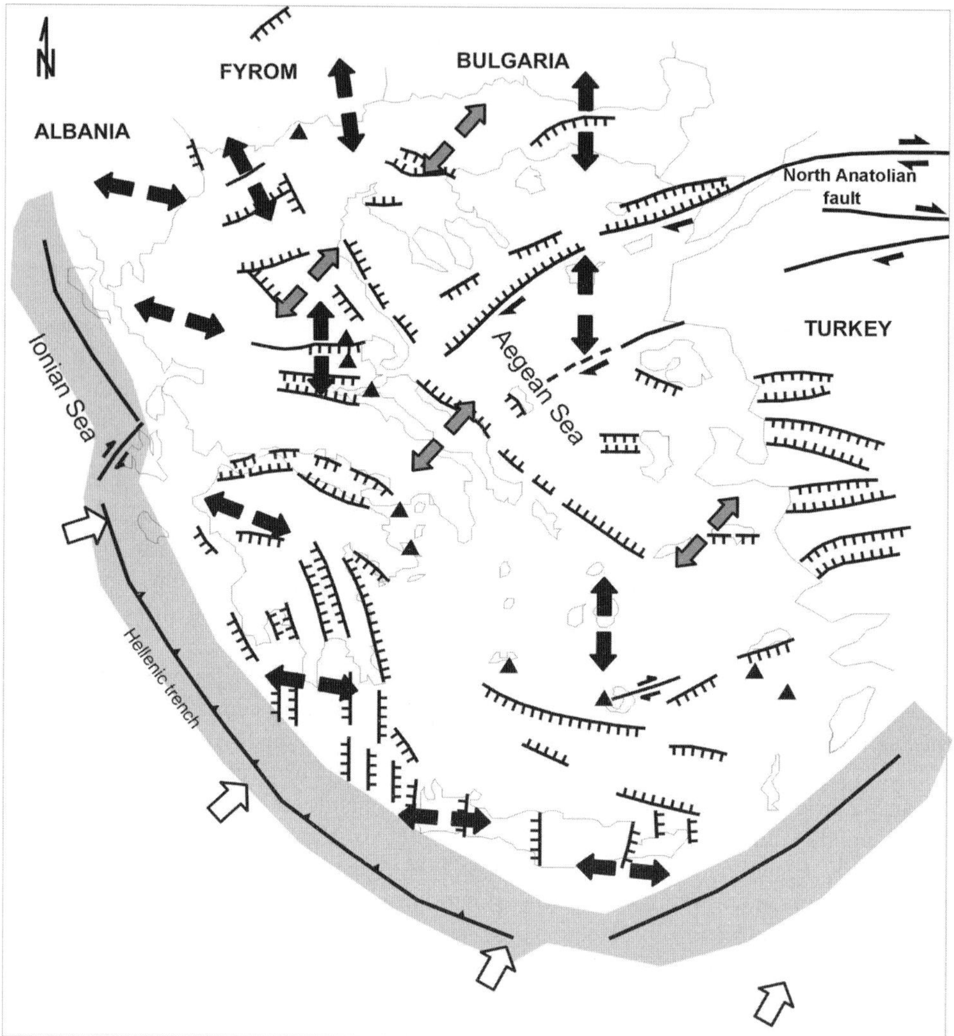

Fig. 5. Sketch-map showing the extensional tectonics of the D_3 event in Aegean and surrounding areas. Grey arrows show the extension in Pliocene times, black arrows the extension in Quaternary to recent times, white arrows the compressional directions outside the Hellenic arc, black lines are indicative normal and strike slip faults of the broader area.

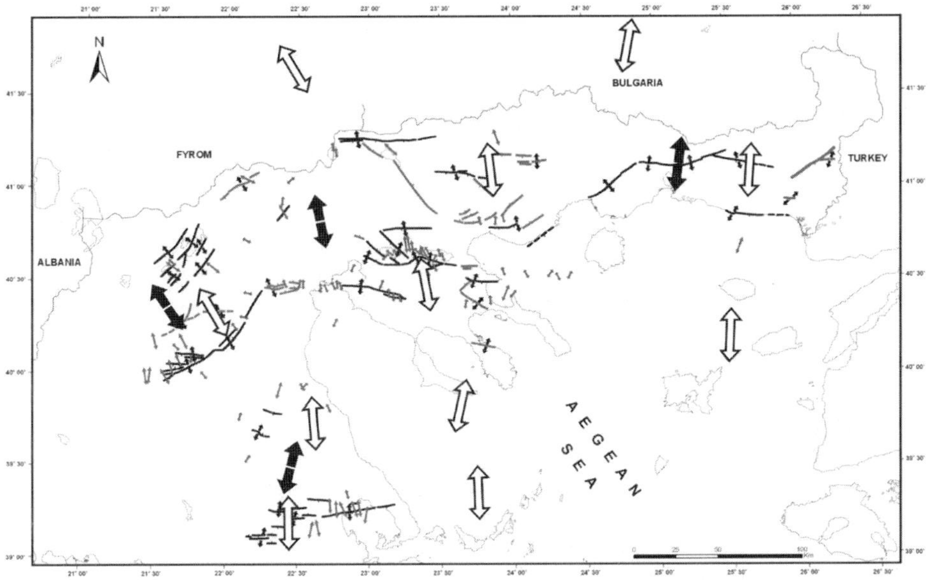

Fig. 6. Map of the main seismic-active faults of Northern Greece. Black lines: seismic-active faults, gray lines: possible active faults. Small arrows show T-extensional axis derived from geological field measurements along the faults (black arrows) and fault plane solutions (gray arrows) of earthquakes M≥4,5 and 3,3≤M≤4,5 (very small arrows). Large arrows show the average field stress determined by geological (black) and seismological (white) data corresponding to large areas.

REFERENCES

Avigad, D. and Garfunkel, Z., 1991. Uplift and exhumation of high pressure metamorphic terrains: the example of the Cyclades blue-schist belt. Tectonophysics, 188: 357-372.

Buick, I.S., 1991. The late-Alpine evolution of an extensional shear zone, Naxos, Greece. Journal of the Geological Society of London, 152: 639-654.

Doutsos, T., Koukouvelas, I., Zelilidis, A. and Kondopoulos, N., 1994. Intracontinental wedging and post-orogenic collapse in the Mesohellenic Trough. Geologische Rundschau, 83: 275-285.

Fassoulas, C., Kilias, A. and Mountrakis, D., 1994. Postnappe stacking extension and exhumation of HP/LT rocks in the island of Crete, Greece. Tectonics, 13: 127-138.

Gautier, P., Brun, J.P. and Jolivet, L., 1993. Structure and kinematics of Upper Cenozoic extensional detachment on Naxos and Paros (Cyclades islands). Tectonics, 12: 1180-1194.

Gautier, P. and Brun, J.P., 1994. Ductile crust exhumation and extensional detachments in the central Aegean (Cyclades and Evia Islands). Geodinamica Acta, 7: 57-85.

Kilias, A. and Mountrakis, D., 1990. Kinematics of the crystalline sequences in the western Rhodope massif. Geol. Rhodop., 2: 100-116.

Kilias, A., Fassoulas, C. and Mountrakis, D., 1994. Tertiary extensional of continental crust and uplift of Psiloritis metamorphic core complex in the central part of the Hellenic arc (Crete, Greece). Geol. Rdsh., 83: 417-430.

Kilias, A. and Mountrakis, D., 1998. Tertiary extension of the Rhodope massif associated with granite emplacement (Northern Greece). Acta Vulcanologica, 10: 331-333.

Kilias, A., Falalakis, G. and Mountrakis, D., 1999. Cretaceous-Tertiary structures and kinematics of the SErbomacedonian metamorphic rocks and their relation to the exhumation of the Hellenic Hinterland (Macedonia, Greece). Int. Journ. Earth Sciences, 88: 513-531.

Lister, G.S. and Forster, M., 1996. Inside the Aegean metamorphic core complexes. Technical publication Australian Crustal Research Centre, 45: 110 p.

Mountrakis, D., 1982. Emplacement of the Kastoria ophiolite on the western edge of the Internal Hellenides, Ofioliti, 7, 213: 397-406.

Mountrakis, D., 1986. The Pelagonian zone in Greece. A polyphase-deformed fragment of the Cimmerian Continent and its role in the geotectonic evolution of the Eastern Mediterranean. Journal of Geology, 94: 335-347.

Mountrakis, D., Tranos, M., Thomaidou, E., Papazachos, C., Karayianni, E. and Vamvakaris, D., 2003. Map of the main seismic-active faults of Northern Greece. (unpubl. Report. Earthquake Planning and Protection Organisation, Athens, Greece).

Papazachos, B.C. and Papazachou, C.C., 2002. The earthquakes of Greece. Ziti publ. Thessaloniki, 315p.

Papazachos, C. and Kiratzi, A., 1996. A detailed study of the active crustal deformation in the Aegean and surrounding area. Tectonophysics, 253: 129-153.

Ring, U.K., Laws, S. and Bernet, M., 1999. Structural analysis of a complex nappe sequence and late-orogenic basins from the Aegean Island of Samos, Greece. Journal of Structural Geology, 21: 1575-1601.

Robertson, A.H.F., Dixon, J.E. and Brown, S., 1996. Alternative tectonic models for the Late Paleozoic-Early Tertiary development of Tethys in the Eastern Mediterranean region. Geol. Soc. London Sp. Publ., 105: 239-263.

Seidel, E., Kreuzer, H. and Harre, W., 1982. A Late Oligocene-Early Miocene high pressure belt in the External Hellenides. Geologisches Jahrbruch, E23: 165-206.

Vandenberg, L.C. and Lister, G.S., 1996. Structural analysis of basement tectonites from the Aegean metamorphic core complex of Ios, Cyclades, Greece. Journal of Structural Geology, 18: 1437-1454.

A model of the Aegean geodynamic zone

B. Ranguelov

Geophysical Institute, Bulgarian Academy of Science, Sofia, Bulgaria, e-mail: boyko.ranguelov@jrc.it

ABSTRACT

The problems of the geodynamic regime in the Aegean active zone are under investigation many years now. A number of authors (Papazachos, 1966; Caputo, 1970; Artemjev, 1984; Ranguelov, 1987, etc.) have suggested different hypotheses on the development of this complicated - from a geodynamic point of view – regional zone. Our hypothesis is based on the generated two models, for the North part (so called NASDM) and for the South subducted part (called SAASZM) and their interaction newly compiled and presented by us. The combined actions of the forces of the South subducted zone (SSZ) and the main element of the North part – North Anatolian Transform Fault (NAF) and their interactions, generate, in our opinion, the main geodynamic and geomorphologic peculiarities of the whole zone. The extensional areas are followed by areas with the compression regime, which suggests a very complicated recent geodynamic regime. All assumptions are based on the observed phenomena reflecting the recent development of the investigated zone (such as seismicity – expressed by the strong and week earthquakes, GPS measured plate movements, geomorphologic expressions observed on the relief, etc.). The main conclusion is that such a model can explain most of the observed structures and geodynamic facts in the region: strong earthquakes occurrence and their complicated positions and mechanisms, the alternation of the compressional and extensional areas from south to north, aseismic zone north of Crete, the recent volcanism expressed in the volcanic arc, the big curvature of the subducted part of the Aegean zone between the African and the Eurasian plates, the locations of many geomorphologic features, etc.

Keywords: Aegean zone, geodynamic model, subduction, transform faults

1. INTRODUCTION

The specific behavior of the Aegean arc subduction zone and its surroundings (mainly to the north, is under investigation. The specificity of this geodynamic zone is dominated by several factors:

- sharp curve outlining the subduction zone;

- practically aseismic zone existence to the inner (northern) part of the subduction zone;
- low volcanic activity on one side and very high seismic activity on the other, reaching depths in the subducted part up to 200 km.
- existence and alternation of the zones of extensional and compressive geodynamic regime;
- existence and active action of the clearly expressed transform fault - NAF and its branches located north to the subduction zone;
- high horizontal displacement velocities measured by GPS and connected mainly to the North Anatolian fault and its western satellites;
- clear normal faults generated by the earthquakes in the extensional regions and clear strike-slip faults connected with the NAF earthquakes;
- complicated earthquake's mechanisms located in the medieval active zones;
- specific behavior of the younger and recently expressed grabens, their position and development, etc.

All these peculiarities were under the focus of our attention to summarize the observed phenomena, using different recent geodynamics data - from the recent GPS measurements, local and regional seismicity, the geomorphologic evidence of the relief expressions, etc. As a result of the comprehensive analysis we suggested a common general geodynamic model, to explain all observed facts in this very interesting and complicated geodynamic zone of the Aegean Sea and its surroundings.

2. NORTH AEGEAN SEA GEODYNAMIC MODEL (NASGDM)

The North Aegean Sea geodynamic model considers the geodynamic peculiarities to the North of the west branches of the North Anatolian Fault system. It is characterized by the dominant influence of the recent North Anatolian Fault movements, established very reliably by the GPS measurements. The direction of these movements is to the Southwest and their amplitudes are about 20-25 mm/y. Due to these relatively fast movements, the surface block elements of the earth crust are moving in the same direction. As a result, the openings (clearly expressed grabens fulfilled by the rivers going to the south – such as Strymon, Nestos, Vardar - Axios and Maritza - Evros) show a clear extension to the North-South. The grabens started their recent development during the Neogene, so their position and shape have a relatively long lasted formation during the geological times. The recent GPS measurements show relatively small amplitudes of the movements – up to 2 mm/y, mainly to the North-South direction. The "dragging" effect of the North Anatolian Fault movements is confirmed as well by the surface relief expressions of the assymmetrical openings of the grabens. The dominant direction of the assymmetry is to the west. The "three fingers" location of the smaller peninsulas of the Halkidiki peninsula is due to the same effect. All other smaller riverbeds located between Vardar and Maritza riverbeds show similar behavior. The seismic regime (as a most recent expression of the stress distribution and redistribution) is very active there. Several large destructive earthquakes occurred during the previous century:

1902 – an earthquake NE of Thessaloniki (M~6.6).

1904 – two very strong earthquakes (M=7.2 and 7.8) near Kresna village.

1931 – Valandovo earthquake – M= 6.7.

1963 – Skopje earthquake M=6.1.

1978 – Thessaloniki earthquake M=6.4.

Most of the strong earthquakes show normal faulting (Ranguelov et al., 2000) expressed by the surface coseismic cracking with vertical movements from tens of centimeters (Thessaloniki earthquake, 1978) up to a few meters (Kresna - Kroupnik earthquakes, 1904). The only possibility to explain all these observations is to accept the dominant role of the North Anatolian recent fault movements and their "dragging" effect, which is responsible for the stress accumulation and redistribution by the strong earthquakes occurring in this space-time domain. There is no other possible explanation of the normal faulting observed for the strongest events there. Thus, the presented model of the dominant geodynamic regime for this area is well accepted and the regime can be considered as the main reason for the formation of the recent, so called Balkan-Aegean Graben System (BAGS), where these geodynamic forces and movements are acting. The complicated earthquake's mechanisms (Ritsema, 1974) can be also explained by this stress accumulation and release.

3. SOUTH AEGEAN ARC SUBDUCTION ZONE MODEL (SAASZM)

The subduction zone itself, as a main geotectonic structure, dominates this model of the south part of the area signalizing the collision of the Northeast part of the African plate to the Southeast part of the European plate. The total length of the zone is about 1500 km. The zone has a big sharp curve trajectory expressed most clearly near the region of Crete Island. Many authors indicate the existence of the Benioff zone here (McKenzie, 1972; Papazachos, 1973, etc.), dipping to the north with an average of 35 degrees due to the presence of the intermediate earthquakes occurring at the depths of 100-160-200 km. Some previous investigations (Ranguelov, 1987) show the depth penetration of the different segments of the Earth crust of this zone as well as the most significant areas of the bigger seismic energy emission (Ranguelov and Gospodinov, 1982). Using simple geometric calculations the dip angles have been calculated about the four different subducted plaques of the local Earth crust (Ranguelov, 1987). They are presented in Table 1.

It's important to note that the recent volcanism is located to the north of the subducted part and outlined the area of the volcanic islands - Cyclades and Sporades (Nichols, 1971). The zone with lower seismicity exists near the north of the subduction zone. This zone has been frequently connected by the different authors to a "mantle dome" of low velocity asthenosphere (McKenzie, 1972; Artemjev, 1984). The bottom relief shows clear evidences typical to all worldwide presented subduction zones – deep trenches, divided by transverse faults. Using the previous knowledge (Ranguelov and Gospodinov, 1982; Ranguelov, 1987; Dimitrova and Ranguelov, 2002) and the recent

Table 1. Main elements (lengths and dipping angles) of the subducted plaques of the South Aegean subduction zone (Roman numbers indicate them from east to west on Fig. 1).

Dimensions	No of plaques			
	I	II	III	IV
Length [km]	135	200	160	320
Depth [km]	160	130	100	160
Dip angle [°]	52	33	32	26

image of the subducted zone, a model of the recent positions of the main Earth's crust elements in the Aegean area and the forces acting on them is constructed, and presented on Fig. 1. Due to the stress distribution and redistribution, the whole Aegean zone is a seismically active region. Very often the submarine earthquakes generated tsunamis (for example – the most significant recent one, that of 1952). Several zones of extension follow to the north after the presence of a subduction zone with compression direction to the south. The volcanism is expressed to the frontal (northern) part of the zone, which makes this area a classic case. The location to the north of the North Anatolian transform fault makes the situation more complicated, which is indicated by some questionable zones, with no clearly expressed geodynamic regime. They are marked on the scheme by "?" marks – Fig. 1.

4. GENERAL INTEGRATED SCHEME OF THE MODEL

Combining the North Aegean Sea model (NASDM) and the South Aegean arc model (SAASZM), an integrated geodynamic scheme is constructed. To the north, the dominant structure is the North Anatolian Fault zone and its satellites. They are responsible for the recent geodynamics, expressed by:

- fast "everyday" relative movements to the both sides of NAF (up to 20-25 mm/y) in SW-NE direction are observed on the west branches of NAFZ and registered by the recent GPS measurements. Almost all of them are outlined by the smaller seismic events. These movements are an expression of the most recent forces acting there continuously;
- strong normal fault earthquakes far from the subducted part (1902, 1904, 1931, 1963, 1978). These actions show the middle-term trend of the recent geodynamics expressed by the "points (areas) of destruction" of the Earth crust. They are due to the stress-strain distribution and redistribution among the local blocks of the earth crust;
- recent position and asymmetry of the many elements of the earth relief (mainly grabens and riverbeds) located on the north part of the Aegean zone, as well as the deviation of the three smaller peninsulas of the Halkidiki peninsula. Here the dominant geodynamic regime is on extension phase, and this is also measured by GPS. These phenomena show the long-term trend of the geodynamics started long time ago (beginning of the Neogene).

All the facts described above explain the existence of the complicated extensional-compression zones located on mosaic and irregular way.

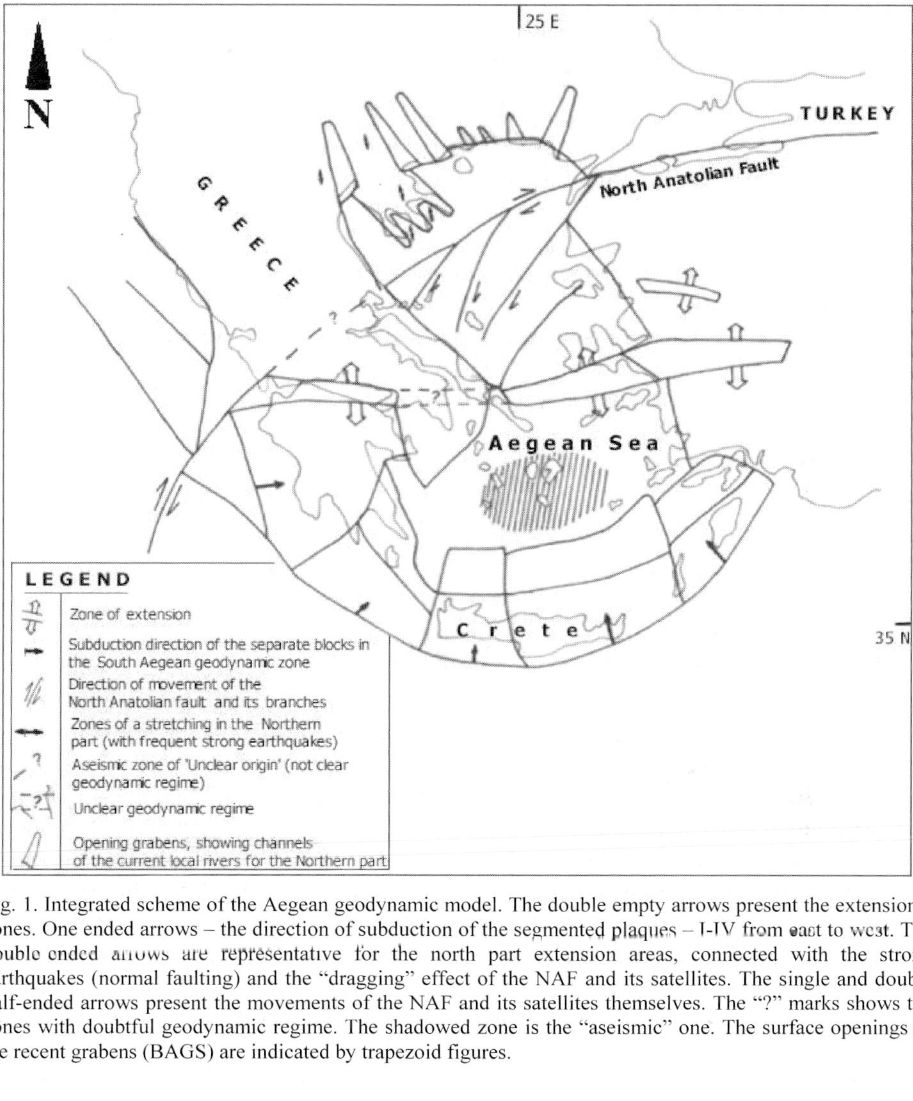

Fig. 1. Integrated scheme of the Aegean geodynamic model. The double empty arrows present the extensional zones. One ended arrows – the direction of subduction of the segmented plaques – I-IV from east to west. The double ended arrows are representative for the north part extension areas, connected with the strong earthquakes (normal faulting) and the "dragging" effect of the NAF and its satellites. The single and double half-ended arrows present the movements of the NAF and its satellites themselves. The "?" marks shows the zones with doubtful geodynamic regime. The shadowed zone is the "aseismic" one. The surface openings of the recent grabens (BAGS) are indicated by trapezoid figures.

To the south, the main element of the recent geodynamics is the Aegean Subduction zone. It also has no typical regular behavior of the subduction zones observed for example in the Pacific – long elongation and smooth curvatures, deeper seismic events, very clear expressed volcanic arcs, etc. The zone has the following peculiarities, not typical of the larger subduction zones:

- very big curvature, looking like "smashed sheet of paper";
- "aseismic zone" to the north, may be due to the "dome" of asthenosphere;
- several plaques dipping in different angles and with different lengths;
- irregular space-time seismic emission by strong deep earthquakes;

- not very active volcanic arc, characterized by very strong but irregular eruptions;
- clear expressed segmented structure observed on the bottom relief and the deep trough, etc.

As it was shown the main "actors" of this geodynamic drama are The North Anatolian Transform fault and the Aegean Subduction zone. The simultaneous actions of these big tectonic units leads to the complicated and irregular picture of the region. Areas of extension follow areas of clear compression to the north. Transform (strike-slip), normal and complicated faulting is often observed. The main expressions, important for the society, about the hazard assessment of the recent geodynamic activity of these structural units are:

- many strong earthquakes (sometimes generating tsunamis) and much more smaller ones, indicating areas of recent active geodynamic structures,
- submarine and surface landslides and rockfalls,
- recent volcanism and
- the fast block movements and relief changes due to the inner Earth forces.

5. CONCLUSIONS

The suggested geodynamic model reflects almost all the recently observed geodynamic phenomena in the Aegean region. The inner parts of the Aegean subduction zone reflect the complicated structure by different elements and geotectonic units. They are due to the structures with different orientation and different (extensional, compressional and transform) geodynamic regime. The location of the fault structures, and observed seismicity (with its spatial and temporal specific behavior), is an important element of the Aegean zone. The subducted earth crust parts, the surface and deep earth crust movements (some of them detected by the GPS measurements, another by more complicated non direct techniques) are the other components of these complicated regional geodynamics. The recent relief on the land and submarine geomorphologic units also give an image of the complex geodynamics from the long-term point of view. All these facts are current elements of this very specific boundary transition zone between the European and the African plates. The observed volcanism is also a clear expression of this recent geodynamic activity in the Aegean geodynamic zone.

Acknowledgments

The contracts with TIG, DVO (RAN) - GFI and a Bulgarian-Greece bilateral contract – NZ-1209/02 supported this study.

REFERENCES

Artemjev, M., 1984. Identification of Mantle and Lithospheric components by isostatic anomalies. Marine Geoph. Res., 7: 65-71.

Caputo, M., 1970. Deep structure of the Mediterranean Basin. JGR 74: 235-261.
Dimitrova, S. and Ranguelov, B., 2002. A Geodynamic model and seismic danger for SW Bulgaria and surroundings. Proc. Int. Conf. "VSU2002". Sofia, pp. 1-6. (in Bulgarian)
McKenzie, D., 1972. Active tectonics of the Mediterranean Region, Geoph. J. R. Astr. Soc., 30, 55-71.
Nichols, I., 1971. Petrology of Santorini volcano - Cyclades. Journal of Petrology, 12: 43-61.
Papazachos, B., 1966. Preliminary results of an investigation of crustal structure in SE Europe, BSSA 56: 1836-1849.
Papazachos, B., 1973. Distribution of seismic foci in the Mediterranean area and its tectonic implication. Geoph. J. R. Astr. Soc., 33: 1123-1132.
Ranguelov, B., and Gospodinov, D., 1982. On some peculiarities of the seismicity and its spatial distribution for the Balkan Peninsula region. Bulgarian Geophysical Journal, VIII, 2: 64-71.
Ranguelov, B., 1987. Depth structure characteristics of the Aegean Arc. Bulgarian Geophysical Journal, XIII, 3: 57-62.
Ranguelov, B., Rizhikova, S. and Toteva, T., 2001. The earthquake (M 7.8) source zone (South-West Bulgaria). Acad. Publ. House "M.Drinov", Sofia.
Ritsema, A., 1974. The earthquake mechanisms of the Balkan region. UNDP (REM(70), 172, UNESCO Report, Skopje.

The South Aegean Active Volcanic Arc
M. Fytikas and G.E. Vougioukalakis

Upper mantle structure of the Aegean derived from two-station phase velocities of fundamental mode Rayleigh waves

I. Kassaras[1,*], K. Makropoulos[1], E. Bourova[2,3], H. Pedersen[2] and D. Hatzfeld[2]

[1] Department of Geophysics-Geothermy, University of Athens, Panepistimioupolis, Zografou 15784, Greece.

[2] Laboratoire de Géophysique Interne et Tectonophysique, Université Joseph Fourier, Grenoble, France.

[3] Institute of Physics, Saint-Petersburg State University, Russia.

Fundamental mode Rayleigh waves generated by 380 teleseismic events were analyzed over the period range 10-100 s, in order to study the structure of the lithosphere and upper mantle of the Aegean region. Using the two-station method, 255 reliable phase velocity dispersion curves were calculated over 35 profiles and further inverted to obtain a new model of S-wave velocity with depth. S-wave velocities are resolved to a depth of 180 km. Important features are defined, such as a not completely amphitheatric geometry for the western ($\approx 25^0$ dipping angle) and eastern segments ($\approx 35^0$ dipping angle) of the subducted slab. In north Aegean, high velocities associate with the North Aegean Trough, which westernmost tip correlates with a high velocity anomaly in eastern continental Greece. This zone of high velocity contrast is extended in depth, dips southwards with an angle $\approx 35^0$ and intersects with the subducted slab at an area where the direction of major tectonic axes changes from ENE-WSW to NNW-SSE towards the continental massif. In Central and North Aegean, where back arc extension and crustal thinning occur, the predominant low velocities observed could be interpreted by upper mantle high thermal flow and partial melting.

Key words: Aegean region, Rayleigh wave, dispersion, inversion.

1. INTRODUCTION

The highest deformation rate along the Africa/Eurasia convergence zone is documented in the Aegean area, being >4 cm/yr (McClusky et al., 2000). This high rate

* Corresponding author: email: kassaras@geol.uoa.gr

is due to the westward motion of the Anatolian plate around a Heulerian pole located in eastern Mediterranean (Le Pichon et al., 1995) and the subduction of the eastern Mediterranean lithosphere beneath the Aegean (McKenzie, 1978; Le Pichon and Angelier, 1979). The deformation pattern of the Aegean region is complex, characterized by high extensional strain rates, as active faulting, shallow seismic activity and geodesy indicate. The complexity of the region reflects to its seismic velocity structure, where strong velocity contrasts have been reported. A large number of geophysical and seismological studies on the seismic velocity structure of the area have been carried out, with the majority of those constrained within the crust and uppermost mantle (Papazachos et al., 1966; Panagiotopoulos and Papazachos, 1985; Voulgaris, 1991; Chailas et al., 1993; Tsokas and Hansen, 1997; Karagianni et al., 2002). Several studies, focused on the lithosphere and upper mantle structure are mainly based on body wave information (Spakman et al., 1988; Christodoulou and Hatzfeld, 1988; Papadimitriou, 1988; Granet and Trampert, 1989; Ligdas et al., 1990; Drakatos and Drakopoulos, 1991; Ligdas and Lees, 1993; Spakman et al., 1993; Papazachos et al., 1995; Papazachos and Nolet, 1997; Tiberi et al., 2000), while only a few are based on surface waves, been however of limited resolution (Papazachos et al., 1967; Papazachos, 1969; Ezen, 1983, 1991; Kalogeras and Burton, 1996). In this paper, using the dispersion properties of surface wave we study the lithosphere and upper mantle seismic velocity structure of the Aegean area. A number of 380 teleseismic events were utilized and 255 reliable two-station phase velocity dispersion curves of the fundamental Rayleigh wave mode were estimated for 35 profiles in the Aegean. A linear stochastic inversion procedure of phase velocity dispersion curves yielded the S-wave velocities distribution down to the depth of 180 km. The observed shear velocity contrasts lead to the recognition of several characteristics of the Aegean upper mantle i.e. the Hellenic subduction, the back-arc spreading area, the North Aegean Trough.

2. DATA AND DISPERSION MEASUREMENT

In 1997, a temporary network of 30 broadband continuous recording seismological stations was installed across the Aegean area for 6 months (January-July) in the framework of the E.U. project "*SEISFAULTGREECE, Active Faulting and Seismic Hazard in Attiki, (GREECE)*". Active members of the project were the Institutes LGIT, and IPGP from France, University of Athens and Thessaloniki from Greece and University of Cambridge from the U.K. A brief description of the experiment can be found at other sources (e.g., Hatzfeld et al., 2001). In this study, recordings from 15 stations, namely 6 stations equipped with Guralp CMG-40T and CMG-3T 60 s sensors, 6 stations equipped with Guralp CMG-3T 100 s sensors and 3 stations equipped with STS-2, 100 s sensors of the Geofon broadband network in southern Aegean (Fig. 1) were used.

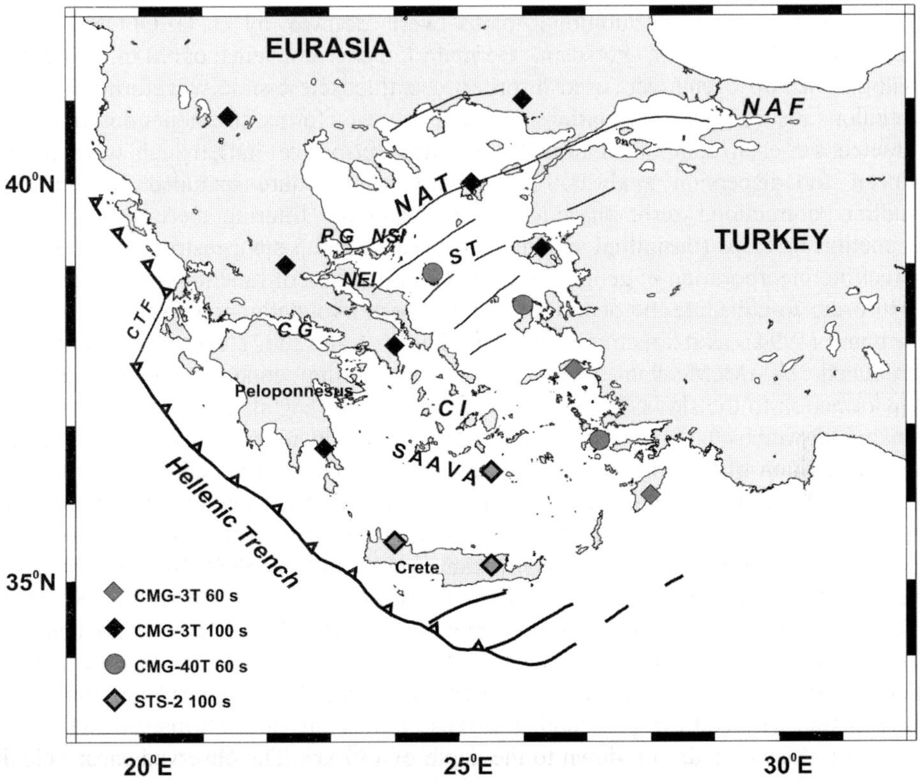

Fig. 1. Map of seismological stations used in this study and main active tectonic features of the Aegean after Hatzfeld et al., 2001. Abbreviations for geographic names discussed in text: CG, Corinth Gulf; PG, Pagassitikos Gulf; NSI, North Sporades Islands; NEI, North Evia Island; NAT, North Aegean Trough; NAF, North Anatolian Fault; ST, Skiros Trough; SAAVA, South Aegean Active Volcanic Arc; CI, Cyclades Islands; CFT, Cephalonia Transform Fault.

For the determination of the elastic properties of the lithosphere and upper mantle of the Aegean, two-station phase velocity measurements were performed over multiple profiles. The basic concept of the two-station method is that the great circles connecting each station with the epicenter ought to have the same azimuth. Therefore, the common source effect, or other uncertainties related to the source are eliminated and subsequently the phase velocity measured between the stations, is only affected by the response of the interstation elastic medium. The usual limitation adopted for each profile in the study is an angular difference between the two stations great circles less than 3^0 (Badal et al., 1992).

Following Levshin (1998), a typical teleseismic magnitude cut-off threshold for the surface wave period range studied, should be of the order of M=5.0. For the time period January-July 1997, the global ISC seismicity catalogue included 423 teleseismic events with M≥5.0, from which 374 fulfilled the criterion that the angular difference between the two stations great circles was less than 3^0, defining 59 profiles in the Aegean. In southernmost Aegean 2 additional paths were defined by 3 GEOFON stations (GeoForschungsZentrums, Potsdam, Germany) using 6 events of M≥6.0. Thus, a catalogue of 380 events was used in order to extract teleseismic waveforms from the continuous recordings of the stations. About 1100 waveforms of 2 hours duration were extracted for each component (about 3300 waveforms in total), which were further reduced for dispersion analysis. The reduction procedure included baseline and trendline correction, zero phase antialias bandpass filtering between 0.01-3Hz, decimation to 1 sps (the initial sampling rate was 20-62.5 sps), instrumental response correction, incorporation of geographical and modification of functional parameters.

In order to calculate the phase velocity between two-stations, we used the code of Herrmann (1994), as it was modified by Baumont et al. (2002). Based upon a concept introduced by McMechan and Yedlin (1981), the code performs wavefield transformation to the slowness-frequency domain, involving slant-stacking of the input signals followed by 1-D Fourier transformation. The main advantage of this technique is the participation of the whole signal in the analysis. The most important disturbance is the coexistence of various propagation modes; therefore the isolation of a single-mode signal is required. Thus, before the implementation of phase velocities measurements, phase-matched filtering in the frequency-time domain was applied to the input signals (Herrmann, 1973; Baumont et al., 2002). Using this approach, a total of 382 phase velocity dispersion curves of the fundamental mode Rayleigh wave were estimated.

The range of the studied periods (10-100 s) with respect to the events location uncertainties, the relatively low relief of the study area, the adequate quality of the experimental data and the good signal-to-noise ratio used, ensured that the systematic errors of phase velocities measurements were almost negligible (Dziewonski and Hales, 1972; Snieder, 1986; Badal et al., 1992; Lana et al., 1997; Pillet et al., 1999).

On the other hand, there are cases where if not properly handled, could affect the reliability of the measurements. The three most commonly encountered effects are:

1. The interstation distance at long periods is smaller than or approximately equal to the wavelength, thus the phase differences between the two stations is small and consequently induce large phase velocity errors. The solution we adopted to minimize this effect was a very careful manual data processing and the use of a large amount of data, to facilitate a statistically reliable estimate.
2. The rays deflection from the great circle path due to lateral velocity variations (multipathing effect), which sometimes can reach an angle of 20^0 (Levshin et al., 1994) and therefore degrade phase velocity calculations, as the interstation distance is different than the one inferred from the great-circle path. The solution we adopted to solve this problem, was to define a statistically reliable variability range for the measured phase velocities. It was based on a thorough investigation of Rayleigh waves polarization properties, evaluation of discrepancies due to the source-receiver propagation effect at symmetric (were applicable) and adjacent great-circle paths.
3. Contamination of Rayleigh waves from other seismic waves propagating with similar group velocities, especially Love waves in the period range 25-50 s. The common solution to solve this problem was firstly careful manual multiple

filtering of the seismograms and secondly the correlation between measurements performed in the vertical and radial component (R>0.9).

In this way, we rejected any of the outlying data likely related to those effects and adopted a total of 255 dispersion curves as the most representative of the elastic medium properties. Those 255 dispersion curves concern 176 teleseismic events (Fig. 2a) and 35 profiles across the Aegean (Fig. 2b). For each path the average phase velocities and the respective standard deviation (σ) were estimated. Individual standard deviation may reach values as high as 0.09 km/s at low periods (<25 s) and 0.07 km/s at periods >25 s, implying though for a realistic peak velocity error estimate ±2.6% for the upper layers and ±1.7% for the lower lithosphere and upper mantle. Such results definitely support the routine procedure followed in this study to constrain dispersion properties.

3. INVERSION FOR S-WAVE VELOCITIES DETERMINATION

3.1. The problem of surface wave linear inversion

In most geophysical applications the velocity model is a continuous function of space coordinates and therefore has unlimited freedom degrees. If we ignore the above and consider that the model has finite number of parameters, a generalized method with a wide range of applications is described below.

Focusing on inversion of surface wave dispersion data, let us consider the phase velocity c. Small perturbation of S-wave velocities β_i of the initial model, corresponds to variations of c with respect to the period accordingly to the following formula, which due to the small magnitude of perturbation is expected to be linear:

$$\delta c_i = \sum_{i=1}^{N} \frac{\partial c_j}{\partial \beta_i} \delta \beta_i = \frac{\partial c_j}{\partial \beta_1} \delta \beta_1 + \ldots + \frac{\partial c_j}{\partial \beta_N} \delta \beta_N \qquad (1)$$

where N is the number of the model layers, δc_j the difference between observed and calculated phase velocity for a period j, $\partial c_j / \partial \beta_i$ is the first order of the Taylor expansion of velocity differences and $\delta \beta_i$ is the resulted S-wave velocity variation for the layer i of the model.

If d is the data vector with M components, defined by the variations of phase velocities and m the model vector with N components, which includes the small S-wave velocity perturbations (unknowns) that define the initial model (1) can be written as:

$$d = Am + e \qquad (2)$$

Therefore, the observation data d are connected to the initial or true model m, through the matrix A (Jacobian), and e is the error of observations. Those equations allow for an iterative procedure which calculates phase velocities by resolving the direct problem for m. The method involves modification of m by $\delta \beta_i$ and calculation of the respective response of the phase velocities δc_j. The procedure restarts with the new model $\tilde{m}$, which results from the substitution of β_i with $\beta_i + \delta \beta_i$ and repeats until the differences between experimental and calculated phase velocities tend to be very small,

Figure 2a

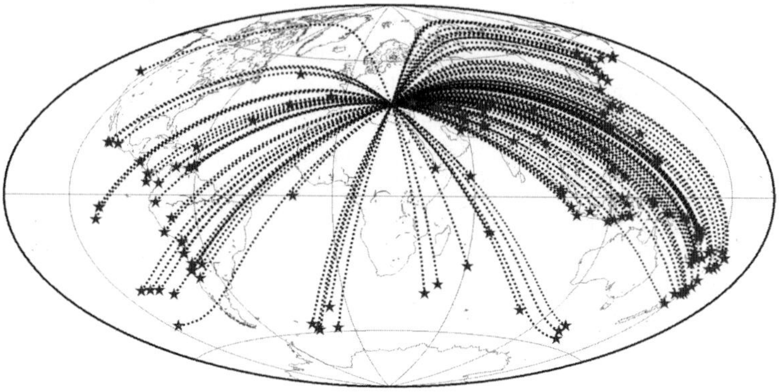

Figure 2b

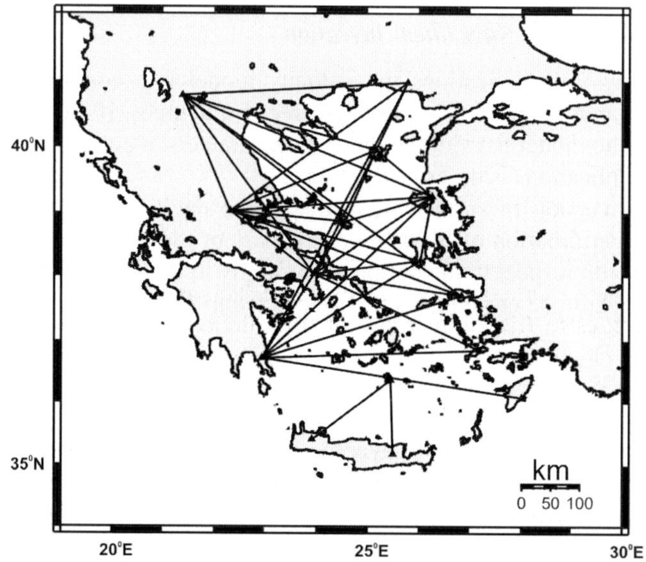

Fig. 2. a) 176 used teleseismic events and propagation great circles. b) 35 profiles across the Aegean for which reliable phase velocity dispersion curves were estimated.

i.e. until the model converges to a solution that provides adequately small residuals.

3.2. The method of damped least-squares solution (stochastic inverse solution)

The method is based in finding a model $\tilde{m}$ that yields the best consistency with the input data, in terms that the length difference between the data vector d and the calculated vector $A\tilde{m}$ is as small as possible. Thus, the damped least-squares solution is given by the model that minimizes the following function:

$$\|A\tilde{m} - d\| + \|\sigma\tilde{m}\| = MIN \tag{3}$$

where $\| \ \|$ denotes the L^2 norm
or

$$(A\tilde{m}-d)^T(A\tilde{m}-d)+\sigma^2\tilde{m}^T\tilde{m}=MIN \tag{4}$$

This typical least squares problem can be expressed in terms of eigenvectors and eigenvalues of matrix A, hence $A=U\Lambda V^T$, $A^T=V\Lambda U^T$ and $UU^T=I=VV^T$. The columns of matrix V are the singular vectors which correspond to the rows of matrix A and the rows of matrix U^T are the singular vectors which correspond to the columns of A. Λ is a diagonal matrix, which diagonal elements are the singular values of matrix A. The above lead to the resolving vector $\tilde{m}$, the variance-covariance matrix C and the resolution matrix R (Herrmann, 1994).

$$\tilde{m}=V(\Lambda^2+\sigma^2 I)^{-1}\Lambda U^T d=Hd \tag{5}$$

$$C=HH^T=V(\Lambda^2+\sigma^2 I)^{-1}\Lambda^2(\Lambda^2+\sigma^2 I)^{-1}V^T \tag{6}$$

$$R=V(\Lambda^2+\sigma^2 I)^{-1}\Lambda^2 V^T \tag{7}$$

Parameter H, is the inverse stochastic operator and $\sigma^2=\sigma_e^2/\sigma_m^2$ is the parameter of Lavenberg-Marquardt (Franklin, 1970) or the damping factor.

Resolution is the most important concept distinguishing inverse theory from parameter estimation. The resolution matrix R is a $P\times P$ square not symmetric array. It relates the estimate to the true model m. Resolution is perfect when $R=I$ and there is no difference between the true model and the estimate. This is true when $\sigma=0$. The degree by which R converges to I is the measure of the solution resolution, i.e. the more the values of diagonal elements of R approximate unity and low values are elsewhere, the more the solution approximates uniqueness. Non-zero off-diagonal elements of R indicate a trade-off between the model parameters. Non-zero R_{ij} indicates failure to separate the effect of the ith ant j^{th} model parameters. R does not depend directly on data errors, but it does depend on A through the data kernels i.e. In other words it depends on where the data were measured rather than on how well they were measured.

The damping factor plays an important role as it allows controlling the instability of the inversion process. The increase of σ reduces the resolution, but simultaneously reduces the error of the solution and stabilizes it. Stochastic inversion requires the selection of an adequate damping factor, as low values result in unstable solution with high energy of the parameter changes, while large values cause very slow convergence and stable but overly-damped low amplitude solution.

On the other hand, the errors e of the experimental data, are not deterministically known, otherwise they should have been eliminated. Ideally, both perfect resolution and lack of experimental errors would be desirable, however, as this can not be practically achieved, what is expected is the definition of an acceptable balance between the errors distribution and the resolution limitations.

The monotonic trade-off curve of $T(\sigma)$, asymptotes to the minimum value of E at the one end and the minimum value allowed for the model norm N on the other. Given the practical lack of confidence about the accuracy of the experimental data, we chose the "knee" of the trade-off curve as a compromise for the damping factor.

3.3. Inversion of experimental data

The inversion method described above was applied to obtain S-wave velocities with depth. An algorithm written by Herrmann (1994) was used. A significant restriction of surface wave inversion is the a priori information given as an initial S-wave velocity model, which has to be relevant to the natural properties of the under study elastic medium. Therefore, we considered four S-wave average velocity models obtained for the Aegean area and tested them in terms of resolving the forward problem for the average dispersion curve of the area. The models, proposed by Papazachos et al., (1967), Papazachos (1969), Papadimitriou (1988) and Papazachos et al., (1995), were deduced from different methods: From group and phase velocity of Love waves, from phase velocity of Rayleigh wave, from teleseismic body wave modeling and from body wave traveltime tomography of local and regional earthquakes respectively. Best consistency between theoretical and experimental dispersion curves was observed for the first and third models and relative consistency with the second (Fig. 3). Relative discrepancy is observed for the latter model of Papazachos et al., (1995) and one reason for that could be that it includes information from western continental Greece, which is not sampled by our data.

An important restriction of surface wave inversion is the measure of dependence of the solution from the a priori information. It is well established that surface wave inversion is very sensitive to large velocity gradients included in the initial model. Apparently, potential large velocity gradients resolved by the inversion, ought to be due to the distribution of reliable experimental data and not due to the reproduction of a priori information. The solution of the inverse problem for the average dispersion curve using the 4 above initial models, resampled where appropriate, apparently confirmed this dependence for the first three models. The latter model (Papazachos et al., 1995), also used as a background model to recent detailed tomography studies (Papazachos and Nolet, 1997; Tiberi et al., 2000; Karagianni et al., 2002) allowed for a solution smoothly influenced by its parameters. Thus it was picked as the most appropriate initial model available. It was moreover chosen since it includes a larger number of parameters with respect to the other 3 models. We modified the crustal thickness of this model from 40 to 30 km, as the solutions using all 4 initial models for the average dispersion curve converged to that depth.

The number of parameters of the initial model is also a critical restriction of surface wave inversion process. Those should be relevant with the amount of the experimental data and their corresponding errors distribution. A small number of initial parameters could lead to underestimated solutions in the sense of a potentially more sufficient data density, while a large number of parameters to limited resolution of the obtained solutions (non uniqueness of the solutions). To estimate the appropriate number of the initial model parameters that our dataset could adequately resolve, we modified the S-wave velocity model proposed by Papazachos et al., (1995), following a manual iterative pro-

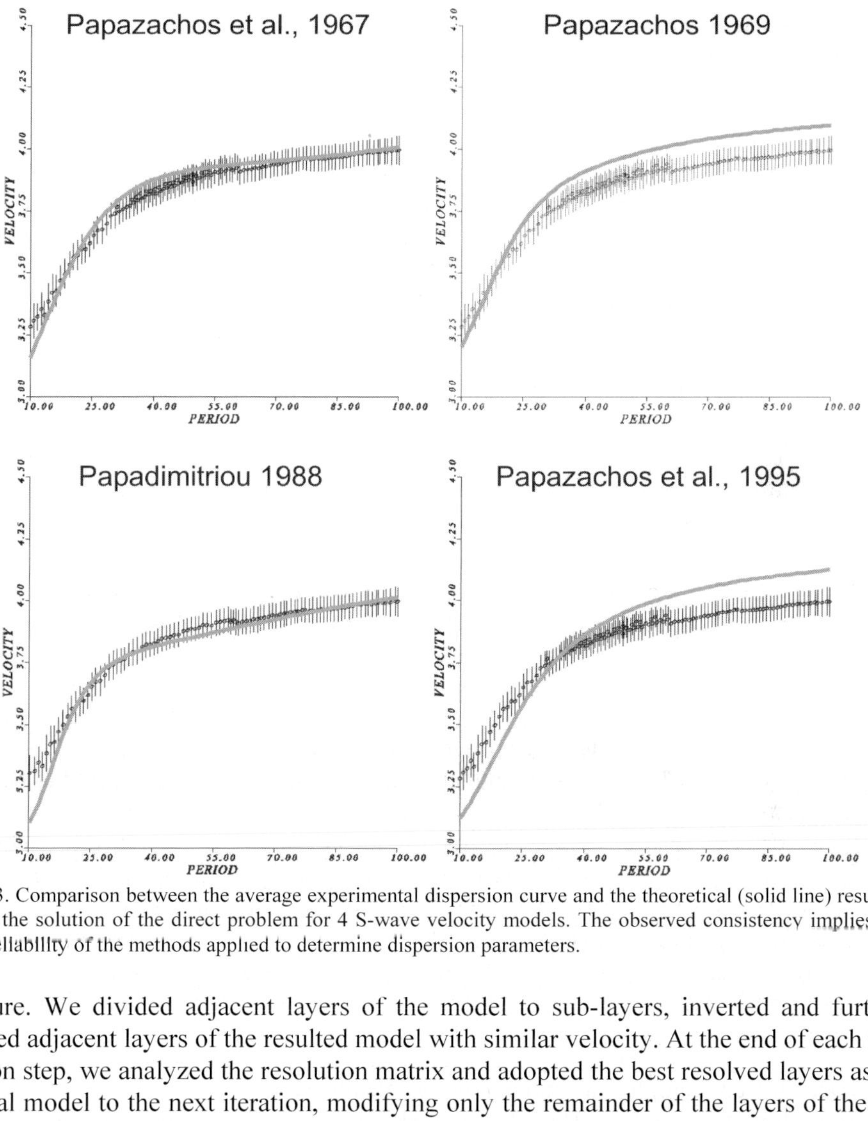

Fig. 3. Comparison between the average experimental dispersion curve and the theoretical (solid line) resulted from the solution of the direct problem for 4 S-wave velocity models. The observed consistency implies for the reliability of the methods applied to determine dispersion parameters.

cedure. We divided adjacent layers of the model to sub-layers, inverted and further joined adjacent layers of the resulted model with similar velocity. At the end of each iteration step, we analyzed the resolution matrix and adopted the best resolved layers as an initial model to the next iteration, modifying only the remainder of the layers of the resulted model. In this way, the most appropriate number of parameters was defined to 12 homogeneous horizontal layers overlying a half-space at 180 km depth. The depth of 180 km was derived from the analysis of the resolution matrix and the partial derivatives matrix or sensitivity matrix (Kassaras, 2002). The concluded initial model yielded a relatively smooth S-wave velocity variation with depth and an adequate resolution for the resolved layers. As expected, two thin layers included in the initial model as representative of the average relief of the studied area (a water and a sediment layer of 0.3 km and 0.7 km thickness respectively), did not influence the inversion results.

Table 1. The parameters of the average S-wave velocity model (MEAN-MOD) and its error estimates.

Thickness (km)	Vs (km/sec)	Error estimates
9.6	3.39	
9.9	3.54	
10.0	3.78	
13.4	4.05	Velocity Dispersion: Mean = 0.000165
13.5	4.28	Std. Err. = 0.065
14.2	4.40	Average SUM = 0.00473
14.2	4.42	Weighted RMS(velocity) = 0.06
14.9	4.42	Model RMS = 0.065
15.1	4.40	Number of observations = 96
20.0	4.35	
20.2	4.39	
20.5	4.49	
∞	4.51	

To estimate S-wave velocities with depth for the Aegean region we inverted the dispersion curves (we adopted an error of 2σ to phase velocities) of each individual profile using 3 initial models. The modified model of Papazachos et al., (1995); the average model of the area (MEAN-MOD) resulted from the inversion of the average dispersion curve (Table 1); a coarse model of 5 layers resulted from incorporating adjacent layers of similar velocity to thicker layers. All models included a half space interface at 180 km depth. The initial crustal thickness information was taken from the tomographic images of Karagianni et al., (2002).

We reduced contributions from noisy model vectors by choosing values of σ, appropriate for the noise level in the data, considering the knee of the empiric trade-off curve between the remaining residual variance after inversion and the squared model length. By this method the rigid minimum norm criterion is replaced with a more flexible one that minimizes a combination of error *(E)* and solution norm (*N*).

$$T(\sigma) = \mathrm{E} + \sigma^2 \mathrm{N} \qquad (8)$$

Where σ is the damping factor, or the trade-off parameter.

For each profile, the final 1D velocity model was derived from the average of the 3 individual solutions. Furthermore, we reinverted the dispersion curve for each profile using the respective final velocity model as initial. Through the output resolving kernels (Kassaras, 2002) we defined that for the depth range 20-180 km the resolution is efficient for more than 80% of the profiles, ≈60% for the depth range 10-20 km, while the topmost layer (0-10 km) is not resolved by the data (Table 2).

Table 2. Resolution (% of all profiles) and error estimates for each layer.

Layer	Resolution (% of 35 profiles)	S-wave velocity error (km/sec)
1	25	±0.08
2	60	±0.06
3	90	±0.06
4	100	±0.06
5	100	±0.06
6	85	±0.06
7	100	±0.04
8	95	±0.04
9	90	±0.05
10	100	±0.03
11	90	±0.03
12	80	±0.03

4. RESULTS

4.1. Average phase velocities distribution pattern

The average dispersion curve demonstrates low velocities in the Aegean, compared to published continental lithosphere dispersion models (Fig. 4). The greatest discrepancy occurs as expected for cratonic environments, such as North Africa (Hazler et al., 2001) and Canada (Brune and Dorman, 1963) at periods >30 s, since those regions reveal significantly faster lithosphere compared to tectonic environments (Toksöz and Anderson, 1966; Kanamori, 1970). Better agreement exists with models derived from Tectonic Europe (Badal et al., 1992; Stange and Friederich, 1993). Furthermore, the Aegean shows lower velocities compared to those models, in agreement with tomographic results for the uppermost mantle (Hearn, 1999) who found an important velocity contrast between Tectonic Europe (low velocities) and the fast sub-African lithosphere of the Adriatic Sea. Moreover, they reported extremely low mantle velocities for the Apenninic, Dinaric and Hellenic arcs and Tyrrhenian, Hellenic back arcs, namely the most active features within Tectonic Europe.

Fig. 5 shows the average dispersion curves over continental and offshore profiles. Offshore profiles present systematically higher phase velocities over the entire period range. This pattern is doubtful due to the composition of the shallow layers or the relief of the region, while it is not restricted at lower periods. The significantly thick crust in continental Greece inferred from published information, may affect the measurements at periods up to 20-30 s, however the extend of the contrast over the entire period range is liable to continuous discrepancy in depth, probably linked to different composition and evolution of the lithosphere and upper mantle of the two environments.

4.2. Phase velocity perturbations

Lateral surface wave variations reflect heterogeneities of the elastic medium. In order to investigate differences in velocity, however small they are (as long as they are greater than the order of the statistical experimental error), we proceeded to a mapping of experimental phase velocities perturbations with respect to the average dispersion curve ($dc/c_0(T)\%$), for periods ('depths') 20, 30, 40, 50, 60, 70, 80, 90 s. As a rough

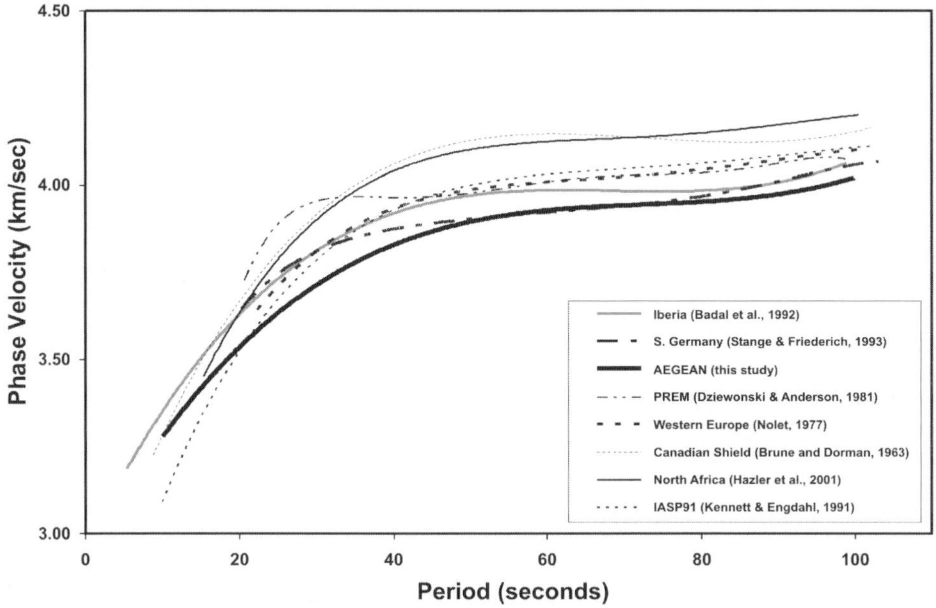

Fig. 4. Comparison of the average phase velocity dispersion curve in the Aegean region with reference dispersion curves of fundamental Rayleigh waves inferred from various regions.

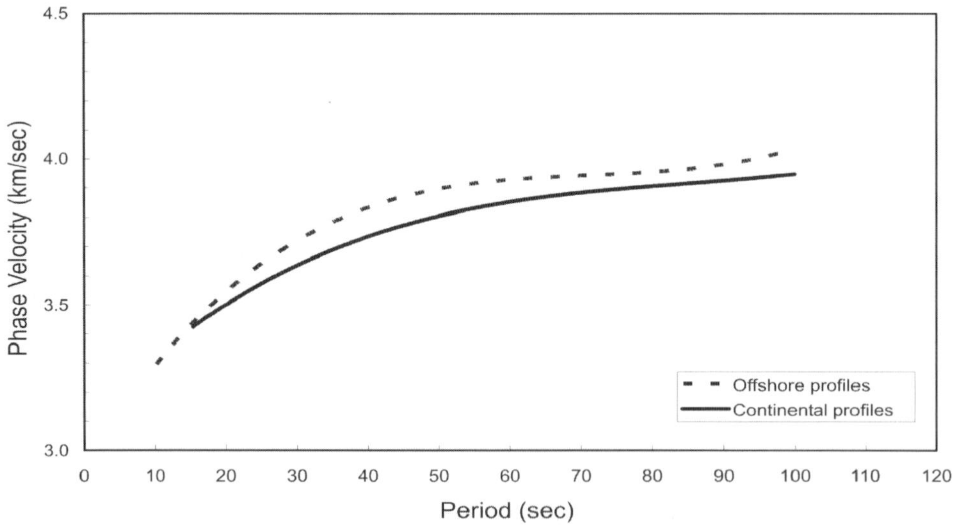

Fig. 5. Comparison between the average dispersion curves over continental and offshore profiles.

approximation inferred from the analysis of sensitivity kernels of phase velocities with respect to the shear velocities, phase velocities at 20, 30, 40, 50, 60, 70, 80, 90 s, sample depths at 40, 70, 100, 120, 140, 160, 180, 200 km respectively. At 100 s, apparently large phase velocity discrepancies introduced high biases to the perturbation values with respect to adjacent periods; hence the 100 s map is omitted. In south Aegean, no phase velocity measurements are available for periods >60 s, due to the response of the sensors installed there, except for three GEOFON stations, which however do not adequately sample the area. Consequently this part of the Aegean has been clipped from the perturbation images of 70, 80 and 90 s.

To construct phase velocity perturbation maps, for each one of the periods investigated, we: a) assigned a perturbation value of each profile to several points along the profile (sampling rate 0.1sample/km), in order to obtain a better sampling, b) applied to the whole region a rectangular grid with nodes of 80 km dimensions, c) for each non-blank node we calculated its center coordinates *(x.y)* and the average perturbation value *z*. Lastly, we interpolated the perturbation values, in order to enlarge the volume of the dataset and be able to draw smooth perturbation contours. The latter was done by solving the equation (Wessel and Smith, 1995):

$$(1-T)*L((L(z)+T*L(z))=0 \tag{9}$$

where T is a tension parameter and L is the Laplacian operator.

In order to smooth the phase velocity function we considered resolution analysis from Bourova et al., (2003) and set the mesh dimensions equal to 80 km (scalar increment of the averaging nodes dimensions up to 120 km length, which is larger than the averaging length estimated by Bourova et al., hardly affects the perturbation images shown in Fig. 6). Apparently, by applying the above restrictions, the mostly preferred E-W and NW-SE orientations of the profiles do not affect the shape of the velocity anomalies contours and moreover a reasonable sensitivity of velocity anomalies variations is not lost.

The range of velocity perturbations is restricted within -3%-6%, consistently with surface wave studies of regional (Levshin et al., 1999) and global scale (Laske and Masters, 1996). Those imply that phase velocity anomalies do not exceed ±7%, with their peak amplitude pertaining to subduction zones. It is moreover in good agreement with results deduced from body wave tomography in the Aegean area (Papazachos and Nolet 1997). The pattern of velocity anomalies does not significantly differ as we traverse the elastic medium, revealing a smooth in depth variation. Moreover, the amplitude of the observed anomalies reduces as we pass from low to large periods, implying for a definitely realistic pattern of larger heterogeneity existing in the upper layers of the lithosphere.

One could possibly assume that phase velocity anomalies are an artifact, as the order of perturbation values might in cases be greater than the experimental error. Fig. 7 shows phase velocity perturbations maps for the periods 20, 40, 60 and 70 s, constructed using only perturbation data greater than an uncertainty of 2σ. Undoubtedly they do not considerably differ from the respective maps of Fig. 6. The latter allows us to adapt the

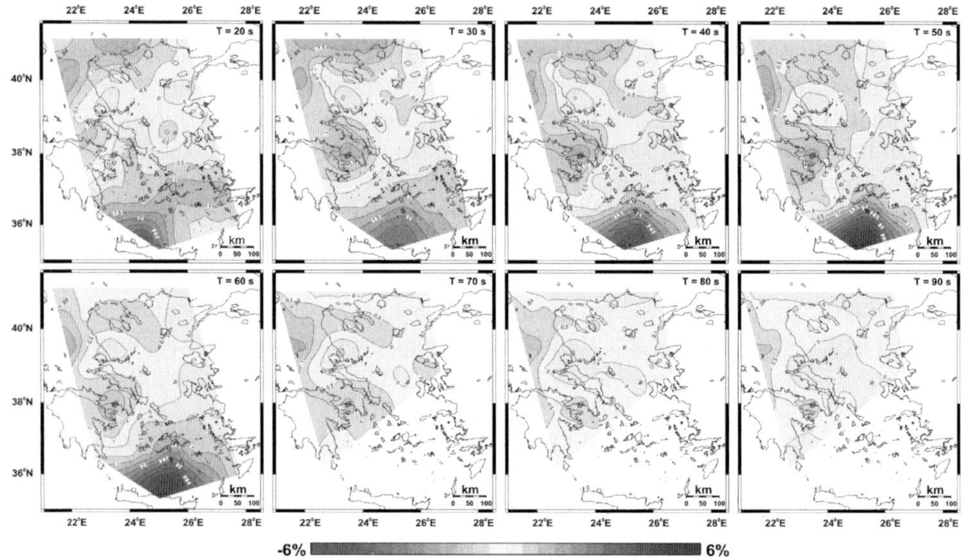

Fig. 6. Rayleigh wave phase velocity perturbations for periods 20, 30, 40, 50, 60, 70, 80, 90 s.

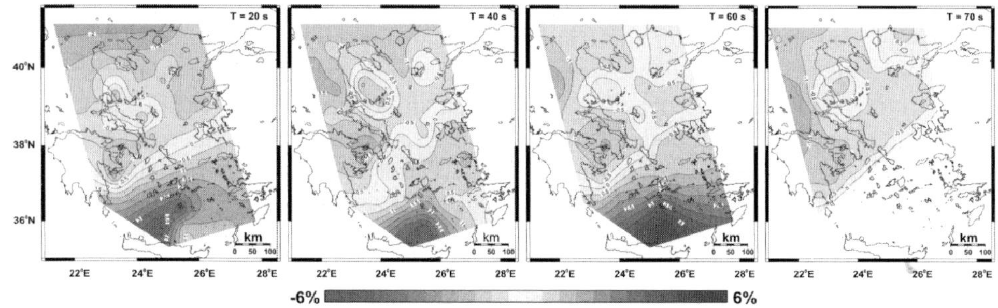

Fig. 7. Rayleigh wave phase velocity perturbations for periods 20, 40, 60, 70 s, constructed using only profiles with perturbations beyond a statistical error of 2σ (confidence level $\approx$95%).

following most obvious characteristics of the velocity perturbation images to a valid lateral heterogeneity of the elastic medium:

1. Highest perturbations of maximum amplitude +6% are observed in southernmost Aegean. Those are reasonably attributed to the northward dipping African subducted lithosphere.
2. Low velocities predominate in continental Greece.
3. North Aegean is characterized by low velocities.
4. A high velocity contrast is observed at the westernmost tip of North Aegean Trough.

Phase velocity anomalies agree quite well with P and S-wave anomalies (Papazachos and Nolet, 1997) down to a depth of 50 km. They further agree with P wave anomalies around the depth of 100 km, however they do not agree with the S-wave perturbations deeper than 50 km. Although local scale traveltime inversion provides enhanced information on absolute velocities, the resolution reduces with depth due to the limited intermediate depth foci available. Especially in the area of north Aegean, which is characterized by the lack of intermediate depth seismicity, the resolution of P and S-wave tomography results are expected to be of limited resolution within the lower lithosphere and upper mantle.

4.3. Comparison of phase velocity anomalies with structural geology, geodesy and seismicity

Successive orogenic episodes since the Eocene followed by rotations since the Lower Cenozoic, clockwise for the western and counterclockwise for the eastern Aegean (Kissel and Laj, 1988) result in the formation of major structures oriented NNW-SSE and ENE-WSW in western and eastern Aegean respectively. Moreover, the successive collisions resulted a shortening of several hundreds of kilometers and therefore likely involve not only the crust but also the upper mantle (Aubouin et al., 1976). A possible relation of phase velocity anomalies with structural and deformation characteristics of the Aegean should reasonably imply that kinematics deformation is not restricted within the upper brittle crust, but has deeper roots located at least in the upper mantle. Therefore, results from this study should be of great importance particularly for the northern Aegean, where due to the lack of deep seismicity, the available geophysical information is only adequately sampling the shallower layers of the lithosphere.

In Fig. 8, the phase velocity perturbation image of 20 s is compared with deformation characteristics, such as main active faults, geological units, measured

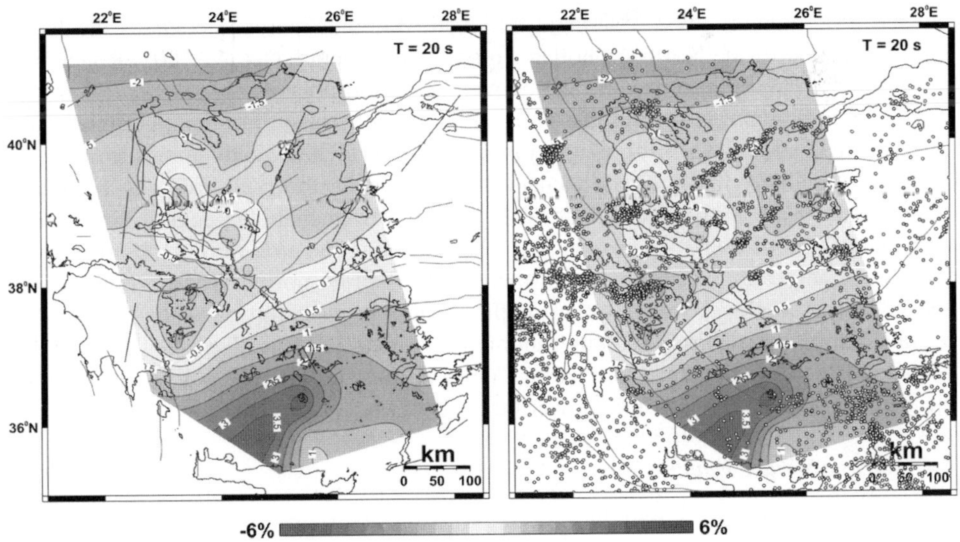

Fig. 8. Left: map of phase velocity perturbations at 20s period superimposed on the active faults of the Aegean (red lines) and strain (black bars) deduced from GPS measurements (McClusky et al., 2000; Hatzfeld et al., 2001). Right: map of phase velocity perturbations at 20s period superimposed on the geological map of the Aegean (Robertson and Dixon, 1984) and earthquake epicenters with M>4 (Makropoulos et al., 1989; ISC, 2001).

horizontal strain and seismic energy distribution. The comparison could be extended to greater periods 'depths', as the vertical variation of velocity anomalies is not significant. Since the averaging dimensions we adopted to construct phase velocity anomalies maps (by decreasing the averaging dimensions we gain a lot to resolution, however we might be lead to artifacts due to the preferred orientations of the profiles) does not allow for detailed investigation, we only discuss below the most evident features.

A high velocity zone in the area of the western tip of North Aegean Trough (WTNAT) correlates with the transition boundary of the major geological units and tectonic features direction from ENE-WSW to NNW-SSE towards the west. Furthermore, the direction of measured horizontal displacement (McClusky et al., 2000; Hatzfeld et al., 2001), changes from NE-SW to almost N-S in the same area. Apparently, the NNW-SSE horizontal strain orientation is related to low velocities, while the almost N-S orientation with high velocities, or with the transition from low to high velocities throughout the Aegean.

North Aegean is characterized by high shallow seismic energy release, implying high rates of brittle extensional deformation. Epicenters are delineated along the branches of North Anatolian Fault in the Aegean in an ENE-WSW direction. This pattern of seismicity correlates with an extended zone of low phase velocities between 38^0-40^0N, which is further correlated with observed high thermal flow throughout the area. The high velocities in the WTNAT area coincide with an abrupt interruption of the linear distribution of earthquakes. Towards the west, the deformation regime changes from dextral strike-slip to E-W normal faulting in PG area (Papadimitriou et al., 1994). On the other hand, the focal mechanism of the 26 July 2001 M=6.5 Skiros earthquake, showed a surprising NW-SE sinistral strike-slip faulting, perpendicular to the ENE-WSW direction of the North Anatolian Trough (Papadopoulos et al., 2002; Roumelioti et al., 2003). Taking into account the above, one could presumably speculate that this high velocity zone, also present to P and S-wave traveltime tomographic maps (Papazachos and Nolet, 1997), is of an important role for the geodynamics of northern Aegean. It is likely that the extended in depth anomaly observed provides a short of explanation to the traditional debate whether the NAF penetrates central Greece and reaches the area of CTF, or terminates at a point around the area were the high velocity anomaly is inferred. We propose that this zone, which is not an artifact, is likely the barrier to the southwestward motion of the Anatolian plate through the Aegean. The overall pattern of phase velocity anomalies, since associated with the complex geotectonic setting deduced from surface observations, implies for a deep deformation mechanism, involving at least the lower lithosphere and upper mantle.

4.4. Average S-wave velocity

The average over 35 individual 1D S-wave velocity profiles (MEAN-MOD) is remarkably consistent with the S-wave velocity model derived by the inversion of the average dispersion curve of the area using as a starting model the one of Papazachos et al., (1995), (Fig. 9). Moreover, the average S-wave velocity when compared to the starting model is systematically lower below the depth of 30 km, consistently with the

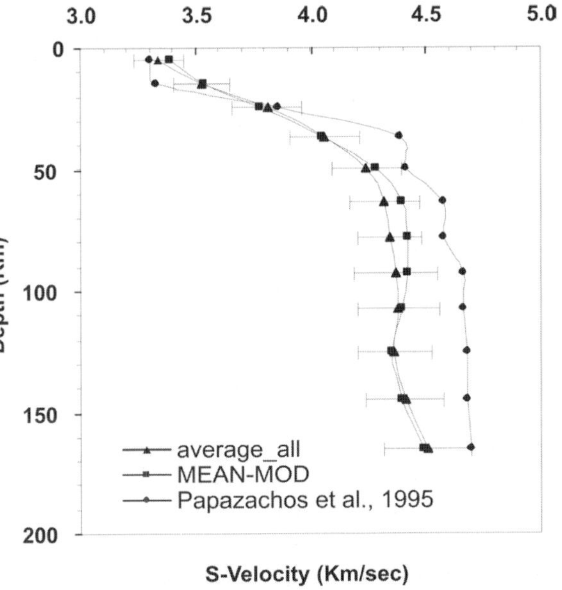

Fig. 9. The average absolute S-wave velocity model over the individual solutions is in good agreement with the model resulted from the inversion of the average dispersion curve. Compared with the starting model of Papazachos et al. (1995) discrepancy is observed for depths >30 km, which is greater than the error bars.

average phase velocities distribution pattern. This discrepancy, which is bigger than the error bars, could be reasonably attributed to several reasons (or some combination), that we will attempt to approach.

A first reason could be that surface wave dispersion inversion provides S-wave velocities of limited resolution when compared to local body wave traveltime inversion. A second reason is that the starting model of Papazachos et al., (1995), does not sample exactly the same area as our data. A third reason could be that the starting model deduced mainly from P-wave information. Subsequently, their S-wave velocity model is partly influenced from this information. Discrepancies between P and S-wave velocity patterns could likely be explained by a high pore fluid pressure related to high heat flow within the Aegean lithosphere. Spencer and Nur (1976), experimentally showed that under the same temperature and pressure conditions, but with a high pore fluid pressure (>1 kbar), a decrease of shear modulus μ and consequently S-wave velocity occurred, while P-wave velocity remained relatively unchanged. The fourth reason could be the wavelength of the seismic waves used in both studies (the starting model derived from 1-2 sec body waves). The currently unknown anelastic properties of the Aegean upper mantle, may likely not affect short period body waves, but may strongly impact the long period surface wave propagation, producing serious traveltime delays (Liu et al., 1976; Kanamori and Anderson, 1977). A fifth reason could be the intrinsic constitution of the Aegean upper mantle. Two types of rocks are typically common in the upper mantle of the Earth, peridotite and eclogite. Both rock types have similar P wave velocities, but quite different Poisson ratios. Experimental Poisson ratios extracted for peridotite and eclogite are 0.24-0.27 and >0.30 respectively (Badal et al., 1992). By arbitrarily assuming that eclogite participates equally as peridotite within the upper mantle, the S-wave velocities (damped by P-wave velocities using a typical Poisson ratio of 0.25) should increase by an order of 5%-10%. By such an increase, the average 1D S-wave velocity models of both studies appear similar.

4.5. Uppermost mantle structure

In Fig. 10 we map the crustal thickness lateral variations, inferred from the distribution of individual 1D S-wave velocity profiles. Following Karagianni et al., (2002), for each profile we assumed a minimum value of 4.0±σ km/sec, as the transition from lower crustal to mantle velocities. In a few cases we relaxed the above criterion and adopted as the depth of the Moho interface, the depth of abrupt increase of S-wave velocity. Given that lower crustal and uppermost mantle velocities appear an average deviation of 2σ=±0.12km/sec, the average error for the Moho depth estimation should range between ±2km for the shallower depths and ±1.5km for the larger depths. Fig. 10 is in remarkable agreement with the 20 s phase velocity map (Fig. 6), implying that low period phase velocities are strongly influenced by crustal thickness variations.

Consistently with recent studies (Kalogeras and Burton, 1996; Chailas et al., 1993; Papazachos 1994; Tsokas και Hansen 1997; Makris et al., 1998; Karagianni et al., 2002), the crust does not appear thicker than 30 km throughout the Aegean. Thin crust is observed in the area of Cyclades islands (22-25 km) and along the North Aegean Trough (NAT) (25-28 km). Thicker crust is observed in the area of southernmost Aegean near Crete (≈30 km) and between N. Aegean and Skiros Troughs (≈30 km). In continental Greece the crust appears thicker (35-40 km).

4.6. Shear wave velocity distribution

As the maximum wavelength of our dataset (450 km) and the resolution analysis indicate, the S-wave velocity distribution is resolved down to a depth of 180 km. To distinguish S-wave velocity contrasts that could be attributed to various structures within the Aegean upper mantle, we constructed four cross-sections along the Aegean (Fig. 11). To construct cross-sections we adapted a 1D S-wave velocity model at several points along the 35 profiles and calculated the average, by applying a 3D mesh with 80x80x15 km node dimensions for the x, y, and z axes respectively. The dimensions of the structures distinguished in the cross-sections should vary between 80 and 200 km. As the limited resolution of our dataset at lower periods (<20 s) does not allow for a detailed crustal structure investigation, we focus only at differences in the mantle structure. Evident is that resolution is better in the N-S than in the E-W direction, due to the preferred orientation of the resolved profiles.

Several S-wave velocity anomalies can be seen at the cross sections. The distribution of high velocity anomalies is in most cases likely related to the subducted slab (cross sections A_1A_2, C_1C_2 and D_1D_2). Moreover, it is consistent with relocated

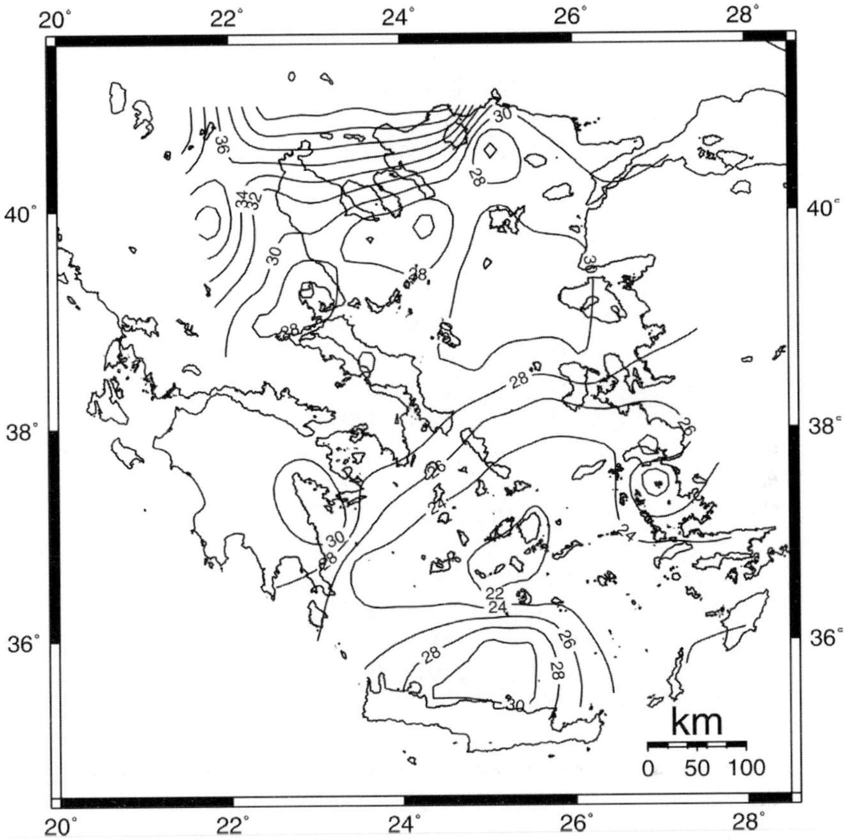

Fig. 10. Map of lateral crustal thickness variations in the Aegean area.

intermediate depth seismicity (Engdahl et al., 1998). Both seismicity and S-wave velocity distribution imply for a non symmetrical geometry for its western and eastern segments. The western segment (cross sections A_1A_2 and C_1C_2) appears larger and shallow dipping ($\approx 25^0$) compared to the smaller and steeper dipping eastern segment ($\approx 35^0$). The exhibited pattern of a not completely amphitheatric geometry of the subducted slab is coherent with Papazachos and Nolet (1997) and may well be interpreted by the combined effect of slab pull of the African lithosphere and the fast spreading of the Aegean (Hatzfeld, 1994). This combined effect is further supported by assuming that the $\approx 25^0$N dipping high velocity zone observed in section D_1D_2 beneath CI possibly corresponds to a detached segment of the African lithosphere.

In cross section B_1B_2, a high velocity anomaly could be associated with a slab beneath north Aegean observed by Spakman et al., (1993), down to a depth of 600 km, possibly related with Pleistocene volcanism observed in the region of PG and NEI (Pe-Piper and Piper, 2002). Two high velocity anomalies observed in cross section B_1B_2, are related with NAT and the western termination of NAF in the area of NSI. This

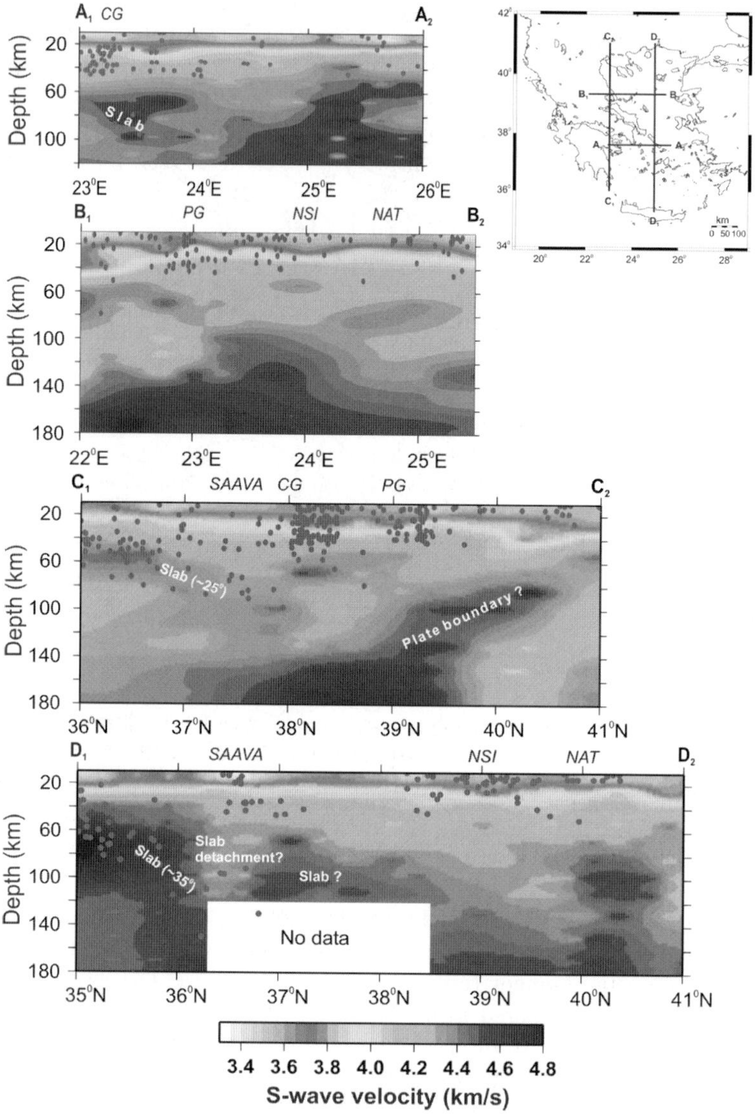

Fig. 11. Absolute S-wave velocity cross sections along four profiles shown on the map at the top right corner. At each cross section we superimpose relocated seismicity (Engdahl et al., 1998). For geographic names discussed in text see Fig. 1.

configuration is in good agreement with GPS constraints (McClusky et al., 2003), who describe the area about NSI as the northwestern boundary of the Anatolian Plate.

A remarkable feature is observed in section C_1C_2, consistently with Papazachos and Nolet (1997). A high velocity zone dipping with an angle of about 25^0S is located beneath northern Greece. This zone is characterized by high S-wave velocity (4.6-4.8 km/s), comparable with velocities observed for the eastern segment of the Aegean subduction. At a depth >100 km, the Aegean subducted slab intersects with this anomaly beneath ~38^{0}N. This configuration is related with a synthetic velocity model based on the conversion of an initial thermal structure (De Jonge et al., 1994) and the tomographic results of Spakman et al., (1993). We could possibly interpret this anomaly as a plate boundary, consistently with McClusky et al., (2003), who describe this area as the southernmost boundary of the stable Eurasian plate. However, the lack of intermediate earthquake foci, or other additional geophysical information constitutes any attempt to interpret this anomaly rather ambiguous. Certainly, additional information is required to resolve this ambiguity.

South of NAT, low velocities predominate in central and north Aegean. This pattern, furthermore associated with back arc extension and crustal thinning, could be interpreted by the presence of high thermal flow in the upper mantle, possibly due to a combination of slab dehydration and upper mantle depletion. All above are consistent with the observed spatial distribution of high velocity anomalies, which likely demonstrates the present subduction process in the Aegean. In south Aegean, low velocities above (slab detachment area?) and beneath the subducted slab are related with the SAAVA, implying for both dehydration process of the subducted slab and mantle upwelling (thermal flow from the asthenosphere).

5. CONCLUSIONS

The results presented in this study reveal important characteristics of the S-wave velocity structure of the Aegean lithosphere and upper mantle. Consistently with Papazachos and Nolet (1997), the subducted slab exhibits a strong velocity contrast in the area of south Aegean (up to 6%) with the background model. Its geometry is not completely amphitheatrical, as the larger shallow dipping western segment and the smaller and steeper eastern one indicate. The dip of the slab varies from $\approx 25^0$ for its western part to $\approx 35^0$ to its eastern part. The back arc low velocity anomaly associated with both mantle upwelling and slab dehydration, is well documented beneath the south Aegean active volcanic arc.

In north and central Aegean, the distribution pattern of velocity anomalies appear quite complex, however it could deduce reasonable explanation for several observed peculiarities related to structural geology, strain rates and magmatism.

1. A high velocity anomaly beneath central Greece and north Aegean could likely be related with a subduction zone observed by in this area by Spakman et al., (1993), as well as with recent volcanism reported in the area of central Macedonia and Thessaly (Pe-Piper and Piper, 2002).
2. A high velocity anomaly beneath NAT intersects in depth with the above zone, beneath the area of NSI. This pattern could be associated with the transition of the main geotectonic features direction from ENE-WSW to NNW-SSE toward the west. The intersection region almost coincides with the western termination of the North Anatolian Fault (NAF system) and the northwestern boundary of the Anatolian Plate proposed by McClusky et al. (2003.).

3. A high velocity anomaly beneath northern and central Greece could be interpreted as the southernmost boundary of the stable Eurasian lithosphere, consistently with McClusky et al. (2003).
4. Low velocities in north and central Aegean could be attributed to high thermal flow in the upper mantle, possibly due to a combination of slab dehydration and thermal flow from the asthenosphere possibly through detachement areas of the African slab.

The above described high velocity anomalies distribution almost coincides with the boundaries between Eurasian, Anatolian and Aegean Plates proposed by McClusky et al. (2003). A slight difference between the results of both studies concerns the segment of the northern boundary of the Aegean Plate between NAT and CG. Our results show that this boundary should likely be located somewhat southern, consistently with the recent 7/9/1999 Athens earthquake, which showed that CG is propagating further east from the Alkyonides region, (Papadimitriou et al., 2002) and the surprising NW-SE sinistral focal mechanism of the recent Skiros earthquake of 26/7/2001.

Even though several features of the Aegean upper mantle have been adequately resolved, our dataset does not include information for the whole Aegean region, as it does not sample western and northernmost Greece. Future incorporation of additional data from those areas should greatly contribute to enlighten ambiguously interpreted structures of great importance, due to the lack of sufficient supporting geophysical information.

Concluding, despite the contribution of the revealed features in this study (ambiguous or not) in connecting the present deep seismic structure with the tectonic evolution of the Aegean, debate as to whether plate/block or continuum descriptions (or some combination) explains better the shallow deformation in the area, still remains.

Acknowledgments

This work was implemented in the framework of the first author's PhD thesis elaboration (December 2002). The "SEISFAULTGREECE" project was supported by contract ENV4-CT96-0277 of the EU. We thank B. Papazachos, H. Lyon-Caen and K. Priestley, lead participants of the project and all our colleagues who participated in the field experiment and data manipulation. We thank GFZ, Potsdam, Germany, for the disposal of the GEOFON data from SANT, KRIS and SKD stations. Lastly, we are grateful to D. Baumont for providing the codes for surface wave dispersion analysis.

REFERENCES

Aubouin, J., Bonneau, M., Davidson, J., Leboulanger, P., Matesco, S. and Zambetakis, A., 1976. Esquisse structurale de l'arc Egeen extreme: Des Dinarides aux Taurides. Bull. Soc. Geol. Fr., 7, 385-401.

Badal, J., Corchete, V., Payo, G., Canas, J.A., Seron, F.J. and Pujades, L., 1992. Deep structure of the Iberian Peninsula determined by Rayleigh wave velocity inversion. Geophys. J. Int. 198, 71-88.

Baumont, D., Paul, A., Beck, S., Zandt, G. and Pedersen, H., 2002. Lithospheric structure of the central Andes based on surface waves dispersion. Geophys. J. Int., 107, B12, 103-115.

Bourova, E., Kassaras, I., Pedersen, H., Yanovskaya, T., Hatzfeld, D. and Kiratzi, A., 2003. Surface-wave tomography in the Aegean region, submitted in Geophys. J. Int.

Brune, J.N. and Dorman, J., 1963. Seismic waves and Earth structure in the Canadian Shield. Bull. Seism. Soc. Am., 53, 167-210.

Chailas, S., Hipkin, R.G. and Lagios, E., 1993. Isostatic studies in the Hellenides. Proc. of the 2nd Congress of the Hellenic Geophysical Union, Florina, 5-7 of May, Greece, Vol.2, 492-504.

Christodoulou, A. and Hatzfeld, D., 1988. Three-dimensional crustal and upper mantle structure beneath Chalkidiki (Northern Greece). Earth Planet. Sci. Lett., 88, 153-168.

De Jonge, M.R., Wortel, M.J.R. and Spakman, W., 1994. Regional scale tectonic evolution and the seismic velocity structure of the lithosphere and upper mantle: The Meditarranean region. J. Geophys. Res., Vol. 99, No B6, 12091-12108.

Drakatos, G. and Drakopoulos, J., 1991. 3-D velocity structure beneath the crust and upper mantle of the Aegean Sea region. PAGEOPH, 135, 401-420.

Dziewonski, A.M. and Hales, A.L., 1972. Numerical analysis of dispersed seismic waves, In: B.A. Bolt (Editor), Seismology: Surface Waves and Earth Oscillations (Methods in Computational Physics), Vol.11, New York, Academic Press.

Engdahl, E.R., Van der Hilst, R.D. and Buland, R.P., 1998. Global teleseismic earthquake relocation with improved traveltimes and procedures for depth determination. Bull. Seism. Soc. Am., Vol. 98, 722-743.

Ezen, U., 1983. An interference phenomenon in Rayleigh wave trains associated with the earthquakes in and around the Aegean sea. Bull. Int. Inst. of Seism. Earthq. Eng., 20, 33-62.

Ezen, U., 1991. Crustal structure of western Turkey from Rayleigh wave dispersion. Bull. Int. Inst. of Seism. Earthq. Eng., 25, 1-21.

Franklin, J.N., 1970. Well-posed stochastic extension of ill-posed linear problems. J. Math. Anal. Appl., 31, 682-716.

Granet, M. and Trampert, J., 1989. Large scale P-velocity structure in the Euro-Mediterranean area. Geophys. J. Int., 99, 583-594.

Hatzfeld, D., 1994. On the shape of the subducted slab beneath Peloponese, Greece. Geophys. Res. Lett., 21, 3, 173-176.

Hatzfeld D., Karagianni, E., Kassaras, I., Kiratzi, A., Louvari, E., Lyon-Caen, H., Makropoulos, K., Papadimitriou, P., Bock, G. and Priestley, K., 2001. Shear wave anisotropy in the upper mantle beneath the Aegean related to internal deformation. J. Geophys. Res., 106, No. 12, 30737-30753.

Hazler, S., Pasyanos, M., Walter, W.R. and Sheehan, A., 2001. Two-station phase velocity determination for structure in North Africa. Proc. of the 21st Seismic

Research Symposium on Technologies for Monitoring the Comprehensive Nuclear-Test-Ban Treaty, September 21-24, Las Vegas, Nevada, USA, pp. 102-110.

Hearn, T.M., 1999. Uppermost mantle velocities and anisotropy beneath Europe, J. Geophys. Res., 104, B7, 15123-15139.

Herrmann, R.B., 1973. Some aspects of band-pass filtering of surface waves. Bull. Seism. Soc. Am., 62, 129-139.

Herrmann, R.B., 1994. "Computer programs in Seismology", User's manual. St. Louis University, Missouri.

International Seismological Centre, 2001. On-line Bulletin. http://www.isc.ac.uk/Bull, Internat. Seis. Cent., Thatcham, United Kingdom.

Kalogeras, I.S. and Burton, P.W., 1996. Shear-wave velocity model from Rayleigh wave dispersion in the broader Aegean aerea. Geophys. J. Int., 125, 679-695.

Kanamori, H., 1970. Velocity and Q of mantle waves. Phys. Earth Planet. Int., 2, 259-275.

Kanamori, H. and Anderson, D.L., 1977. Importance of physical dispersion in surface wave and free oscillation problems: Review. Rev. Geophys. Space Phys., 15, 105-112.

Karagianni, E.E., Panagiotopoulos, D.G., Panza, G.F., Suhadolc, P., Papazachos, C.B., Papazachos, B.C., Kiratzi, A., Hatzfeld, D., Makropoulos, K., Priestley, K. and Vuan, A., 2002. Rayleigh wave group velocity tomography in the Aegean area. Tectonophysics, 358, 187-209.

Kassaras, I., 2002. Upper mantle structure of the Aegean region derived from the dispersion of seismic surface waves. PhD Thesis, University of Athens, Greece.

Kennett, B.L.N. and Engdahl, E.R., 1991. Traveltimes for earthquake location and phase identification. Geophys. J. Int., 105, 429-465.

Kissel, C. and Laj, C., 1988. The tertiary geodynamical evolution of the Aegean arc: A paleomagnetic reconstruction. Tectonophysics, 146, 183-201.

Lana, X., Fernandez Mills, G., Badal, J., Canas, J.A., 1997. Objective regionalization of Rayleigh-wave dispersion data by clustering algorithms. Geophys. J. Int., 129, 421-438.

Laske, G. and Masters, G., 1996. Constraints on global phase velocity maps from long-period polarization data. J. Geophys. Res., 101, 16059-16075.

Le Pichon, X. and Angelier, J., 1979. The Hellenic Arc and trench system: a key to the neotectonic evolution of the Eastern Mediterranean area. Tectonophysics, 60, 1-42.

Le Pichon, X., Chamot-Rooke, N., Lallemant, S., Noomen, R., and Veis, G., 1995. Geodetic determination of the kinematics of Central Greece with respect to Europe: Implications for Eastern Mediterranean Tectonics. J. Geophys. Res., 100, 12675-12690.

Levshin, A., Ritzwoller, M and Ratnikova, L., 1994. The nature and cause of polarization anomalies of surface waves crossing Northern and Central Eurasia. Geophys. J. Int., 117, 577-591.

Levshin, A., 1998. Surface Wave Analysis and Tomography. Lecture presented in the

4th Workshop on Three Dimensional Modeling of Seismic Wave Generation Propagation and their Inversion, 28 Sep.-9 Oct. 1998, Trieste, Italy.

Levshin, A., Ritzwoller, M. and Resovsky, J., 1999. Source effects on surface waves group travel times and group velocity maps. Phys. Earth Planet. Int., 15, 293.312.

Ligdas, C.N. and Lees, J.M., 1993. Seismic velocity constraints in the Thessaloniki and Chalkidiki areas (northern Greece) from a 3-D tomographic study. Tectonophysics, 228, 97-121.

Ligdas, C.N., Main, I.G. and Adams, R.D., 1990. 3-D structure of the lithosphere in the Aegean Sea region. Geophys. J. Int., 102, 219-229.

Liu, H.P., Anderson, D.L. and Kanamori, H., 1976. Velocity dispersion due to anelasticity : implications for seismology and mantle composition. Geophys. J. R. Astr. Soc., 47, 41-58.

Makris, J., Harje, P., Papanikolaou, D. and Stavrakakis, G., 1998. The island of Crete and on shore seismic experiments. Paper presented in the "First workshop I.E.D.P.", Chanea, Greece, October 15-19, 1998.

Makropoulos, K.C., Drakopoulos, J.K. and Latoussakis, J.B., 1989. A revised and extended earthquake catalogue for Greece since 1900. Geophys. J. Int., 98, 91-394.

McClusky, S., Reilinger, R., Mahmoud, S., Ben Sari, D. and Tealeb, A., 2003. GPS constraints on Africa (Nubia) and Arabia Plate motions. Geophys. J. Int., 155, 126-138.

McClusky, S., Balassanian, S., Barka, A., Demir, C., Ergintav, S., Georgiev, I., Gurkam, O., Hamburger, M., Hurst, K., Kahle, H., Kastens, K., Kekelidze, G., King, R., Kotzev, V., Lenk, O., Mahmoud, S., Mishin, A., Nadariya, M., Ouzounis, A., Paradissis, D., Peter, Y., Prilepin, M., Reilinger, R., Sanli, I., Seeger, H., Tealeb, A., Toksöz, M.N. and Veis, G., 2000. Global Positioning System constraints on plate motions and dynamics in the eastern Mediterranean and Caucasus. J. Geophys. Res., 105, No B3, 5695-5719.

McKenzie, D.P., 1978. Active tectonics of the Alpine-Himalayan belt: the Aegean Sea and surrounding region, Geophys. J. R. Astron. Soc., 55: 217-254.

McMechan, G.A., and Yedlin, M.J., 1981, Analysis of dispersive waves by wave field transformation. Geophysics, 46, 869-874.

Nolet, G., 1977. The upper mantle under Western Europe inferred from the dispersion of Rayleigh modes. Z. Geophys., 43, 265-286.

Panagiotopoulos, D.G. and Papazachos, B.C., 1985. Travel times of Pn waves in the Aegean and surrounding area. Geophys. J. Royal. Astr. Soc., 80, 165-176.

Papadimitriou, P., 1988. Etude de la structure du manteau superieur de l' Europe et modelisation des ondes de volume engendrees par des seismes egeens, PhD Thesis, Paris University, France.

Papadimitriou, P., Makropoulos, K. and Drakopoulos, J., 1994. Large earthquakes, rupture process and microseismicity in central Greece. Proc. of the XXIV General Assembly of the ESC, Athens, pp. 952-961.

Papadimitriou P., Voulgaris N., Kassaras I., Kaviris G., Delibasis N. and Makropoulos K., 2002. The Mw = 6.0, 7 September 1999 Athens earthquake. Natural

Hazards, October 2002, 27, Issue 1-2: 15 – 34.
Papadopoulos, G., Ganas, A. and Plessa, A., 2002. The Skyros earthquake (Mw 6.5) of 26 July 2001 and precursory seismicity patterns in the North Aegean Sea. Bull. Seism. Soc. Am., 92, 1141-1145.
Papazachos, B.C., 1969. Phase velocities of Rayleigh waves in southeastern Europe and eastern Mediterranean Sea. PAGEOPH, 75, 47-55.
Papazachos, B.C., Comninakis, P. and Drakopoulos, J., 1966. Preliminary results of an investigation of crustal structure in southeastern Europe. Bull. Seism. Soc. Am., 56, 1241-1268.
Papazachos, B.C., Polatou, M. and Mandalos, N., 1967. Dispersion of surface waves recorded in Athens. PAGEOPH, 67, 95-106.
Papazachos, C.B, 1994. Contribution to the study of crustal and upper mantle structure in Northeastern Europe by the inversion of seismological and geophysical data. PhD Thesis, University of Thessaloniki, Greece.
Papazachos, C. and Nolet, G., 1997. P and S-wave velocity structure of the Hellenic area obtained by robust nonlinear inversion of travel times. J. Geoph. Res., 102, B4, 8349-8367.
Papazachos, C.B., Hatzidimitriou, P.M., Panagiotopoulos, D.G. and Tsokas, G.N., 1995. Tomography of the crust and upper mantle in southeast Europe. J. Geophys. Res., 100, 405-422.
Pe-Piper, G. and Piper, D.J.W., 2002. The igneous rocks of Greece, Gebruder Borntraegen, Berlin-Stuttgart, p. 573.
Pillet R., Rouland, D., Roult, G. and Wiens, G.A., 1999. Crust and upper mantle heterogeneities in the southwest Pacific from surface waves phase velocity analysis. Phys. Earth Planet. Inter., 110, 211-234.
Robertson, A.H.F. and Dixon, J.E., 1984. Aspects of the geological evolution of the eastern Mediterranean, in "The Geological evolution of the eastern Mediterranean. Geol. Soc. Spec. Publ., 17, 1-74.
Roumelioti, Z., Kiratzi, A. and Dreger, D., 2003. The source process of the July 26, 2001 Skyros Island (Greece) earthquake. Geophys. J. Int., in press.
Snieder, R., 1986. Phase speed perturbations and three-dimensional scattering effects of surface waves due to topography. Bull. Seism. Soc. Am., 76, 1385-1392.
Spakman, W., Wortel, M.J.R. and Vlaar, N.S., 1988. The Hellenic subduction zone: a tomographic image and its geodynamical implications. Geophys. Res. Lett., 15, No1, 60-63.
Spakman, W., Van der Lee, S. and Van der Hilst, R.D., 1993. Travel-time tomography of the European-Mediterranean mantle down to 1400 Km. Phys. Earth Planet. Inter., 79, 3-74.
Spencer, J.W. and Nur, A.M., 1976. The Effects of Pressure, Temperature and Pore Water on Velocities in Westerly Granite. J. Geophys. Res., 81, 899-904.
Stange, S. and Friederich, W., 1993. Surface wave dispersion and upper mantle structure beneath Southern Germany from joint inversion of network recorded teleseismic events. Geophys. Res. Lett., 20, No 21, 2375-2378.
Tiberi, C., Lyon-Caen, H., Hatzfeld, D., Achaeur, U., Karagianni, E., Kiratzi, A., Louvari, E., Panagiotopoulos, D., Kassaras, I., Kaviris, G., Makropoulos, K.

and Papadimitriou, P., 2000. Crustal and upper mantle structure beneath the Corinth rift from a teleseismic tomography study. J. Geophys. Res., 105, 28159-28171.

Toksöz, M.N. and Anderson, D.L., 1966. Phase velocities of long period surface waves and structure of the upper mantle - 1. Great-circle Love and Rayleigh wave data. J. Geophys. Res., 71, 1649-1658.

Tsokas, G.N. and Hansen, R.O., 1997. Study of the crustal thickness the subducting lithosphere in Greece from gravity data. J. Geophys. Res., 102, 20585-20597.

Voulgaris, N., 1991. Crustal structure investigation in Western Greece (Zante-NW Peloponnesus). PhD Thesis, University of Athens, Greece.

Wessel, P., and Smith, W.H.F., 1995. The Generic Mapping Tools (GMT) version 3.0 Technical Reference & Cookbook, SOEST/NOAA.

Deep structure and active tectonics of the southern Aegean volcanic arc

B.C. Papazachos[1], S.T. Dimitriadis[2], D.G. Panagiotopoulos[1], C.B. Papazachos[1] and E.E. Papadimitriou[1]

[1] Geophysical Laboratory, Dept. of Geophysics, School of Geology, Aristotle University of Thessaloniki, Greece

[2] Department of Mineralogy, Petrology and Economic Geology, School of Geology, Aristotle University of Thessaloniki, Greece

ABSTRACT

Data on spatial distribution of intermediate focal depth earthquakes, fault plane solutions and deep velocity structure have been used to further investigate active tectonics related to the deep structure of the southern Aegean volcanic arc. It is observed that the top layer of the subducted east Mediterranean lithospheric slab is seismically very active at depths 60 – 110 km and 140 – 170 km and that its low seismicity part (110 – 140 km) is under the volcanic arc. This observation, geochemical data and tomographic results suggest that the primary magma reservoir of the Hellenic volcanic arc is in the mantle wedge between the subducted Mediterranean and the overriding Aegean slabs, at depths 60 – 90 km. The genesis of earthquakes at the shallow part of the subducted Mediterranean slab is attributed to dehydration embrittlement of basalt, the low seismicity at intermediate depths is due to increase of temperature and confining pressure and the increase again of seismicity in the lower active part of the slab to a second dehydration embrittlement of hydrous eclogite.

Keywords: Aegean Sea, volcanic arc, active tectonics, deep structure, magma's genesis

1. INTRODUCTION

Southern Aegean is one of the most active tectonically regions of the whole western Eurasia. It is there where fast convergence of the Aegean microplate and the eastern Mediterranean lithosphere (front part of the African plate) occurs. The eastern Mediterranean lithosphere subducts under the Aegean and the Aegean microplate overrides the eastern Mediterranean (Papazachos and Comninakis, 1971; McKenzie, 1972). This physical process results in the generations: of shallow earthquakes with magnitudes up to 8.3 in the convergence boundary of the two plates (Hellenic trench), of intermediate focal depth earthquakes with magnitude up to 7.6 in the upper part of the subducted Mediterranean slab, and of shallow earthquakes with magnitudes up to 7.5 in the overriding Aegean microplate (Papazachos and Papazachou, 2003). It also results in the generation of volcanic activity along the Hellenic volcanic arc in southern Aegean (Georgalas, 1962; Fytikas et al., 1985).

Much work has been already done on active tectonics of the southern Aegean by the use of seismological data (Papazachos and Comninakis, 1969/70, 1971; McKenzie, 1970, 1978; Hatzfeld et al., 1989; Jackson, 1994) as well as of geodetic data (Straub and Kahle, 1994; Le Pichon et al., 1995; Reilinger et al., 1997; Papazachos, 1999; McClusky et al., 2000). However, some recent mainly seismological data can be used to better understand the deep structure and active tectonics in southern Aegean. Such data concern the spatial distribution of intermediate depth earthquakes, fault plane solutions, deep velocity structure and volcanic activity. For this reason, an attempt is made in the present work to use such data for further studying of the active tectonics setting in the southern Aegean. Moreover, in this attempt we also focus on the processes definition of primary magma's genesis in southern Aegean and propose a model for that.

2. SPATIAL DISTRIBUTION OF INTERMEDIATE DEPTH SHOCKS

The spatial distribution of intermediate focal depth earthquakes in southern Aegean forms an amphitheatrical surface (Benioff zone) that dips from the convex to the concave part of the Hellenic arc. This zone was first identified by Papazachos and Comninakis (1969/70) and later investigated in detail (Papazachos and Comninakis, 1971; Hatzfeld et al., 1989; Papazachos et al., 2000). Additional new information is now available to better study the intermediate depth earthquakes in southern Aegean.

In the present work two complete samples of data for intermediate depth earthquakes in southern Aegean (34^0 N – 39^0 N, 20^0 E – 29^0 E) are used. The first sample includes 22 intermediate depth earthquakes (h = 60 – 155km) that have magnitudes M= 6.0 – 7.6 and occurred in the period 1911 – 1964 and the second sample includes 10 such earthquakes with M= 5.6 – 6.8 which occurred in the period 1965 – 2002. The parameters of these earthquakes are given in Papazachos et al. (2003). Fig. 1 shows a map of the epicenters of these earthquakes which are denoted by open or solid triangles for earthquakes with focal depths 60 – 100 km or 101 – 155 km, respectively. In the same figure the isodepth of 100 km and the five main volcanic centers are also shown. Fig. 2 shows a plot of the distance, x (in km), of the epicenters from this isodepth of 100 km (positive to the concave and negative to the convex part of the arc) as a function of the focal depth, h (in km). Least squares suggest a mean dip angle equal to ~25 degrees for x > 0.0 km (depths >100 km). Separate calculations for the western and eastern part of the arc give similar dip angles for the two parts, although tomographic profiles (Papazachos and Nolet, 1997) and spatial distribution of small recent shocks (Papazachos et al., 2000) show a steeper subduction for the eastern part of the slab and depths >100 km. However, the dip angle gradually decreases for smaller depths (Fig. 2) in agreement with previous results.

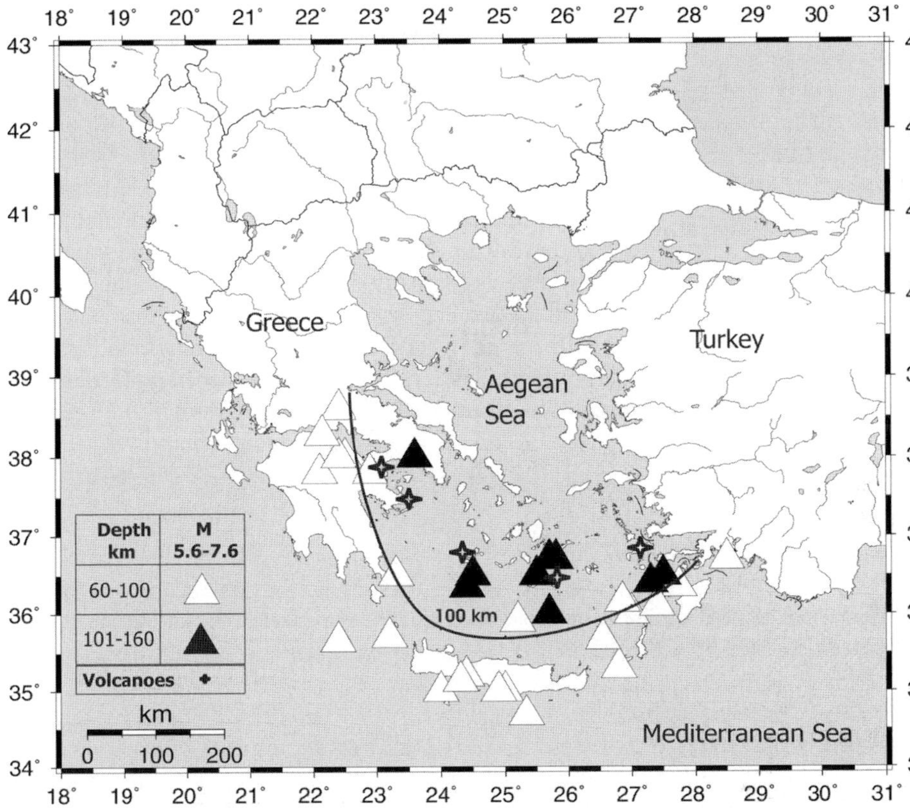

Fig. 1. Epicenters of strong intermediate depth earthquakes, which occurred in southern Aegean since 1911 ($M^3$6.0) and since 1965 ($M^3$5.6). White and black triangles denote depths smaller and larger than 100 km, respectively. The isodepth of 100 km and the five volcanic centers are also shown.

An interesting feature observed in Fig. 2 is the decrease of strong earthquake seismicity in the section of the subducted Mediterranean slab that is under the volcanic arc. To further test this observation, data for accurately located 104 intermediate depth shocks with magnitudes 4.0 – 6.3, which occurred in the southern Aegean in the period 1964 – 1979 (Comninakis and Papazachos, 1980) have been used. Fig. 3 shows a plot of the frequency of shocks, n (percentage number of shocks per 10 km depth), as a function of the focal depth. This plot clearly shows a minimum of the frequency of shocks in the depth range of 110 km – 140 km. Based on the fact that both the sample of strong (M = 5.6 – 7.6) and the sample of smaller (M = 4.0 – 6.3) intermediate depth earthquakes in southern Aegean show a clear decrease of the frequency of shocks, as well as of the maximum magnitude (M_{max}= 6.7 in depths 110 – 140 km), we may assume that this phenomenon is due to a change of the geophysical conditions in this part of the subducted oceanic crust. This change is discussed later in detail, in conjunction with results from tomography and geochemical data.

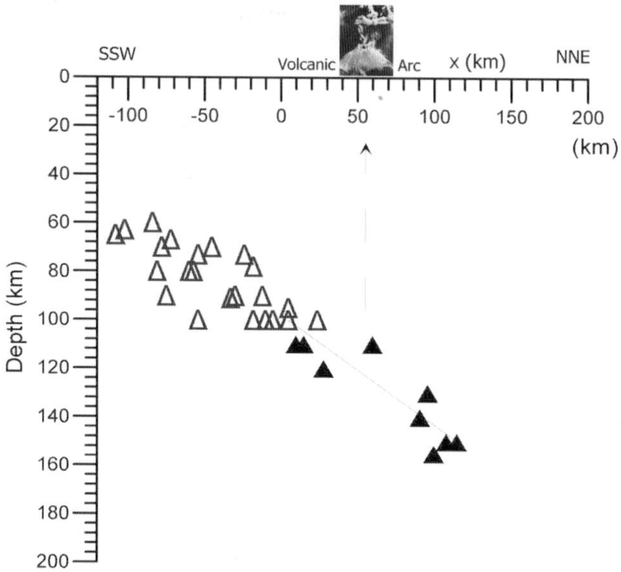

Fig. 2. Vertical profile of the Benioff seismic zone in southern Aegean as defined by the location of strong earthquakes (M ^{3} 6.0 since 1911, M ^{3} 5.6 since 1965).

3. THE STRESS FIELD

The orientation of the tectonic stress components and the corresponding faulting which are of interest for the present work are derived from fault plane solutions: a) of shallow earthquakes that occur in the southern Aegean volcanic arc, b) of shallow earthquakes that occur in the convex part of the Hellenic arc (Hellenic trench) where the Mediterranean and Aegean lithospheres converge and c) of intermediate depth earthquakes that are generated in the Benioff zone along the top oceanic crustal layer of the subducted Mediterranean slab. Such fault plane solutions have been published by several workers (Papazachos, 1969; McKenzie, 1972; Kiratzi and Langston, 1989; Taymaz et al., 1990; Papadimitriou, 1993, among many others). The most reliable of these solutions which concern strong ($M \geq 5.5$) earthquakes are included in a catalogue published by Papazachos and Papazachou (2003) that has been the source of relative data used in this work.

Reliable fault plane solutions for three strong shallow earthquakes (9.7.1956 M=7.5, 4.7.1968 M=5.5, 5.12.1968 M=6.0) in the volcanic arc are available. From these the representative solution has been derived by the method suggested by Papazachos and Kiratzi (1992) and is shown in the first line of Table 1, where ζ, δ, λ are the strike, dip and rake of the representative fault and ξ and θ are the trend and plunge of the tensional axis. That is, the crust in the volcanic arc is dominated by a tensional field with a NW – SE direction which results in normal faults which strike NE – SW. Fig. 4 shows the

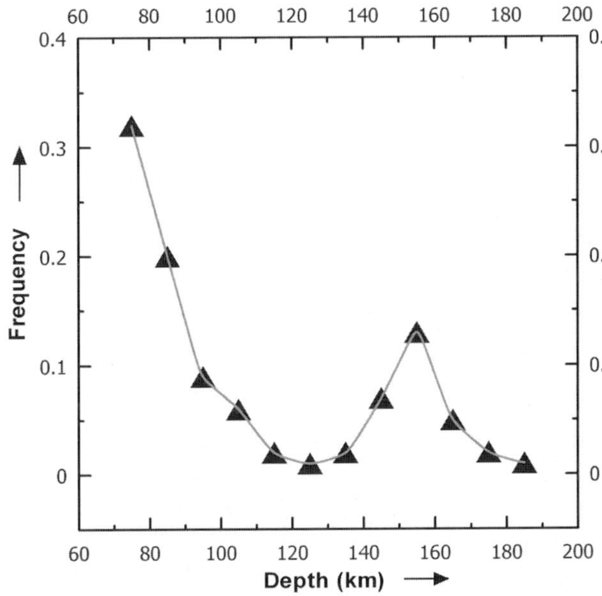

Fig. 3. Frequency, n, of intermediate depth shocks (M ^{3} 4.0, per 10 km depth) as a function of the focal depth (data for 104 accurately located shocks of the period 1964 – 1979, published by Comninakis and Papazachos (1980). A minimum for 110 – 140 km focal depths is observed.

strikes of five such faults which are located in the five volcanic centers of the southern Aegean volcanic arc where a clustering of earthquake epicenters also occur (Papazachos and Panagiotopoulos, 1993).

A typical fault plane solution for the shallow earthquakes in the convex part of the Hellenic arc (Hellenic trench), where the Mediterranean and Aegean lithosphere converge, is shown in the second line of Table 1 (Papazachos and Papazachou, 2003). That is, horizontal compression directed SSW – NNE dominates in the Hellenic trench, due to the convergence of the two plates in this direction. This results in the generation of shallow earthquakes ($h < 60$ km) on low dip angle thrust faults, which strike NW – SE and dip NE.

Reliable fault plane solutions are available for sixteen intermediate depth earthquakes that occurred in the shallower part (60 – 110 km) of the descending oceanic crust. Four of these earthquakes occurred in the western part of the subducted slab and have typical solution shown in the third line of Table 1. Eight of these earthquakes occurred in the central part of the descending slab and have a typical solution which is shown in the fourth line of Table 1. The rest four of the earthquakes occurred in the eastern part of the slab and have a typical solution which is shown in the fifth line of Table 1. Fig. 5 shows on a map the horizontal projections of the P (compression) and of the T (tension) axes in the three parts of the slab. It is shown that the maximum tension is directed to the inner (concave) part of the arc and dips almost parallel to the dip direction of the descending Mediterranean slab, while the maximum compression is

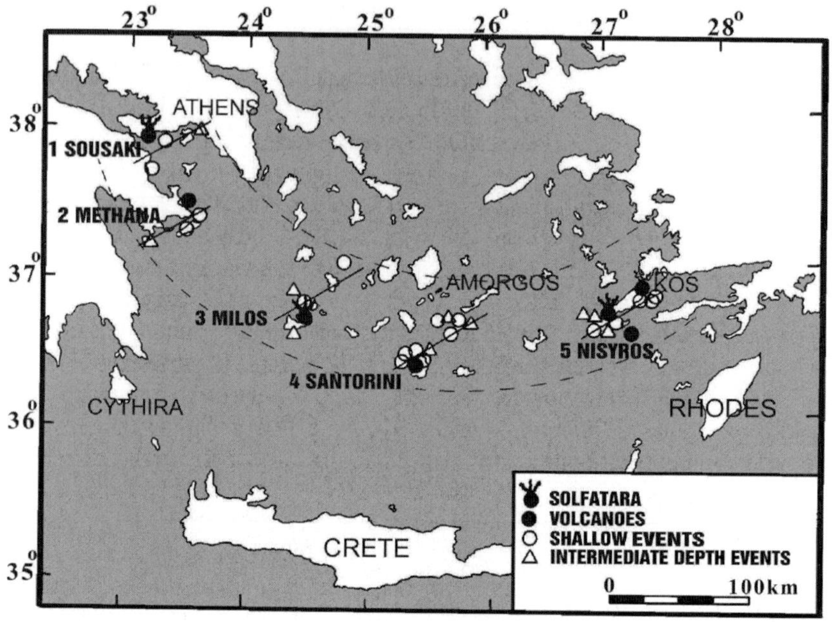

Fig. 4. The five seismovolcanic centers in the southern Aegean volcanic arc (after Papazachos and Panagiotopoulos, 1993).

Table 1. Parameters of the typical fault plane solutions for the shallow earthquakes in the volcanic arc and the Hellenic trench and for the intermediate focal depth earthquakes in the western, southern and eastern descending slab, respectively.

Region	Fault Parameters			Tension		Compression	
	ζ	δ	λ	ξ	θ	ξ	θ
Volcanic Arc	2310	400	-1040	1520	00	-	-
Hellenic Trench	3090	230	1010	2110	230	-	-
W. Descending Slab	-	-	-	840	630	1960	120
S. Descending Slab	-	-	-	200	490	2790	90
E. Descending Slab	-	-	-	3190	630	1780	210

almost parallel to the strike of the arc, as it has been previously observed (Kiratzi and Papazachos, 1995). There are only two available fault plane solutions for the lower active part (140 – 180 km) of the subducted slab, which also seem to show down-dip tension.

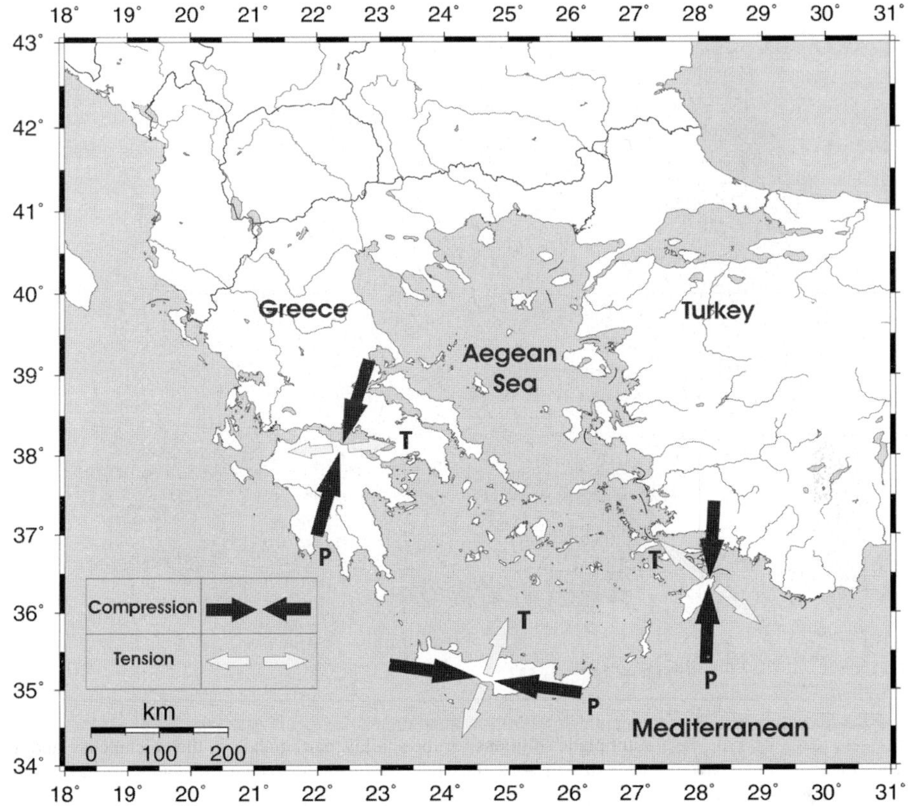

Fig. 5. Stress field in the shallow (60 – 110 km) part of the subducted Mediterranean oceanic crust, beneath the western, central and eastern part of the Aegean.

4. DEEP VELOCITY STRUCTURE

Velocity structure concerns the crust and the upper part of the mantle where active tectonics take place. Information for such structure comes from recent seismic tomographic studies (Sapkman, 1986; Ligdas et al., 1990; Drakatos and Drakopoulos, 1991; Papazachos et al., 1995). Thus, such studies led to the conclusion that the subducted lithospheric eastern Mediterranean slab can be followed further to the north – northeast down to 600 km or more (Spakman et al., 1988).

P-wave travel time tomography images of the crust and upper mantle for the depth range 60 – 90 km show a clear low velocity layer along the volcanic arc of the southern Aegean, while no such layer is observed for depths 40 – 60 km and 90 – 120 km (Papazachos et al., 1995). This distribution is observed in Fig. 6 where very low P velocities under the volcanic arc show that high temperature/partial melt material is found in this layer (60 – 90 km). Since this layer lies directly under the volcanic arc, it

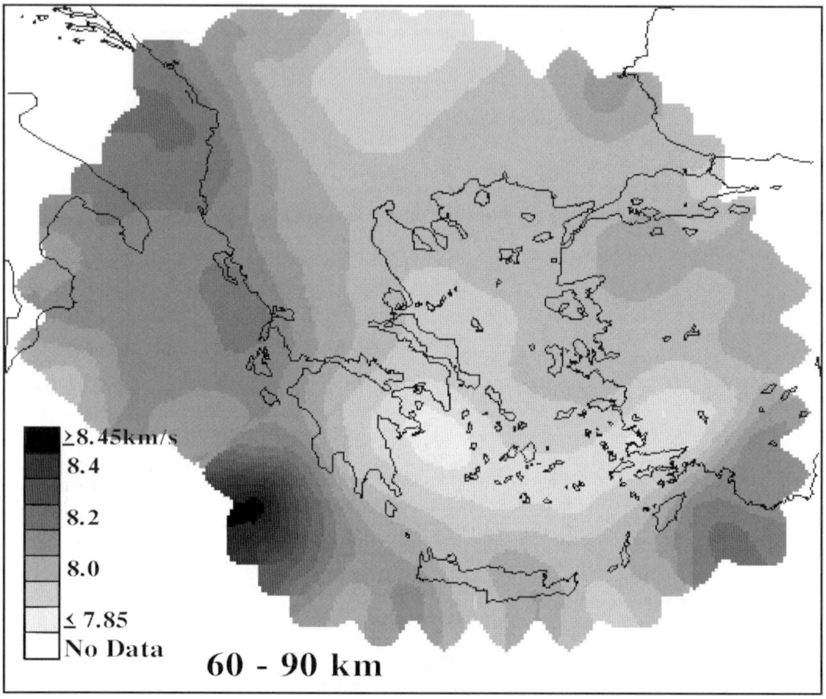

Fig. 6. P-velocity distribution at the depth of 60-90 km in southern Aegean (Papazachos et al., 1995). A low velocity layer along the volcanic arc is clearly identified.

may reasonably be concluded that this area corresponds to the primary magma source of the southern Aegean volcanic arc.

This idea is clearly supported by the results presented in Fig. 7, where three cross-sections in the west, central and eastern part of the subduction are shown. P-wave velocities are derived from the model of Papazachos and Nolet (1997) and the earthquakes of the western and eastern parts of the Benioff zone are superimposed on the corresponding tomographic images. In all sections we can easily identify the lack of strong earthquakes under the volcanic arc, as well as the low P-velocities at the depth of ~70-80 km in the mantle wedge under the volcanic arc. This low velocity anomaly is less clear in the eastern section, however the corresponding mantle wedge velocities are still much lower than the corresponding slab velocities for the same depth range.

5. GEOCHEMICAL CONSTRAINTS

As previously discussed, the main characteristic feature of the Aegean subduction-arc system is the vertical superposition of three of its main components: a) active volcanic centers at the surface, b) a major low velocity, relatively flat body (Fig. 7) below them, at a depth of 60 to 90 km within the overriding mantle wedge and, c) an

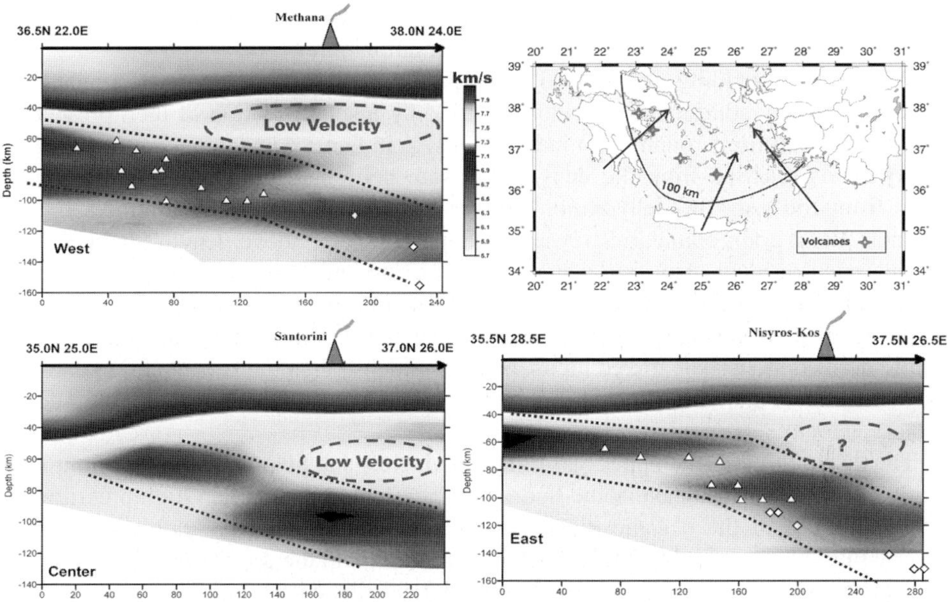

Fig. 7. P-velocity distribution along three cross-sections (west, central and east part) of the southern Aegean subduction, using the results of Papazachos and Nolet (1997). The intermediate depth seismicity of figure (2) is also superimposed for the western and eastern section of the slab. Notice the lack of seismicity in the Benioff zone and the low velocities in the mantle wedge under the volcanic arc.

110 to 140 km depth segment of the subducting slab which is characterized by lower seismicity in comparison to the shallower and deeper segments of the same slab (see Figs. 2, 3 and 7). This vertical superposition probably implies genetic links, since the presence of volcanoes above the low velocity body shows that the latter is most likely the main magma reservoir from which other higher lying magmatic chambers and finally the volcanoes themselves are periodically feeded with magma. However, the presence of the intermediate segment of lower seismicity in the subducted slab, straight below the main magma reservoir, is not easily understood and needs further explanation.

We propose that the critical factor determining the distance from the trench, the depth and the vertical arrangement of the three previous components is the presence of the geophysically observable location of the partial melting zone within the overriding mantle wedge, between the low seismicity segment below and the main magma reservoir above. This location is mainly determined by the subduction rate, the thermal state of the subducting slab, and the thermal structure of the overriding wedge. A partial melting front is actually forming when and where the hydrous fluids ascending from the down-going and dehydrating slab cross the wet mantle lherzolite solidus within the overriding mantle wedge. Further up-slab any watery fluids that were gradually released from the descending slab are either kept in the pores between the mineral grains sustaining a high pore fluid pressure or were stabilized in the overriding mantle in hydrous phases such as serpentine, phlogopite or amphiboles.

Within the overriding mantle wedge the partial melting front (to be imagined as a nearly completely solid mantle peridotite with only a small fraction of melt present as droplets and thin films round the mineral grains) is in fact a demanding water sink. The watery fluids released from the dehydrating slab below are therefore actively driven away from the zone of dehydration and are streaming up via interconnected tiny channels along grain boundaries towards the melting front above, leaving behind a far less hydrated residual. Any free water left in the zone of dehydration is probably also consumed in the partial melting of metasediments that might have been driven at this depth; such melts transport also upwards, reach the melting front and mix with the mantle derived melts. The mixed melts, being less dense than the surrounding minerals, migrate upwards and are collected in the main magma reservoir (the 60 to 90 km depth flat low velocity zone).

The dehydration of the upper part of the subducted slab below the melting front may be not complete. It can leave some hydrous minerals that can sustain high pressures in the metabasites of the oceanic crust. However, the removal of all the available free water will drop considerably the pore fluid pressure within the upper part of the slab at a level far below that necessary to compensate the existing high confining pressure. This may well explain the 110 to 140 km depth low seismicity segment of the down-going slab. Also, the resulted densification of this part of the slab can cause an increase in the subduction angle, a feature identified in the subducting Mediterranean oceanic slab from both tomographic results and the Benioff zone (Fig. 2 and 7, as well as results from Papazachos et al. (2000)).

Further dehydration within the subducted slab, probably driven to completeness, will start again when the remaining high-pressure hydrous phases will no longer be stable under the conditions prevailing at greater depths (more than 140 km). The relatively little water released during this second dehydration stage is less likely to find its way to the previous or form a new zone of partial melting due to removal of solidus to higher temperatures and will probably only rise the pore fluid pressure.

It should be noted that existing geochemical data (Zellmer et al., 2000) are in accordance with the above proposed model, since they show that in the generation of at least the Santorini magma three components are involved: mantle wedge, sediment melts and slab derived fluids. Unfortunately, there are no available data on the thermal structure of either the subducting slab or the mantle wedge above it. However, the thermal structure of the subducting slab is mainly depended on the subduction rate, which can be inferred from the convergence rate between Africa and Europe, considering velocities ranging between 1 cm/yr and 5 cm/yr, with the velocity of 3-3.5 cm/yr as a realistic estimate (e.g. Papazachos, 1999).

Using v=1 cm/yr, v=5 cm/yr and v=3 cm/yr, a length scale (L) of 40 km (~one-half the thickness of the oceanic lithosphere) and a thermal diffusivity (κ) of 1 mm^2/s, from the equation $P_e = (v \cdot L)/\kappa$ the three corresponding Peclet numbers (the lower, the higher and a more probable intermediate one) are 13, 63 and 38, all greater than one. Thus, the advection of heat along with the subducting slab should occur at a much faster rate than the conduction of heat into the subducting slab. The thermal structure of the subducting slab must therefore feature strong depression of the isotherms. Additional factors farther depressing the isotherms are the endothermic intermediate metamorphic reactions, especially the devolatilization ones, and the partial melting of the metasediments. Factors opposing such a depression are the shear heating at the interface between the subducted slab and the overriding lithosphere and the radiogenic heat produced within the subducting sediments.

Since the thermal depression and the resulted inversion of the isotherms near the boundary between the subducted slab and the mantle wedge above it is a necessary prerequisite for melting to occur in the overriding mantle wedge and for the consequent expression of volcanic activity on the surface, we suggest that the depression and inversion of the isotherms characterizing most subduction zones (Anderson et al., 1978, 1980; Furlong et al., 1982) also exist in the Aegean subduction system. Shear heating and radiogenic heat production, therefore must not play a significant role in the case of the Aegean subduction. Given the large amount of sediments that are driven into the trench, a low radiogenic heat production may appear strange; however, due to the thin skin tectonic regime in the eastern Mediterranean (Kastens, 1991) large parts of these sediments may be detached from the slab and may not follow it at great depths.

We cannot go further deducing from the little available data more details for the thermal structure beneath the Aegean arc system. Nevertheless, we can make the realistic assumption that at 100 to 110 km depth, where we suggested that drastic dehydration of the basic oceanic rocks and partial melting of the subducted sediments should occur, the temperature in the upper part of the subducting slab should have reached ~700° C (Fig. 8). The corresponding lithostatic pressure (assuming a column of 20 km of granite and 70 km of peridotite) should be there ~28 kbars (2.8 GPa). These conditions suggest a steep steady state geotherm within the upper part of the subducted slab, which will probably pass through the zeolite, lawsonite-chlorite and lawsonite-blueshist facies, before entering the eclogite facies at pressures higher than 20 Gpa corresponding to depths more than 70 km (see for example Peacock, 1993, Fig. 2).

Taking into account that delays in the completeness of mineral reactions and metastable preservation of metamorphic assemblages is the rule rather than the exception for rocks following steep geotherms, we can make a realistic suggestion that at about 100 to 110 km lawsonite-glaucophane metabasites of the oceanic crust are converted to glaucophane bearing eclogites, at the same time that partial melting of metasediments is taking place. This conversion will drop the water content of the oceanic metabasites from nearly 6% to less than 1%, liberating thus significant amounts of water that escape up to the melting front within in the overriding mantle wedge. The not completely dry eclogite formed may contain zoisite, paragonitic mica and glaucophane, along with garnet and pyroxene. These minerals remain stable until the increasing temperature and pressure further deeper will gradually destabilize them, liberating again small amounts of water and leaving behind a completely dry garnet-pyroxene eclogite. This second dehydration may just increase the pore fluid pressure

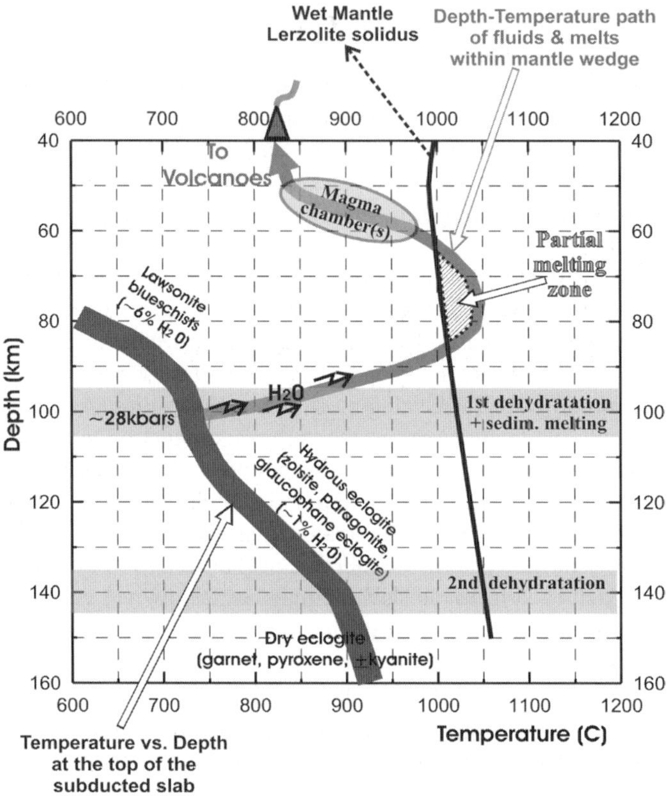

Fig. 8. Schematic depth-temperature diagram and corresponding geochemical information for the top part of the subducted slab and the depth-temperature path of fluids and melts derived from slab dehydration (see text for explanation).

allowing seismicity to appear again (for depths > 130-140 km). It should be noted that the possibility that hydrous phases within the subducted oceanic crust may persist at depths much larger than 100 km, even down to 250 km, is not unlikely; such a possibility is supported by theoretical and geophysical data (Peacock, 1993; Abers, 2000).

6. A SYNTHETIC MODEL FOR THE SOUTHERN AEGEAN SUBDUCTION

Data presented in the present work combined with already published information can be used to propose a synthetic model for the deep active tectonic structure of the southern Aegean. Such a model, in addition to information on the converging lithospheric plates, should also include information on the location and size of the

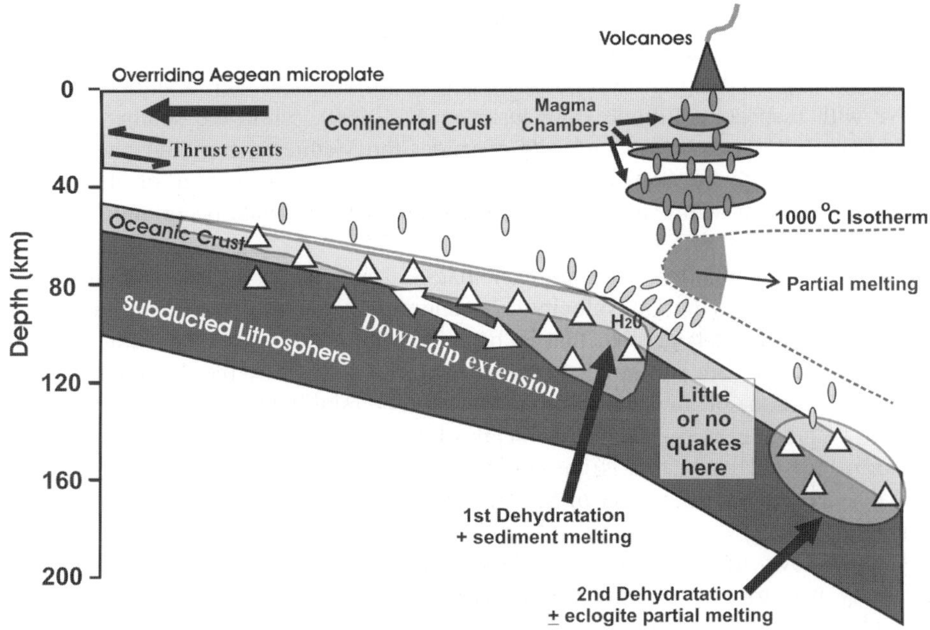

Fig. 9. A synthetic model for the active tectonic structure of southern Aegean.

primary magma zone. Such information and knowledge on the genesis of primary magma in island arcs (Abers, 2000; Yamasaki and Seno, 2003, among others) can be used to propose a reasonable hypothesis on the way that primary magma is generated in the Hellenic arc.

A schematic synthetic model is shown in Fig. 9, which shows a vertical profile of the deep tectonic structure of the Hellenic arc in an ~SSW – NNE direction (approximately normal to the arc). The thickness of the descending eastern Mediterranean slab (front part of the African plate) is taken equal to 70-90 km, which is the thickness of the layers above the top of the low shear wave velocity layer, estimated from phase velocity of Rayleigh waves (Papazachos, 1969), as well as tomographic results (Papazachos and Nolet, 1997). The thickness of the crustal oceanic layer of the slab has been set to ~7 km by wide aperture seismic data (Bohnhoff et al., 2002). The thickness of the overriding Aegean crust varies along the profile, while its lithospheric thickness is considered to be of the order of 50 km in the volcanic arc area (Papazachos, 1969; Makris, 1976, 1978; Papazachos and Nolet, 1997; Bohnhoff et al., 2002; Karagianni et al., 2002). This is also supported by the observations that intermediate depth earthquakes ($h \geq 60$ km) under the concave part of the arc are generated by down – dip tension (Fig. 5), while shallow earthquakes ($h < 60$km) in the convex part of arc are generated by thrust faulting, as it is expected if the interaction between the two lithospheres occurs at a depth of about 40-50 km. Considering that the mean thickness of the crust in the continental slab of southern Aegean is about 20-25 km (Makris, 1976;

Bohnhoff et al., 2002), the crust and lithospheric mantle of the overriding southern Aegean slab have similar thickness.

The eastern Mediterranean lithosphere moves to the north with a velocity of 5-10 mm/yr with respect to Europe (McClusky et al., 2000), while the southern Aegean lithosphere moves southwestwards and overrides the Mediterranean lithosphere with a velocity of 30 mm/yr (Papazachos, 1999). Hence, the subduction of the eastern Mediterranean slab with respect to the Aegean lithosphere should occur at a maximum rate of 4 cm/yr in an NNE direction.

The vertical dimension of the primary magma zone can be considered equal to the vertical dimension of the low velocity layer under the volcanic arc as this dimension is derived by seismic tomography, that is, between 60 km and 90 km. The extent of this zone normal to the arc can be also considered to be defined by the distribution of the vertical projections of the five volcanic centers. The minimum and maximum distance of these volcanic centers from the isodepth of 100 km is 10 km (Sousaki) and 90 km (Santorini), respectively, that is, the horizontal dimension of the primary magma zone normal to the arc is of the order of 80 km, similar to what is found by the tomographic results (e.g. Fig. 7).

Fig. 9 shows that the zone of primary magma generation is just above the low seismicity part of the subducted oceanic crust. Based on this observation we may conclude that its low seismicity part of the oceanic crust is dehydrated at depths less than 110 km and the released water contributes to the formation of primary magma. Information on trace element and isotopic data for the Santorini volcano supports the idea that the sources of magma generation in the Hellenic volcanic arc is the subducted oceanic crust and the mantle wedge between the two slabs (Zellmer et al., 2000).

The generation of intermediate depth earthquakes at the shallow part (60 – 110 km) of the Benioff zone in south Aegean is convincingly explained by the hypothesis of dehydration embrittlement of the subducted slab in these depths. For the generation of earthquakes at the deeper part (140 – 180 km) of these zone several explanations have been proposed worldwide (e.g. Wiens, 2001). It is possible that earthquakes at these depths are also generated by second dehydration embrittlement of basalt because calculations of seismic wave velocities show that the oceanic crust remains distinct to these depths (Abers, 2000). The second possibility is that earthquakes at these depths are generated by dehydration embrittlement of serpentine in case that the dipping slab is of cold type, because in such case the lower part of the dehydration loci of serpentine penetrates the lower seismically active part of the subducted slab (Yamasaki and Seno, 2003). For the southern Aegean subduction we propose that the second dehydration embrittlement of hydrous eclogite-basalt is the most locigal explanation for this seismicity, given the relatively young age and possibly high temperature of the southern Aegean subduction.

7. CONCLUSIONS AND DISCUSSION

The Mediterranean slab, with a thickness of about 80 km and an oceanic crust of 7 km, dips under the Aegean microplate at a mean angle of 25°. The seismically active part of the upper layer of this slab (with earthquakes of $M \geq 4.0$) reaches up to a depth of 180 km, but at depths 110 – 140 km this layer is of relatively low seismicity due probably to its higher temperature. Earthquakes in the shallow part (60 – 110 km) of this dipping Benioff zone are generated by down – dip tension, while maximum compression is horizontal and parallel to the Hellenic arc. Strong evidence is presented that the cause of intermediate depth earthquakes in the shallow part of the Benioff zone of southern Aegean is the dehydration embrittlement of basalt while the earthquakes in the lower part of this zone are caused in the same way or are caused by a second similar dehydration embrittlement of hydrous eclogite.

Shallow earthquakes in the zone of convergence (Hellenic trench) of the eastern Mediterranean and Aegean lithospheres are generated by low dip angle thrust faulting trending NW – SE and dipping NE. Shallow earthquakes in the southern Aegean volcanic arc are generated by normal faulting trending in an about NE – SW direction.

The occurrence under the volcanic arc of both the low velocity mantle layer, at depths of 60 – 90 km and of the low seismicity dipping oceanic crust, at depths of 110 – 140 km, indicates that the primary magma in southern Aegean is located in the asthenospheric wedge between the descending Mediterranean and the overriding Aegean microplates under the volcanic arc, in agreement with geochemical data.

The available data do not allow to test the hypothesis whether there are two almost parallel Benioff zones in southern Aegean, as it occurs in other arcs (e.g.Yamazaki and Seno, 2003). We cannot, for example, exclude the possibility that earthquakes with focal depths 130 – 180 km belong to a different zone than those at depths 60 – 120 km. If this is the case, the shallower seismic zone is due to dehydration embrittlement of basalt and the lower seismic zone to dehydration embrittlement of hydrous eclogite. However, the problem needs further examination by the use of additional new data which will help to clarify the geometrical and geophysical constraints of the southern Aegean subduction.

Acknowledgements

This work has been funded by the e-Ruption project (contract EVR1-2001-00024).

REFERENCES

Abers, G.A., 2000. Hydrated subducted crust at 100-250Km depth. Earth Planet. Scie. Lett., 176: 323-330.

Anderson, R.N., De Long, S.E. and Schwarz, W.M., 1978. Thermal model for subduction with dehydration in the downgoing slab. J. Geol., 86: 731-739.

Anderson, R.N., DeLong, S.E. and Schwarz, W.M., 1980. Dehydration, asthenospheric convection and seismicity in subduction zones. J. Geol., 88: 445-451.

Bohnhoff, M., Makris, J., Papanikolaou, D. and Stavrakakis, G., 2002. Crustal investigation of the Hellenic subduction zone using wide aperture seismic data. Tectonophysics, 343: 239-262.

Comninakis, P.E. and Papazachos, B.C., 1980. Space and time distribution of the intermediate focal depth earthquakes in the Hellenic arc. Tectonophysics, 70: 35-47.

Drakatos, G. and Drakopoulos, J., 1991. 3-D velocity structure beneath the crust and upper mantle of the Aegean sea region. Pure Appl. Geophys., 135: 401-420.

Fytikas, M., Innocenti, F., Manetti, P., Mazzuoli, R., Pecerillo A. and Villari L., 1985. Tertiary to Quaternary evolution of the volcanism in the Aegean region. In: J.F. Dixon and A.H.F. Robertson (Editors), The Geological Evolution of the Eastern Mediterranean. Blackwell Publ., Oxford, p. 848.

Furlong, K.P., Chapman, D.S. and Alfeld, P.W., 1982. Thermal modeling of the geometry of subduction with implications for the tectonics of the overriding plate. J. Geophys. Res., 87: 1786-1802.

Georgalas, G.C., 1962. Active volcanoes in the world including solfatara fields. Intern. Volcan. Assoc., 12: 1-40.

Hatzfeld, D., Pedotti, G., Hatzidimitriou, P., Panagiotopoulos, D.G., Scordilis, M., Drakopoulos, J., Markopoulos, K., Delibassis, N., Latoussakis, J., Baskoutas, J. and Frogneua, M., 1989. The Hellenic subduction beneath the Peloponese: first results of a microearthquake study. Earth Planet. Sci. Lett., 93: 283-291.

Jackson, J., 1994. Active tectonics of the Aegean region. Annual Rev. Earth Planet Sci., 22: 239-271.

Karagianni, E.E., D.G. Panagiotopoulos, G.F. Panza, P. Suhadolc, C.B. Papazachos, B.C. Papazachos, A. Kiratzi, D. Hatzfeld, K. Makropoulos, K. Priestley and A. Vuan, 2002. Rayleigh wave group velocity tomography in the Aegean area. Tectonophysics, 358: 187-209.

Kastens, K., 1991. Rate of outward grouth of the Mediterranean Ridge accretionary complex. Tectonophysics, 199: 25-50.

Kiratzi, A.A. and Langston, Ch.A., 1989. Estimation of earthquake source parameters of the May 4, 1972 event of the Hellenic arc by the inversion of waveform data. Physics Earth and Planet. Inter., 57: 225-232.

Kiratzi, A.A. and Papazachos, C.B., 1995. Active seismic deformation in the southern Aegean Benioff zone. J. Geodynamics, 19: 65-78.

Le Pichon, X., Chamot-Rooke, N., Lallemant, S., Noomen, R. and Veis, G., 1995. Geodetic determination of the kinematics of central Greece with respect to Europe: implication for eastern Mediterranean tectonics. J. Geophys. Res., 100: 12675-12690.

Ligdas, C.N., Main, I.G. and Adams, R.D., 1990. 3-D structure of the lithosphere in the Aegean Sea region. Geophys. J. Int., 102: 219-229.

Makris, J., 1976. A dynamic model of the Hellenic arc deduced from geophysical data. Tectonophysics, 36: 339-346.

Makris, J., 1978.The crust and upper mantle of the Aegean region from deep seismic soundings. Tectonophysics, 46: 269-284.

McClusky, S., Balassanian, S., Barka, A., Demir, C., Ergintav, S., Georgiev, I., Gurkan, O., Hamburger, M., Hurst, K., Kahle, H., Kastens, K., Kekelidze, G., King, R., Kotzev, V., Lenk, O., Mahmoud, S., Mishin, A., Nadariya, M., Ouzounis, A., Paradissis, D., Peter, Y., Prilepin, M., Reilinger, R., Santli, I., Seeger, H., Tealed, A., Toksoz, M.N. and Veis, G., 2000. Global Positioning System constraints on plate kinematics and dynamics in the eastern Mediterranean and Caucasus. J. Geophys. Res., 105: 5695-5719.

McKenzie, D.P., 1970. The plate tectonics of the Mediterranean region. Nature, 226: 239-243.

McKenzie, D.P., 1972. Active tectonics of the Mediterranean region. Geophys. J.R. Astr. Soc., 30: 109-185.

McKenzie, D.P., 1978. Active tectonics of the Alpine-Himalayan belt: the Aegean Sea and surrounding regions. Geophys. J.R. Astr. Soc., 55: 217-254.

Papadimitriou, E.E., 1993. Focal mechanism along the convex side of the Hellenic arc and its tectonic significance. Bull. Geof. Teor. Appl., 35: 401-426.

Papazachos, B.C., 1969. Phase velocities of Rayleigh waves in southeastern Europe and eastern Mediterranean Sea. Pure Appl. Geophys., 75: 47-55.

Papazachos, B.C. and Comninakis, P.E., 1969/1970. Geophysical features of the Greek island arc and eastern Mediterranean ridge. Final Proc. Seances de la Conference Reunie a Madrid , Madrid, Spain, 74-75.

Papazachos, B.C. and Comninakis, P.E., 1971. Geophysical and tectonic features of the Aegean arc. J. Geophys. Res., 76: 8517-8533.

Papazachos, B.C. and Panagiotopoulos, D.G., 1993. Normal faults associated with volcanic activity and deep rupture zones in the southern Aegean volcanic arc. Tectonophysics, 220: 301-308.

Papazachos, B.C., Karakostas, V.G., Papazachos, C.B. and Scordilis, E.M., 2000. The geometry of the Wadati-Benioff zone and lithospheric kinematics in the Hellenic arc. Tectonophysics, 319: 275-300.

Papazachos, B.C. and Papazachou, C.B., 2003. The earthquakes of Greece. Ziti, Thessaloniki.

Papazachos, B.C., Comninakis, P.E., Karakaisis, G.F., Karakostas, B.G., Papaioannou, Ch.A., Papazachos, C.B., Scordilis E.M., 2003. A catalogue of earthquakes in Greece and surrounding area for the period 550BC-2002. Publication, Geophys. Lab. Univ. Thessaloniki, Thessaloniki, Greece.

Papazachos, C.B., 1999. Seismological and GPS evidence for the Aegean – Anatolia interaction. Geophys. Res. Lett., 26: 2653-2656.

Papazachos, C.B. and Kiratzi, A.A., 1992. A formulation for reliable estimation of active crustal deformation and an application to central Greece. Geophys. J. Int., 111: 424-432.

Papazachos, C.B., Hatzidimitriou, P.M., Panagiotopoulos, D.G. and Tsokas, G.N., 1995. Tomography of the crust and upper mantle in southeast Europe. J. Geophys. Res., 100: 12405-12422.

Papazachos, C.B. and Nolet, G.P., 1997. P and S deep velocity structure of the Hellenic area obtained by robust nonlinear inversion of travel times. J. Geophys. Res., 102: 8349-8367.

Peacock, S.M., 1993. The importance of blueschist-eclogite dehydration reactions in subducting oceanic crust. Geol. Soc. America Bull., 105: 684-694.

Reilinger, R. E., McClusky, S. C., Oral, M. B., King, R. W., Toksoz, M. N., Barka, A. A., Kinik, I., Lenk, 0. and Sanli, I., 1997. Global Positioning System measurements of present-day crustal movements in the Arabia-Africa-Eurasia plate collision zone. J. Geophys. Res., 102: 9983-9999.

Spakman, W., Wortel, M. J.R. and Vlaar, N.J., 1988. The Hellenic subduction zone: a tomographic image and its dynamic implication. Geophys. Res. Let., 15: 60-63.

Straub, Ch. and Kahle, H.G., 1994. Global positioning system estimated in the Marmara Sea region, northwestern Anatolia. Earth Planet Sci. Lett., 121: 495-502.

Taymaz, T., Jackson, J. and Westaway, R., 1990. Earthquake mechanisms in the Hellenic trench near Crete. Geophys. J. Int., 102: 695-731.

Wiens, D.A., 2001. Seismological constraints on mechanism of deep earthquakes: temperature dependence of deep earthquake source properties. Phys. Earth Planet. Inter., 127: 145-163.

Yamasaki, T. and Seno, T., 2003. Double seismic zone and dehydration embrittlement of the subtucting slab. J. Geophys. Res., 108: 2212-2233.

Zellmer, G., Turner, S. and Hawkesworth, C., 2000. Timescales of destructive plate margin magmatism: new insights from Santorini, Aegean volcanic arc. Earth Planet. Scie. Lett., 174: 265-281.

A West-East Traverse along the magmatism of the south Aegean volcanic arc in the light of volcanological, chemical and isotope data

L. Francalanci[1,2,*], G.E. Vougioukalakis [3], G. Perini[1], P. Manetti[1,2]

[1] Dipartimento di Scienze della Terra, Università degli Studi di Firenze, via La Pira, 4, I-50121, Firenze, Italy.

[2] C.N.R., I.G.G., Sezione di Firenze, via La Pira 4, I-50121, Firenze, Italy.

[3] I.G.M.E, Mesogeion, 70, Athens, Greece.

ABSTRACT

The volcanic rocks of the South Aegean arc (SAAVA) form a chain from the Gulf of Saronikos (Susaki, Egina, Poros, Methana) at West, to an area close to the Anatolian coast at East (Kos, Nisyros and minor islands), through the central part (Milos and Santorini island groups). The volcanic activity began in the Lower Pliocene at Egina (4.7 Ma) and lasted until present days, with the still active Methana, Milos, Santorini and Nisyros volcanoes. The beginning of volcanism is younger in the central sector of the arc.

Volcanic center location was controlled by large tectonic lineaments, most of them still active, trending E-W to NW-SE for the western part and mainly NE-SW for the central and eastern parts of the arc. Volcanic fields developed along ellipse shaped areas with the longest axis oriented perpendicular to the subduction front.

In the western volcanic fields (Susaki, Egina-Poros-Methana and Milos), volcanic centers are mostly monogenetic and no composite volcanic structures are present. In the eastern sector of the arc, Santorini and Nisyros are important composite volcanoes with caldera structures.

Volcanic rocks belong to the calc-alkaline and high-K calc-alkaline (mainly Pliocenic in age) series. Basalts are mainly present in Santorini island group. Magmas underwent to complex differentiation processes, dominated by crystal fractionation, often associated to crustal contamination and mixing-mingling. Large compositionally

* Corresponding author: e-mail: lorella@unifi.it

zoned magma chambers often fed highly explosive eruptions, especially in the central and eastern sectors of the arc.

In the western and eastern parts, potassium content of erupted magmas decreased with time, probably due to the increase of partial melting degree of the mantle source. Trends of evolution tend to pass from calc-alkaline to tholeiitic from West to the Santorini volcanic field, back again to calc-alkaline toward the Nisyros volcanic field. Incompatible trace element contents are lower in Santorini mafic magmas. From West to East, Sr and Pb isotope ratios decrease, whereas Nd isotope ratios increase.

Partial melting of a MORB-like astenospheric mantle, metasomatised by prevailing subducted sediments, is thought to produce the entire spectrum of parental magmas of SAAVA. Slab-derived fluids are generally reduced. Low Ba/La values suggest the occurrence of even lower fluid contents during magma genesis at Santorini. Total amount of subducted sediments involved in the magma genesis decreases from West to East. Alternatively, the West-East $^{87}Sr/^{86}Sr$ and $^{144}Nd/^{143}Nd$ variation of the Mediterranean sediments leads to suggest a similar variation in the subducted sediment composition. Most of the geochemical characteristics of the Santorini magmas can be explained by higher partial melting degrees of the mantle source, probably triggered by the greater lithosphere extension, inducing adiabatic upwelling of the mantle. The higher lithosphere extension also caused Santorini magmas to stop at shallower levels, thus preventing amphibole crystallisation and allowing a higher amount of mafic magmas to reach the surface.

From the beginning of SAAVA magmatism, occurred in the external parts of the arc, the partial melting degrees of the magma mantle source seem to have increased with decreasing time and going towards the central sector of the arc.

Keywords: Magmatic arc, volcanology, geochemistry, magma genesis, Aegean Sea.

1. INTRODUCTION

The Aegean area is one of the most rapidly deforming parts of the Alpine - Himalayan mountain belt, pointed out by the extremely high number of the seismic events occurring in this area. Geophysical data record a thinned continental crust in all the Aegean area and an anomalous heat flow, implying a complex geodynamic - geotectonic setting (Papazachos et al., this volume, Mountrakis et al., this volume, Pe-Piper and Piper, this volume and references therein). Deformation seemed to be dominated by the effects of the westward motion of the Anatolian block (~20 $^{mm}/_{y}$), the southwestward motion of southern Aegean (~30-35 $^{mm}/_{y}$) and the vertical movements of big lithosphere portions. Alternatively, differential convergence rates between the northeastward directed subduction of Africa relative to the hangingwall disrupted Eurasian lithosphere (faster motion of Greece relative to Cyprus-Anatolia), however, have been recently proposed in order to determine the Aegean extension (Doglioni et al., 2002 and references therein).

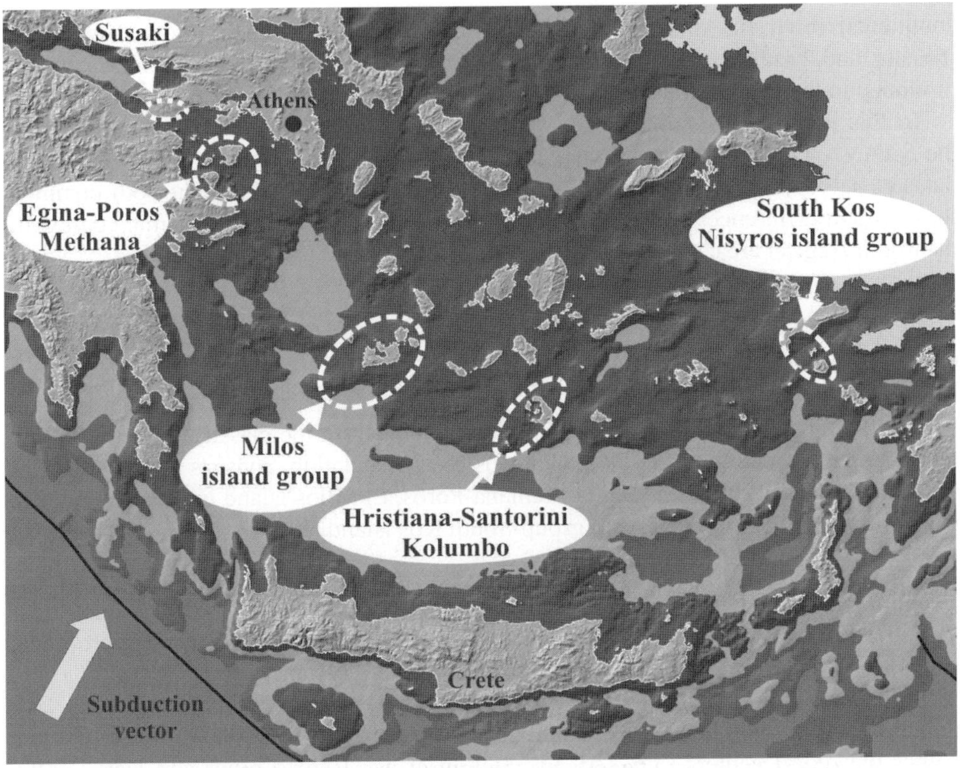

Fig. 1. The South Aegean active volcanic arc fields (dashed white curves).

Calc-alkaline (CA) and high-K (HK) calc-alkaline volcanic activities were manifested in Plio-Quaternary along a restricted belt, which extends in an arc form from Susaki to the west to Nisyros island to the east, the so-called South Aegean Active Volcanic Arc (SAAVA) (Fig. 1). This arc hosts the active (Methana and Santorini) and potentially active (Milos and Nisyros) Hellenic volcanic centers. The correspondence of the active volcanic arc over a Benioff zone at a depth of about 110-130 km is an evidence for the role of subduction in the genesis of the arc (e.g., Innocenti et al., 1981; Manetti, 1997; Pe-Piper and Piper, 2002).

The volcanism is voluminous, with subaerial volumes for the volcanic fields up to 100 km^3. Volcanic field and volcanic center shapes and distribution appear to be associated with big tectonic lineaments and active faults whose trending changes from West to East of the arc. The central and eastern sectors of the arc are characterised by large composite volcanoes, with caldera structures (Santorini and Nisyros), whereas the western sector is mainly characterised by small, often monogenic, eruptive centers.

The products of this volcanism form a typical calc-alkaline association, which displays a continuous evolution from basalts to rhyolites. Their chemical characteristics are closely comparable with those of the volcanics of island arcs sited on thin continental margins. Andesites and dacites are dominant, but less evolved members (basalts and basaltic andesites) are also common (about 25% of the total erupted products) in the central-south part of the arc.

In this paper, we summarise the main geological-volcanological characteristics of the SAAVA fields. We also discuss the available chemical, isotopic and petrological dataset, based on more than 1500 analyses, in an attempt to highlight both common features and differences and try to interpret them in the known geodynamic context of the South Aegean area.

2. VOLCANOLOGICAL AND GEOCHEMICAL CHARACTERISTICS OF THE SAAVA FIELDS

The South Aegean Volcanic Arc Province, intended in the classical meaning of the term as petrologically broadly homogeneous rock association, consists of five different volcanic fields: a) Susaki, b) Egina-Methana-Poros, c) Milos island group, d) Santorini island group and e) Nisyros island group-South Kos island (Fig. 1).

2.1. Susaki

For Susaki area (Fig. 2), the term volcanic field is not probably the most appropriate, as volcanic outcrops are of limited extension (total volume less than 1 km^3) and scattered in a large area. Volcanic centers are monogenetic, comprising dacitic lava domes and associated lava flows. Outcrops are found in two subgroups: the western one gather the oldest centers (3,6-4,05 Ma, Bellon et al., 1979, Collier and Dart, 1991), whereas the eastern one comprise the youngest lavas (2,7-2,3 Ma, Fytikas et al., 1976; Schroder, 1976). Both the vent distribution and the shape of the edifices are controlled by the E-W and the NW-SE tectonic lineaments of the area, which were present since Pliocene and continue to be active up to the present. The NW trending extensional tectonic lines, which are considered also as the youngest ones, mostly seems to control the eastern group (Mettos et al., 1988; Pavlides, 1993; Stiros, 1995). In the coeval lacustrine deposits, which surround the volcanic outcrops, a few layers of tuffs and tuffites are found, documenting a subordinate explosive activity related with the emplacement of the lava domes (Mettos et al., 1988).

The Susaki area host today a low enthalpy geothermal field, with a max temperature of 80 °C at a depth of 200 m, as well as a weak fumarolic activity with low temperature, CO_2-H_2S rich, hot gasses (30-40 °C). Both hydrothermal fluid circulation and hot gas outpouring lead to an extensive argilization-

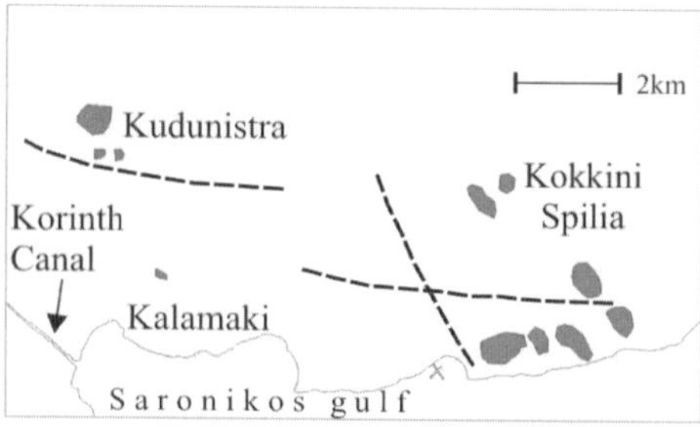

Fig. 2. The Susaki volcanic area. Gray areas: volcanic outcrops. Dashed lines: main tectonic lineaments.

silicification of the rocks in the area.

The volcanic products are HK dacites and rhyolites (Fig. 3) with a glassy groundmass and phenocrysts of plagioclase, biotite and quartz. $^{87}Sr/^{86}Sr$ ratios are high, with values around 0.713 (Fytikas et al., 1986a; Pe-Piper and Hatzipanagiotou, 1997).

2.2. Egina-Poros-Methana volcanic field

This magmatism occurs in the Gulf of Saronikos, a tectonic depression which started to form in the Pliocene. The volcanic activity developed in three limited areas starting from lower Pliocene (Egina) to historical time (Methana) (Fig. 4).

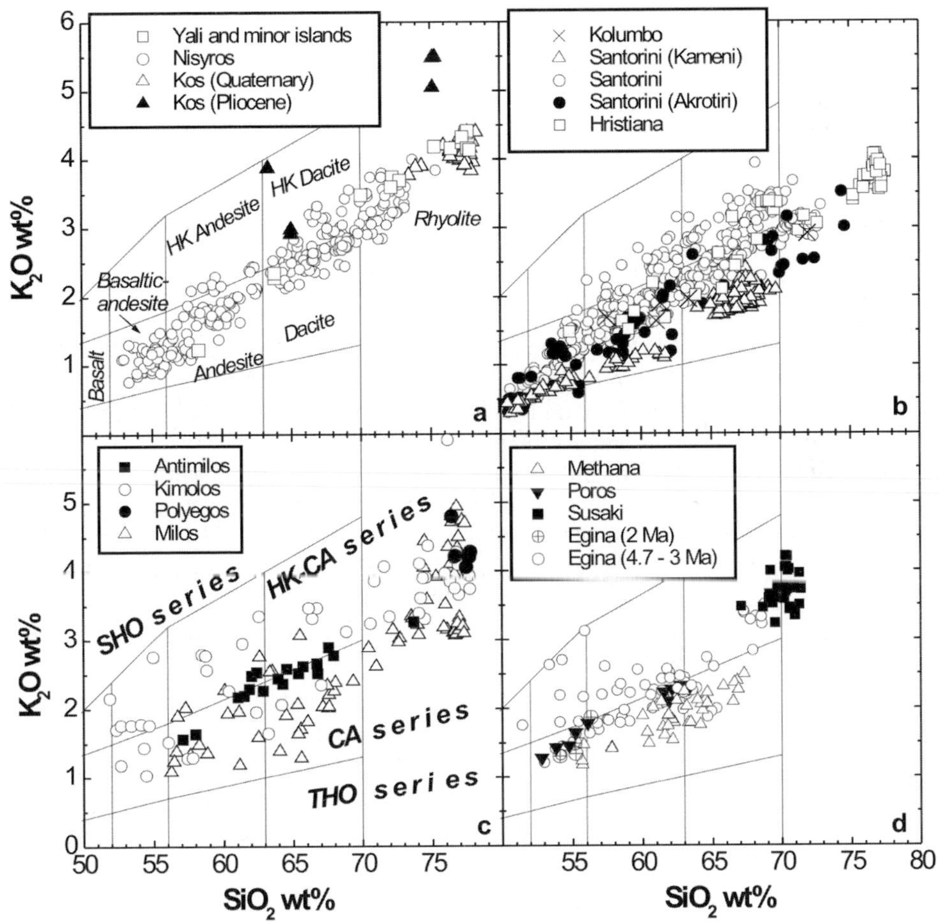

Fig. 3. K_2O versus silica classification diagrams (Peccerillo and Taylor, 1976) for all the SAAVA fields. Data are reported on water-free bases. SHO: shoshonitic, HK-CA: high-K calc-alkaline, CA: calc-alkaline, THO: tholeiitic. See text for the data font.

2.2.1. Egina

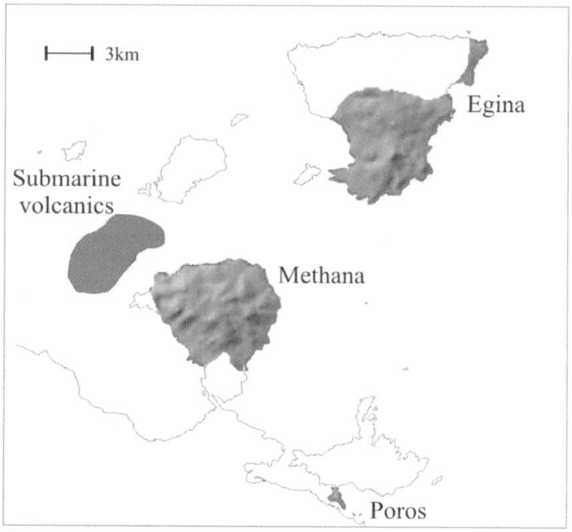

Fig. 4. The Egina-Poros-Methana volcanic field.

Egina island (Fig. 5) is formed by volcanic products for about 2/3 (about 55 km^2) of its area. Volcanic activity started in the Lower Pliocene (Muller et al., 1979; Pe-Piper et al., 1983; Fytikas et al., 1986a). Between 4,7 and 4,3 Ma a variety of volcanic products, from rhyodacitic pumice flows and tuffites to andesitic pillow lavas and hyaloclastites, were deposited by different centers in a shallow submarine environment. From 3,9 to 3,0 Ma the main phase of the volcanic activity was manifested, building up the central and southern Egina island. Extrusion of dacitic to andesitic magmas built up lava domes and dome complexes with associated lava flows, in mainly subaerial environment.

Between 3,0 and 2,1 Ma a low rate eruption period probably occurred: two samples coming from debris flows gave K-Ar ages of 2,5 and 2,2 Ma (Matsuda et al., 1999). At 2,1-2,0 Ma the last eruptive period on Egina occurred, producing subaerial andesitic and basaltic andesitic lava domes and associated flows.

The distribution of the volcanic centers and the shape of the edifices is mainly controlled by the active tectonic lines trending E-W and ENE-WSW In some cases, NW-SE trending lineaments had also an important role.

Egina hosts a hot spring (25 °C) in its north coast, and low temperature (30-40 °C) geothermal reservoirs at a depth of a few hundred meters.

The rocks composition ranges from basaltic andesites to dacites of CA and HK-CA serial affinities (Fig. 3).

Lavas are generally porphyritic; basaltic andesites and andesites have phenocrysts of plagioclase, augite, hornblende, magnetite and minor olivine and hypersthene, whereas dacites contain phenocrysts of plagioclase, hornblende, magnetite and minor clinopyroxene.

Mg-values [molecular Mg/(Mg+Fe)*100] are up to 70. TiO_2 (0.6-0.9 wt%) and K_2O (1.2 – 3 wt%) contents are quite scattered among basaltic andesites. The youngest rocks tend to have lower potassium contents, being CA basaltic andesites (Fig. 3). $^{87}Sr/^{86}Sr$ ratios range from about 0.704 to 0.7067 in rocks with silica less than 56 wt% and remain around 0.706 in the most evolved rocks (Pe, 1975; Pe-Piper and Piper, 1979; Fytikas et al., 1986a; Mitropoulos et al., 1987; Gulen, 1989).

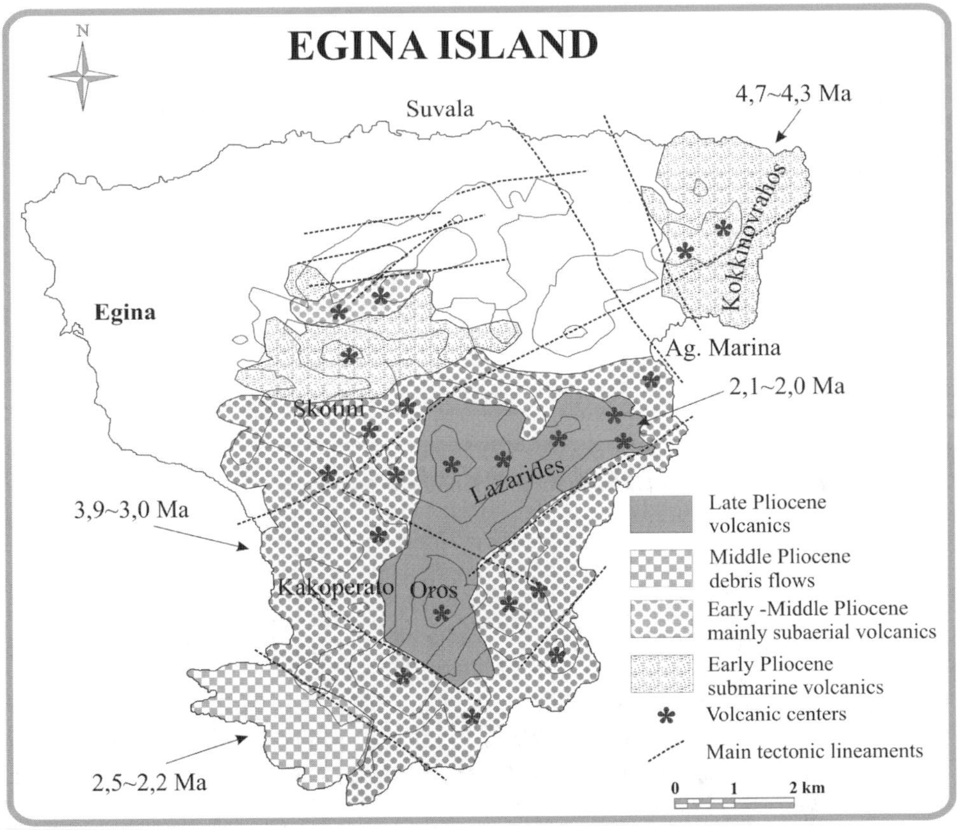

Fig. 5. Egina island geological sketch map (modified from Fytikas et al., 1986a).

2.2.2. Poros

On Poros island one limited outcrop (about 1 Km^2) of a lava dome and associated lava flows, with a K-Ar age of 3,1-2,6 Ma is found (Fytikas et al., 1986a; Matsuda et al., 1999). This volcanic center is located in the eastern edge of the ENE-WSW trending active Trizina graben.

The rocks are evolved andesites with silica between 60.7 - 63.5 wt% and Mg-values around 55. They are porphyritic with phenocrysts of plagioclase and clinopyroxene. The magmatic enclaves found in the lavas are basaltic andesites with Mg-values up to 67, thus a compositional gap (silica between 56 - 61 wt%) is present between the compositions of the host rocks and the enclaves (Fig. 3). The available three data of $^{87}Sr/^{86}Sr$ on andesites are variable and quite high, ranging from 0.7058 to 0.7074 (Pe, 1975; Fytikas et al., 1986a; Mitropoulos et al., 1987).

2.2.3. Methana

Methana peninsula (Fig. 6) is characterised by the youngest volcanic products of the whole Saronikos area: the oldest K-Ar dated products are of 0,9 Ma and older volcaniclastics were also probably deposited in Quaternary. The youngest volcanic event built up the Kameno Vuno andesitic dome and the associated flow at about 230 BC (Strabo, Geographica).

Volcanic edifices are lava domes and dome complexes, with associated lava flows, block and ash flows and related debris flows. The presence of distinctive lithosomes and the absence of guide levels do not permit an accurate reconstruction of Methana volcanic history. Different data agree for a nearly continuous low rate production activity for the whole active period. Vent distribution and edifices shape seems to be

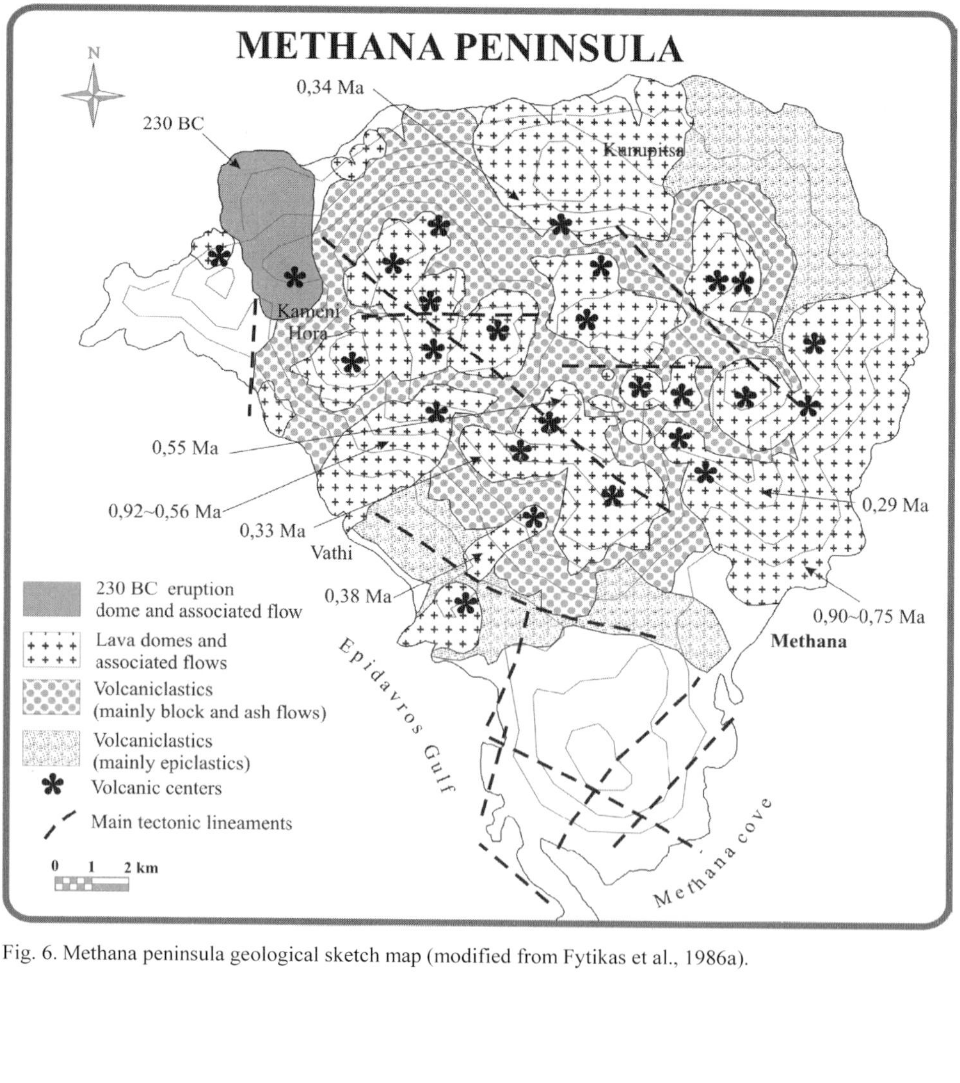

Fig. 6. Methana peninsula geological sketch map (modified from Fytikas et al., 1986a).

controlled by both E-W and NW-SE trending tectonic lineaments.

A few km NW offshore of the Kameno Vuno center, submarine volcanics have been identified (Fig. 4) (Papanikolaou et al., 1988). The age of these volcanics has been estimated at about 1,0 Ma, based on the thickness of the sedimentary cover (Pe-Piper and Piper, 2002).

The volcanic rocks are mainly CA andesites and dacites, with rare basaltic andesites (Fig. 3), and generally show a lower potassium content than those of Egina and Poros.

Plagioclase, clino- and orthopyroxene, and magnetite constitute the phenocryst paragenesis of andesites, whereas in the dacites, plagioclase and magnetite are associated with horneblende and biotite.

Major and trace element contents are generally well correlated with silica and Mg-values are up to 65. TiO_2 contents of the most mafic rocks are around 0.8 wt%. Sr isotope ratios range from about 0.7057 to 0.7066; they do not show a clear correlation with the degree of magma evolution (Pe, 1975; Fytikas et al., 1986a; Mitropoulos et al., 1987; Gulen, 1989).

2.3. Milos volcanic field

Milos, Kimolos, Polyegos and Antimilos islands, as well as Ananes islets, are almost entirely constituted by volcanic products (Fig. 7). The activity started about 3.5 Ma ago in Milos and Kimolos and continued up to recent time with hydrothermal explosions in Milos.

Milos island group is the third SAAVA field with no important central volcano structure. Only Antimilos center could be considered a composite volcano, built up by lava flows and volcaniclastics, as well as acid lava domes and associated flows (Marinos, 1960; Fytikas et al., 1986b).

Milos, Kimolos and Polyegos islands are compound volcanoes constituted by different volcanic edifices, mainly lava domes resulting from limited vent migration, which intrude thick volcaniclastic deposits (Fytikas et al., 1986b; Fytikas and Vougioukalakis, 1993, Francalanci et al., 1994, 2003; Stewart and McPhie, 2003).

The oldest K-Ar date on the lowest volcanic outcrops in the SW coast of Milos island, gave an age of 3,5 Ma (Fytikas et al., 1986b). Biostratigraphic - magnetostratigraphic data and astronomical polarity time scales (APTS) correlations from the marine sediments deposited before the onset of the volcanic activity in South Milos (Van Hinsbergen et al., 2004) gave similar results (4,5-3,7 Ma). It is not clear if the volcanic ash layers of older ages (between 6 and 4,6 Ma) intercalated in these sediments are early volcanic events from Milos centers or they have

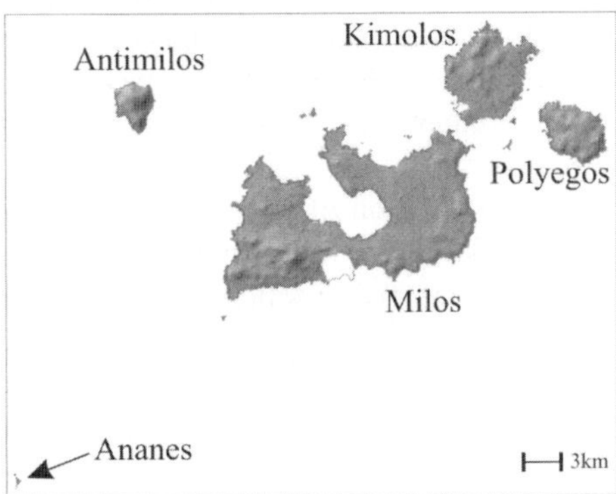

Fig. 7. Milos volcanic field.

been deposited as distal ash from Northern Aegean centers (Antiparos, Patmos).

From 3,5 to 1,6 Ma, alternation of explosive and extrusive activity built up Kimolos, Polyegos and most of the present Milos island (Figs. 8-10). Ananes islets volcanic center is probably one of the earliest in the area. Extensive hydrothermal alteration of the outcropping volcanic rocks does not permit any dating or stratigraphic correlation with the rest of the Milos island group outcrops.

During most of this period, Milos volcanites were deposited in submarine environment, while Kimolos was partly emerged since about the middle of Pliocene. The presence of a large intrusive body in the subsurface of Kimolos area was detected by geophysical exploration, by very limited outcrops, as well as by numerous lithic fragments in the pyroclastic deposits. This is probably the cause of the rapid uplift of this part of the volcanic field.

Volcanic vent distribution and shape of the edifices is clearly controlled in the whole area by NE-SW to E-W trending lineaments, with a subordinate role of the N-S lineaments.

During Quaternary, volcanic activity fed by acid magmas was concentrated in central Milos island and in Antimilos. It developed in three distinctive periods. Between 1,1-0,9 Ma huge rhyolitic extrusive activity was focused in a N-S trending area of central Milos, building up a huge lava field and some lava domes. At about 0,38 Ma the perlitic Trachylas tuff ring and associated rhyolitic flows were built up in the northern extent of previous N-S trending lineament. The rhyolitic domes intruding the southern part of Antimilos volcano have the same age. The last volcanic event on Milos occurred in Holocene (90 Ka according to a K-Ar dating by Fytikas et al., 1986b; 19 Ka according to fission track dating by Principe et al., 2003) building up the perlitic Fyriplaka tuff ring and associated rhyolitic flows, in the South edge of the N-S trending Zephyria active tectonic graben.

The installation in the East-Central Milos area of a huge geothermal field (320 °C in 1 km depth nowadays) and the high seismicity of the area since Holocene, had the result of triggering of many large hydrothermal explosions, both before and after the last volcanic eruption. These events deposited in the central-eastern Milos area a vast mantle of debris and mud flows rich in basement schist fragments which was mapped as the "Green Lahar" formation (Fytikas et al., 1986b). Hydrothermal explosions continued up to historic time (80-200 AD, Traineau and Dalabakis, 1989) in the Agia Kyriaki area, where a fumarolic activity with gas temperatures of 100-101 °C is also present nowadays.

The composition of Milos rocks is mainly CA, with only few samples falling in the HK-CA field of the K_2O versus silica classification diagram (Fig. 3). The most mafic rocks have 56 wt% of silica and Mg-values around 55. They are found together with abundant andesites, dacites and rhyolites. The youngest rhyolites appear to have higher K_2O contents than the oldest rocks.

Most of the rocks are porphyritic with holocrystalline to glassy groundmasses; only some rhyolites are completely glassy (obsidians). Andesites contain phenocrysts of plagioclase, augite, orthopyroxene and magnetite; rare unstable olivine and hornblende

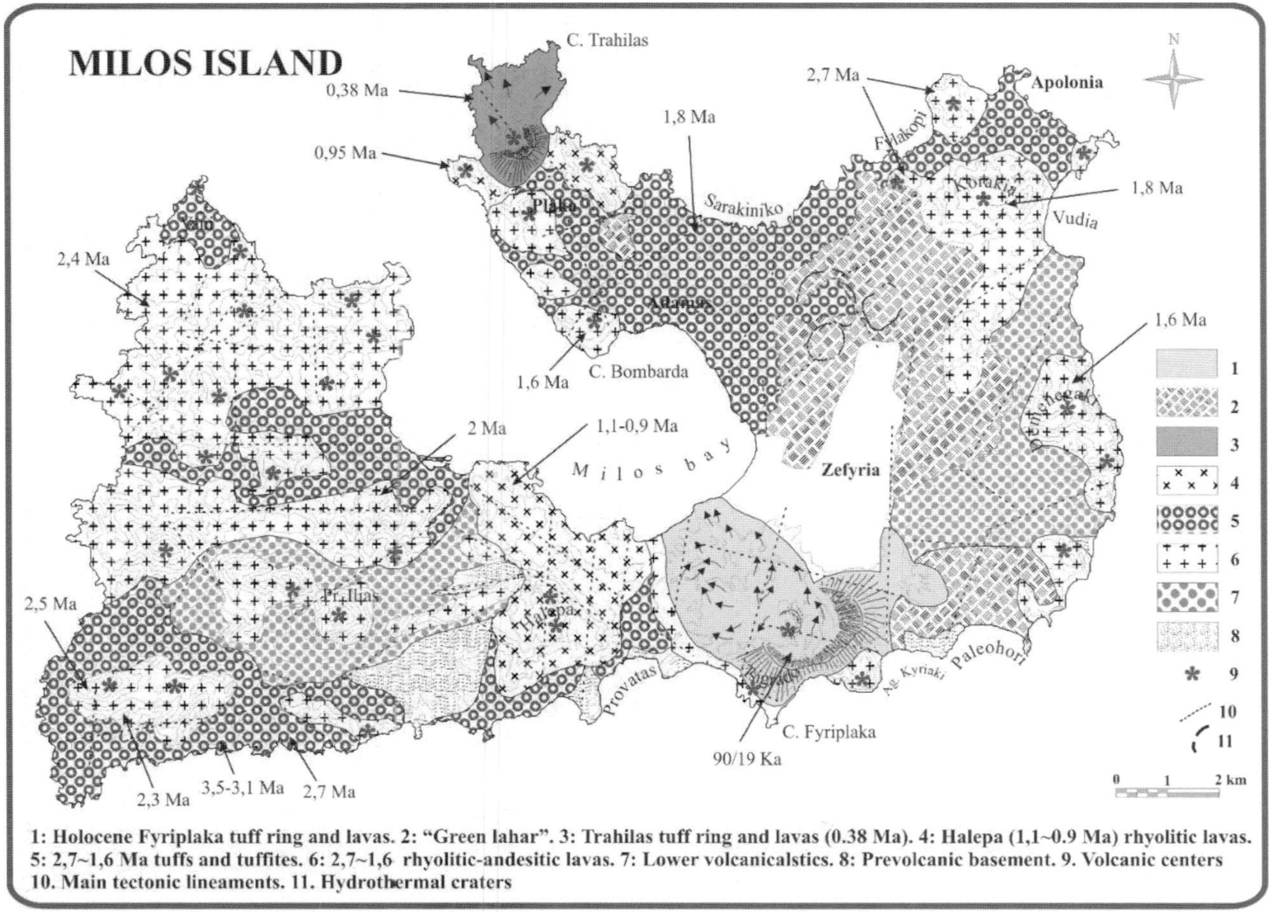

Fig. 8. Milos island geological sketch map (modified from Fytikas et al., 1986a).

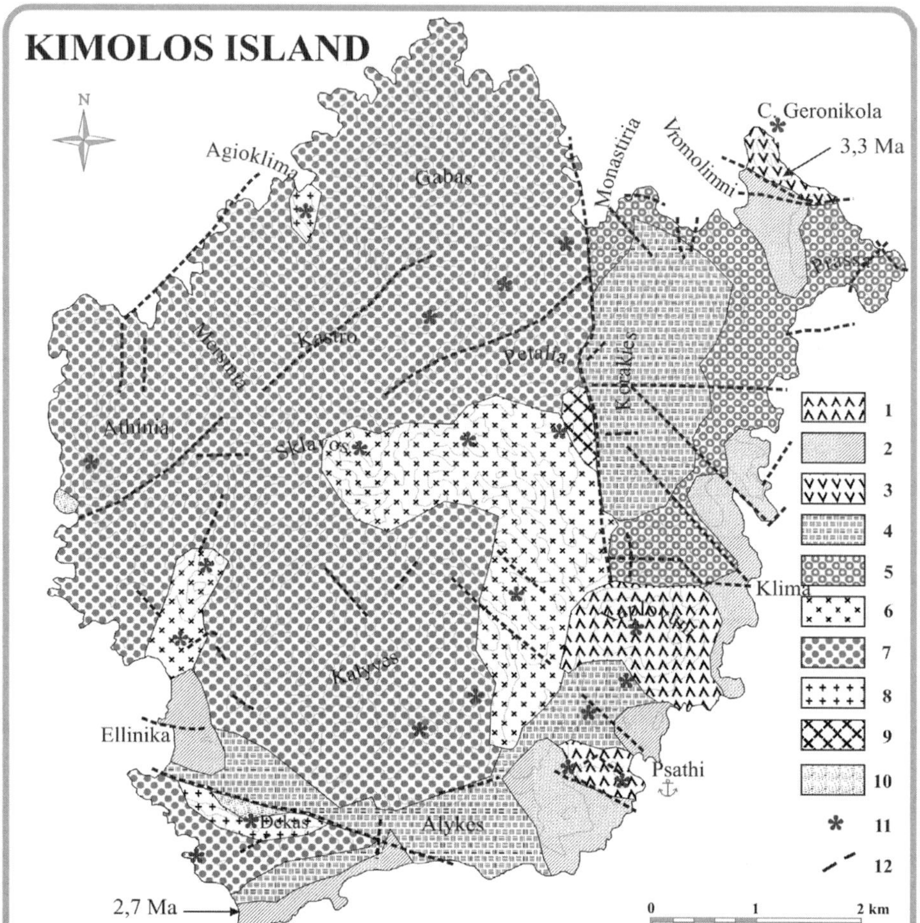

1. Upper Rhyolitic lavas. 2. Upper Rhyolitic volcaniclastics. 3. Geronikola basaltic andesitic to andesitic tuff ring and lava flows. 4. Subaerial (northern) and submarine (southern) epiclastics. 5. Middle rhyolitic volcaniclastics. 6. Middle basaltic to andesitic lavas. 7. Lower volcaniclastics. 8. Lower lavas. 9. Intrusive body. 10. Prevolcanic basement. 11. Volcanic centers. 12 Main tectonic lineaments

Fig. 9. Kimolos island geological sketch map (modified from Francalanci et al., 2003).

are present. In the dacites, the phenocryst paragenesis is constituted by plagioclase, ortho- and clinopyroxene associated to hornblende, rare biotite and corroded quartz. In the rhyolites, plagioclase, biotite and quartz are invariably present, but amphibole and sanidine only sometime occur.

TiO_2 contents are generally lower than 0.9 wt%. The HK-CA samples generally contain higher TiO_2, Rb, Sr and Zr abundances than the CA samples. Sr isotope ratios largely vary, passing from 0.7037 to 0.7076. This large range is found among rocks

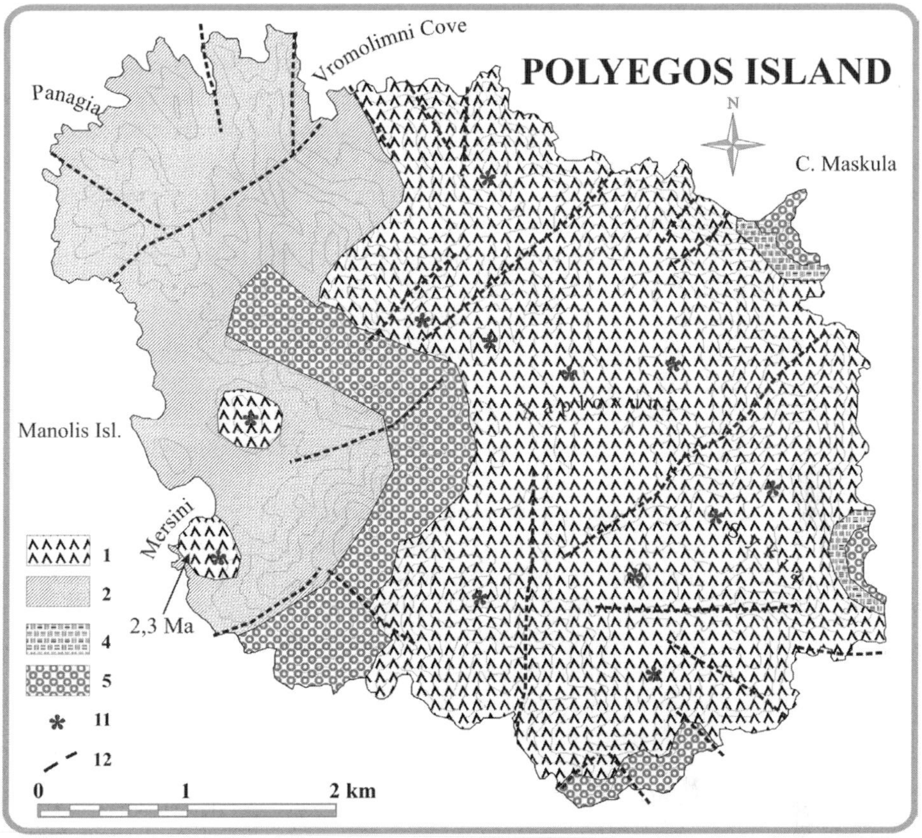

Fig. 10. Polyegos island geological sketch map (modified from Francalanci et al., 2003).

with silica <68 wt%; for rocks with higher silica contents, $^{87}Sr/^{86}Sr$ values generally show a narrower variation, from 0.7051 to 0.7065. $^{143}Nd/^{144}Nd$ values range from 0.51247 to 0.51275 and are negatively correlated with Sr isotope ratios (Innocenti et al., 1981; Barton et al., 1983; Briqueu et al., 1986; Fytikas et al., 1986b, Mitropoulos et al., 1987; Gulen, 1989)

The Kimolos rocks belong to the CA and HK-CA series, ranging from basalts (only few) to rhyolites (Fig. 3). The rocks are variably porphyritic with hypo-crystalline to glassy groundmasses. Olivine is only present in the oldest basalts, where orthopyroxene is lacking. The latter phase is, however, always present in the other rocks becoming accessory in the rhyolites. Clinopyroxene is only lacking in some rhyolites. Plagioclase is ubiquitous and abundant in all the rocks. Hornblende is abundant in dacites and some rhyolites. Biotite, magnetite, sanidine and quartz are also found as accessory minerals in rhyolites.

Mg-values are quite variable also among the most mafic rocks ranging between 65-42. TiO_2 contents of mafic rocks are quite high, varying between 0.9-1.4 wt%. The

Kimolos rocks, except for the youngest ones, generally display higher TiO_2, K_2O and incompatible element contents and lower Sr abundances and Mg-values than the Milos and Antimilos rocks. The youngest products, however, are more similar to the Milos and Antimilos rocks with the same degree of evolution. As for potassium contents, the Kimolos rhyolites are similar to the youngest rhyolites of Milos. The $^{87}Sr/^{86}Sr$ values are between 0.7045-0.7064, displaying a narrower range in respect with the Sr isotopes of Milos (Fytikas et al., 1986b; Francalanci et al., 1994, 2003).

The Polyegos rocks are all rhyolites similar in composition to the Kimolos rhyolites (Fig. 3). Even the Sr isotope ratios are in the range of the most evolved rocks of Kimolos. The Polyegos rocks have glassy, often perlitic, groundmasses with abundant plagioclase and accessory amount of orthopyroxene, hornblende, biotite, magnetite, sanidine and quartz (Fytikas et al., 1986b; Francalanci et al., 1994, 2003).

In the K_2O – silica classification diagram the Antimilos rocks fall at the boundary between the CA and HK-CA series. They are mainly andesites and dacites, with rare rhyolites (Fig. 3).

These rocks are variably porphyritic with hypo-crystalline to glassy groundmasses. Olivine is rarely present in the less evolved members. Orthopyroxene is present only in the evolved rocks becoming accessory in the rhyolites. Clinopyroxene is only lacking in some rhyolites. Plagioclase is ubiquitous and abundant in all the rocks. Hornblende is abundant in dacites and some rhyolites. Accessory abundances of biotite, magnetite, sanidine and quartz are also found in rhyolites.

Major and trace element characteristics of Antimilos rocks more strictly resemble those of Milos rocks (Fytikas et al., 1986b).

2.4. Santorini volcanic field

Santorini volcanic field (Hristiana islets, Santorini island group and Kolumbo submarine volcano) (Fig. 11) is found in the central part of the SAAVA. This volcanic field was built up on the thinnest continental crust (~25 km thick) in comparison with the rest of the SAAVA fields and nearest to the max extension area of the Aegean Sea (Cretan basin).

Santorini volcanic field has developed on the northern margin of a basement horst called the Santorini-Amorgos ridge. If extrapolated, the fault defining the northern margin of the horst passes through the centre of Santorini caldera, where it gives rise to a NE-SW zone of surface faulting, vent

Fig. 11. Hristiana-Santorini-Kolumbo volcanic field.

alignments, and gas emission known as the Kameni Line (Heiken and McCoy, 1984; Druitt et al., 1989; Fytikas et al., 1990; Barberi and Carapezza, 1994). Basement rocks which crop out in the south-eastern half of Thira island lie on a continuation of the Santorini-Amorgos ridge. The north-western half of the volcanic field lies within the Anydros graben. The Kameni Line splits the caldera in two halves and played a fundamental role in the structural development of the volcanic field. Historic subaerial vents of the Kameni Islands lie on the Kameni Line and are confined to a 600m-wide elongate zone trending N65°E parallel to it. Vents for at least six prehistoric plinian eruptions also lay on the Kameni Line. The Amorgos 1956 earthquake (Ms 7,5), the largest shallow seismic event of the Aegean area in the past century, was generated along the same tectonic lineament. Present shallow seismic events, which are registered in the area, are also located along the same lineament, mostly concentrated near to the Kolumbo submarine volcano (Dimitriadis et al. this volume).

2.4.1. Hristiana islands

Hristiana islet group (Hristiani, Askania and Eschati) are constituted by lava domes and flows, with some relevant volcaniclastics intercalations (Puchelt et al., 1977; Aarbourg 1998, Aarbourg and Frechen, 1999). Bathymetric and tectonic data (Mountrakis et al., 1998) suggest that the islets are the remnants of a large volcano that has been dissected and submerged. No dating has been carried out on the Hristiana volcanics. Correlation of the Hristiani in situ pyroclastic tuffs with the oldest ash layers mantling the pre-volcanic basement on Santorini (1 Ma, Seward et al., 1980) suggest that Hristiana volcanic activity is the earliest in the area.

The Hristiana rocks are andesites, dacites and rhyolites, with rare basaltic andesites. They belong mainly to the CA series and only some dacites plot in the HK-CA field (Fig. 3). Few rhyolites have around 72 wt% of silica, but most of the rhyolites show silica between 75-78 wt%.

Mg-values are quite variable with a maximum up to 50. The trace element composition of the Hristiana samples generally overlaps that of the Santorini rocks, a part from a slightly higher Nb content of the former samples. The HK-CA dacites and the least evolved rhyolites display higher incompatible trace element contents than the other Hristiana rocks. The most silicic rhyolites are less enriched in Zr, Nb and Rb and seem to more closely resemble the oldest (Akrotiri) Santorini rocks (Aarburg, 1998; Aarburg and Frechen, 1999).

2.4.2. Santorini island group

Santorini is composed of five islands. Thira, Thirasia, and Aspronisi are arranged in a dissected ring around a flooded caldera containing the post-caldera islands of Palea Kameni and Nea Kameni. The caldera is a composite structure resulting from at least four collapse events (Druitt and Francaviglia, 1992), the last of which was triggered by the Minoan eruption (3,6 ka BP). It is bounded by cliffs up to 300 m high and extends to as much as 390 m below sea level. The islands of Palea and Nea Kameni postdate the

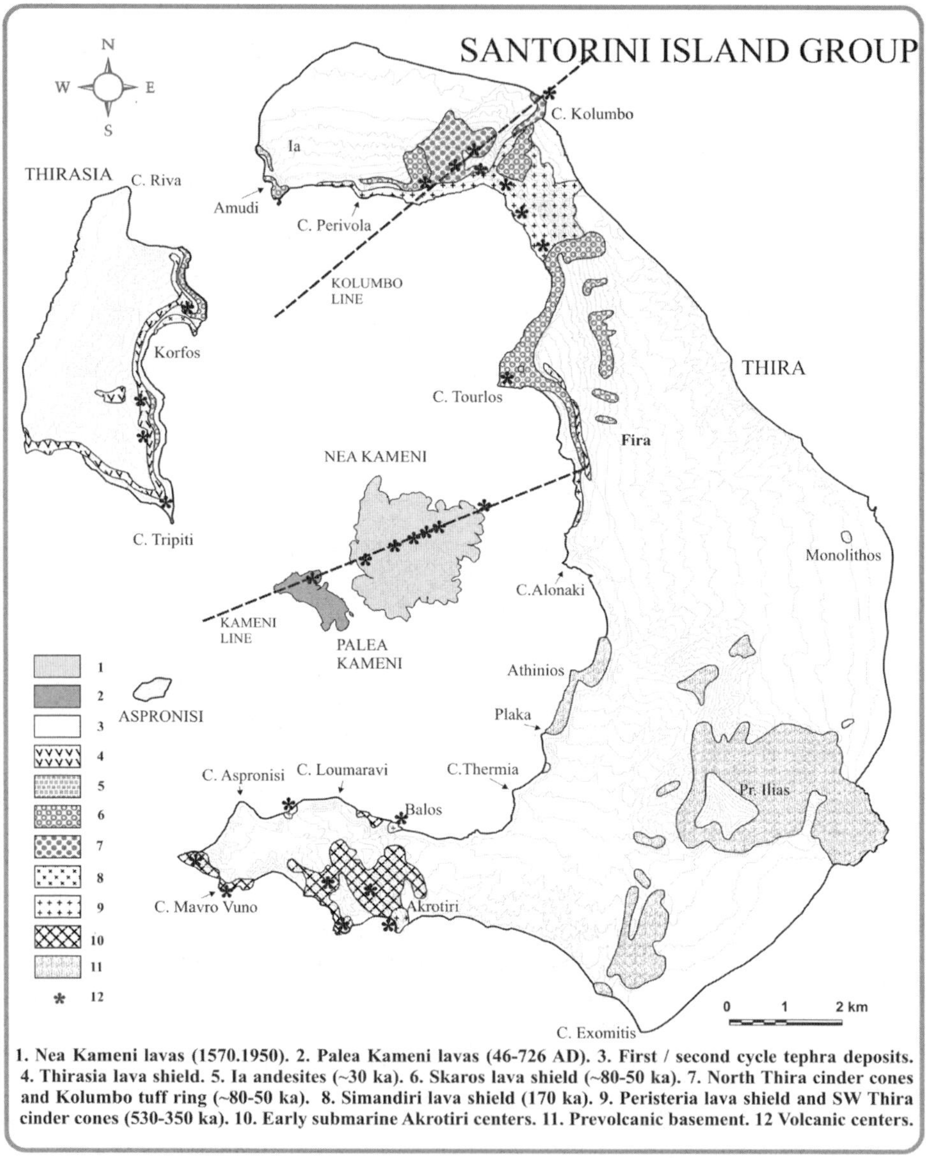

Fig. 12. Santorini island group geological sketch map (modified from Druitt et al., 1999).

Minoan caldera collapse and are the subaerial expressions of a dominantly submarine dacitic shield ~2 km^3 in volume. Complessively, the whole Santorini structure could be defined as a composite volcano (Fig. 12).

Santorini is one of the most violent caldera-forming volcanoes of the world. During the last 360 ka, over a hundred explosive eruptions were manifested. Twelve of these

All the historically registered explosions at Nisyros (1873-1887) are hydrothermal and created three small craters in the west area of the remaining caldera floor (Gorceix, 1873, 1874; Martelli, 1917). The presence of more than 8 older hydrothermal explosion craters in the caldera floor, with a maximum diameter of 300m, indicate that this type of activity was frequent in the past few thousand years. The island is today a site of intense hydrothermal activity which feed many fumaroles in the caldera floor area and hot springs along the coast, hosting a high enthalpy geothermal field (fluids with more than 500°C at 1800m depth) (Chiodini et al., 1993).

Active tectonics is prominent on the Nisyros island group. Some of the active faults have been reactivated during seismic crisis manifested in the last century. During the last 1995-1997 shallow earthquake activity, two of these fractures trending N-S and NW-SE were opened on Nisyros and Yali, respectively (Vougioukalakis et al., 1998).

Nisyros is characterised by a nearly continuous series of volcanic rocks, from basaltic andesites to rhyolites. Basaltic andesites and most of andesites are CA, whereas dacites and rhyolites plot along the boundary between CA and HKCA series (Fig. 3).

Rocks have a variable porphyritic index, ranging from nearly aphyric to highly porphyritic (up to 60 vol%). Plagioclase, clino- and orthopyroxene and opaques are ubiquitous, but present in different amount. Plagioclase is always higher than 70 vol%, whereas the others are always lower than 20 vol% of phenocryst content. Olivine is found up to andesites, whereas amphibole is present in dacitic and rhyolitic lavas and in andesitic enclaves. Biotite is rarely present.

TiO_2 contents of the most mafic rocks range between 0.6-1.0 wt%, but tend to increase up to 1.2 wt% in rocks with about 60 wt% of silica. Mg-values are up to 70. Some elements, like K_2O, Rb and Ba, form single and smooth trends with silica, whereas other elements, such as Sr, Zr, Nb, show more scattered variations, with the samples of the post-caldera domes forming nearly distinct trends. The latter rocks tend to have lower Zr, Nb and LREE and higher Sr and Mg-values than the other Nisyros rocks with the same degree of evolution. $^{87}Sr/^{86}Sr$ ratios are mostly between 0.7034-0.7051, with a higher value of 0.7064. Sr isotope ratios are negatively correlated with Nd isotope ratios (0.51258 –0.51282) (Mitropoulos et al., 1987; Wyers and Barton, 1989; Gansecki, 1991; Vougioukalakis, 1993; Francalanci et al., 1995; Innocenti, 1998).

2.5.3. Yali and Strogyli islets

Yali islet is the youngest volcanic centre of Nisyros island group. It is an Upper Quaternary rhyolitic volcanic edifice. Two small hills, with a maximum height of 165m, are connected with a narrow isthmus of alluvial and seashore deposits. The SW part consists entirely of volcaniclastic deposits. The NE part consists of obsidian lava domes and flows with subordinate pyroclastic deposits. Agios Antonios islet emerge between the two hills of Yali and consist of dacitic lavas.

Two volcanic cycles are distinguished, both characterised by an initial explosive eruption that emplaced rhyolitic pumice fall (Lower Pumice - submarine and Upper Pumice - subaerial), followed by extrusions of obsidian-perlitic lava domes and flows. A marine terrace deposit separate these two cycles. Reliable absolute dating are lacking and chronostratigraphy is not well constrained. In any case, Lower Pumice post-date the Kos Plateau Tuff as pumice of this eruption is found as lithics in the Yali Lower Pumice deposits. Moreover, the probable age of 31 ka for the Yali Upper Pumice and 24 ka for the following obsidian domes suggest a quite young age for the second eruptive cycle of Yali (Wagner et al., 1976). This young age leads us to consider this centre as a potentially active volcano.

Strogyli is a steep andesitic cone, lying on a 500 m deep sea bottom and with a top of 120 m above sea level. The Yali Upper Pumice cover the floor of its small central crater, postdating the Strogyli cone formation to an age >31 ka.

Yali volcanics are all rhyolites with silica between 70-78 wt%, whereas Strogyli rocks are less evolved, being andesites with about 57 wt% of SiO_2. In the potassium-silica diagram all the previous rocks follow the same trend of Nisyros and Quaternary rocks from Kos (Fig. 3).

Agios Antonios dacite is porphyritic, with phenocrysts of plagioclase, clinopyroxene, orthopyroxene and magnetite. The Yali pyroclastites are aphyric with rare microphenocrysts of plagioclase and corroded quartz in the Lower Pumice and of plagioclase, hornblende, clinopyroxene and magnetite in the Upper Pumice. The obsidian lavas have rare phenocrysts similar to the previous minerals, but biotite is also present. Strogyli andesites are porphyritic with phenocrysts of plagioclase, clinopyroxene, orthopyroxene and magnetite.

Strogyli andesites have the highest Mg-value of 58. TiO_2 is always lower than 0.6 wt%. Strogyli, Pahia and Pyrgussa rocks have lower Nb and Zr and higher Sr, Ni and Mg-values than the main Nisyros rocks with the same silica contents. Yali rhyolites show higher Zr, Rb and lower Ba and Ni than the Kos youngest rhyolites.

3. PETROCHEMICAL VARIATIONS ALONG THE SAAVA

In spite of the generally similar calc-alkaline character of all the SAAVA magmas, several systematic mineralogical and geochemical variations in space and time occur.

Mitropoulos and Tarney (1992) found several mineralogical differences between Santorini rocks and those from the other volcanic fields. The most evident are those occurring in olivine, amphibole and opaques. Olivine is more commonly present in Santorini, whereas in the rocks of the other volcanic field, it occurs only in the most mafic basalts. Hornblende is usually found as phenocryst in all the volcanic centres except for Santorini, where it is mainly restricted to the Akrotiri lavas, erupted during the early stages of the volcano history. The opaques in Santorini have higher Ti content and Fe^{2+}/Fe^{3+} values, indicating lower oxygen fugacity.

In the external sectors of the arc a general decrease with time of potassium and incompatible trace element contents is observed, contrarily to what is expected for the volcanic arcs. Indeed, in the western part, among the youngest rocks (2 million years old Egina products and Methana peninsula) only CA compositions are found, whereas the older volcanics are CA and HK-CA (Figs. 3, 17a). At East, the only HK-CA rocks are the Pliocene rocks from Kos, but all the younger samples have lower K_2O contents.

At Milos volcanic field this tendency is not clear, even if the youngest rocks of Kimolos are clearly CA. In the central sector of the arc, on the other hand, Hristiana and Akrotiri rocks follow the same trend of the younger Santorini rocks. Only the more evolved Kameni lavas have lower K_2O contents, probably due to different processes of evolution (Francalanci et al., 1998). It is noteworthy, however, that all the volcanics from this central group are Quaternary in age (Figs. 3, 17a), thus suggesting a general more depleted character for all the youngest rocks of the arc.

The lowest silica contents (47-51 wt%), highest Mg-values and compatible trace element abundances (e.g., Sc and V up to 40 and 400 ppm, respectively) are found in the Santorini volcanic field rocks (Figs. 3, 17b). The most abundant and highest silica rocks (up to about 78 wt%) are present at the Milos and South Kos – Nisyros volcanic fields. Quite high silica samples (> 75 wt%) are, anyway, also observed at Hristiana, whereas in the western part, silica contents remain always lower than 72 wt% (Figs. 3, 17b).

Passing from the western to the eastern sectors of the arc, there is a general decrease of Sr isotope ratios, both in the lowest and highest values. This variation is evident either among all the samples or in the rocks with silica <60 wt% (Fig. 18). In the same direction, among variably evolved samples an increase of Nd isotope ratios is also generally observed. From Santorini to Nisyros island groups, however, Nd isotope ratios tend to slightly decrease again. This leads to a general correlation of $^{87}Sr/^{86}Sr$ versus $^{143}Nd/^{144}Nd$ of the most mafic rocks (silica <60 wt%), with the Santorini samples having the highest Nd isotope ratios but not the lowest Sr isotope ratios (Fig. 19). Fig. 19 also shows that the general correlation of SAAVA samples points to the isotope compositions of marine sediments and, in particular, to those of the Eastern Mediterranean sediments (Weldeab et al., 2002).

Pb isotope ratios also show a general tendency to decrease from the western to the eastern volcanic fields. $^{206}Pb/^{204}Pb$ values range from about 18.65 to 18.95, falling in the field of mid ocean ridge basalts (MORB), whereas $^{207}Pb/^{204}Pb$ (15.62-15.70) and $^{208}Pb/^{204}Pb$ (38.5-39.2) are generally higher than MORB values. Only the Pb isotopes of Nisyros are very close to those of MORB (Fig. 20).

According to the variation of FeO/MgO along the magmatic evolution series, there is an increase of the tholeiitic character of SAAVA magmas from west to Santorini island group, for decreasing again toward the eastern magmas (Fig. 21). However, based on the Al_2O_3 versus alkali index diagram for the mafic rocks (Fig. 22) most of the SAAVA magmas fall in the calc-alkaline field, with the exception of some basalts from Santorini island group.

Incompatible trace element patterns for the most mafic rocks are typical of calc-alkaline magmas, with high LILE/HFSE and REE/HFSE ratios. They have quite similar shapes for all the SAAVA sectors, even if the rocks of Santorini island group have lower incompatible element contents (Fig. 23). Rare earth element patterns are similar, with $La_N/Sm_N > 1$ and Tb_N/Yb_N mostly around 1. La_N/Sm_N is usually > 3, reaching lower values (up to 1.5) only in Santorini rocks.

4. DISCUSSION

The SAAVA rocks show large volcanological, geochemical and petrological variations within each volcanic fields and going from the western to the eastern sectors of the arc.

Some systematic correlations seem to exist with age of volcanism. Indeed, the onset of subaerial activity started at West on Egina island (at about 4.7 Ma), then occurred on Milos island group and in the eastern part and, finally, it appears in the central sector, on the Santorini island volcanic field (Fig. 17a). Thus, in a time span of about 1-2 Ma, subaerial activity appeared shifting from the external to the central area of the SAAVA. The end of volcanism, however, does not occur with a similar systematic time sequence, because active or potentially active volcanoes are present in most of the volcanic fields.

Correlations between age and compositional characteristics of rocks in the volcanic fields of the external sectors of the arc are also found, with a general shift from HK-CA to CA magmatisms.

Most of the volcanological, geochemical and isotopic characteristics follow geographical trends, changing along the arc from the western to the eastern part. At West, the volcanic fields are mainly constituted by small, monogenic eruptive centers, whereas in the central and eastern sectors, large composite volcanoes with caldera collapses dominates. From a geochemical point of view, a decoupling between the behaviour of isotope ratios (especially Sr and Pb isotopes) and the other petrochemical characteristics is observed. Indeed, isotope ratios mostly change from West to East of the arc, whereas element contents and ratios vary passing from the central to the external sectors of the SAAVA (Figs. 3, 17-23).

Several magma differentiation processes can be responsible for the compositional variability in the SAAVA rocks. Many petrochemical variations, especially those occurring within a single volcanic center, are clearly determined by shallow level evolutionary processes, whereas the West-East compositional changes are more probably due to magma source variations.

4.1. Magma differentiation processes at shallow levels

Large compositional variations are present in all the volcanic fields, where magmas generally range from basalts or basaltic andesites to rhyolites (Fig. 3), with highly variable Mg-values ranging between 80-15. Silica contents and Mg-values are also usually well correlated with the other major and trace element abundances. This large spread of compositions is mostly caused by fractional crystallisation (FC) of the main mineral phases found in the variably evolved magmas. Sr, Nd and Pb isotope ratios, on the other hand, also show quite large variations within the single volcanoes (Figs. 18-20) and they are usually positively correlated with the degree of magma evolution. These isotopic characteristics have been usually ascribed to processes of crustal assimilation, which are often associated to fractional crystallisation (AFC). The FC and AFC processes acted at more extent for producing the large amount of rhyolitic magmas erupted from the SAAVA volcanic centers. Most of the SAAVA rhyolites, in fact, can

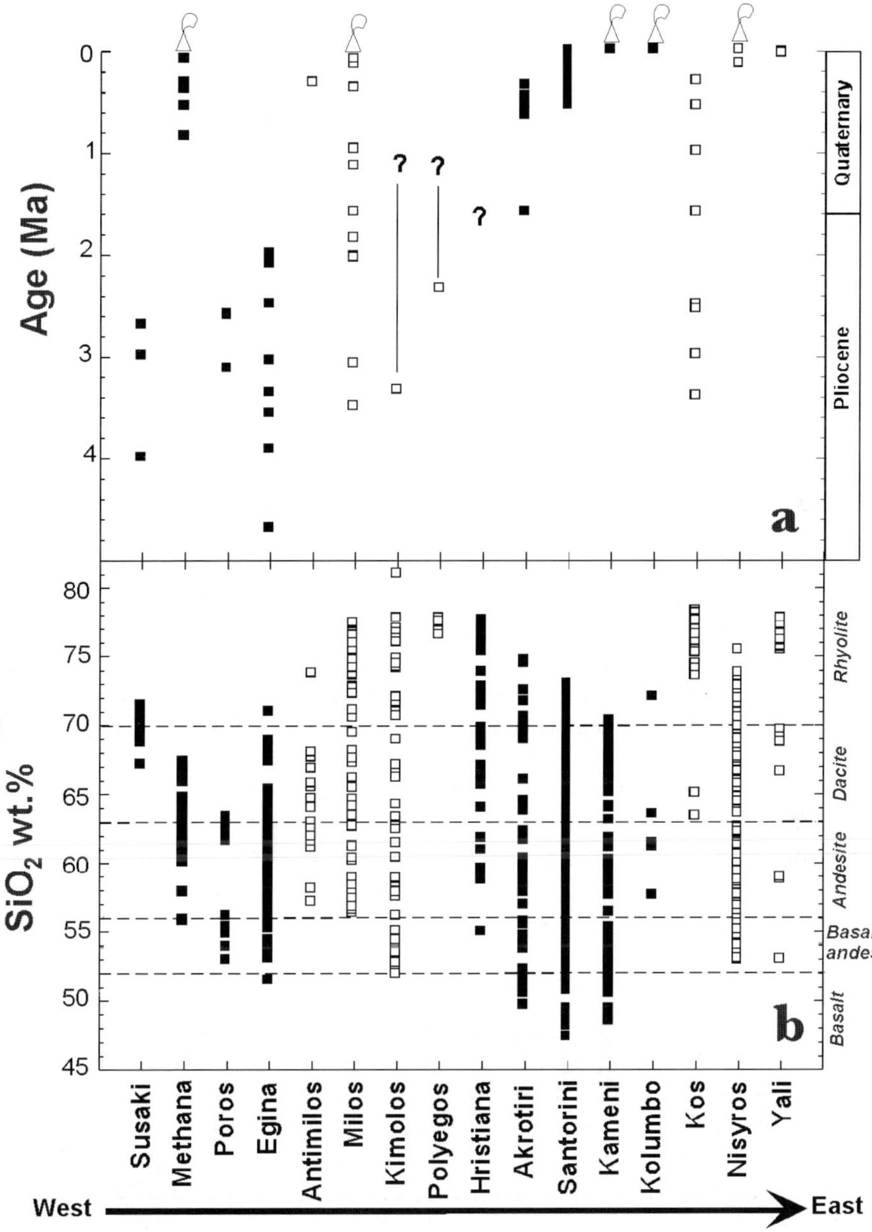

Fig. 17. West - East variations of ages (a) and silica contents (b) of rocks along the SAAVA. Volcanoes indicate the still or potentially active volcanic centers. See text for the data font.

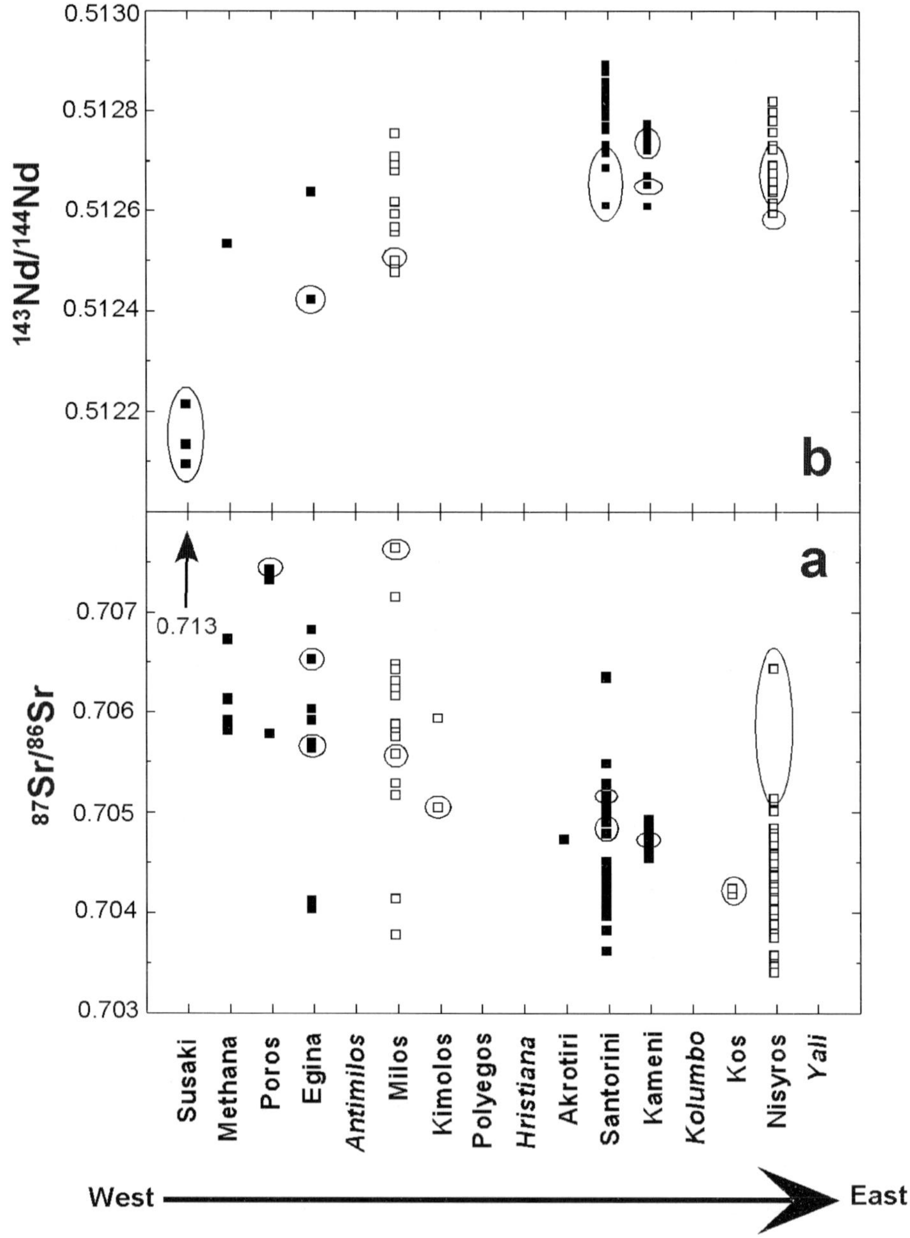

Fig. 18. West - East variations of Sr (a) and Nd (b) isotope ratios of rocks along the SAAVA. The encircled data are referred to samples with silica higher than 60 wt%. See text for the data font.

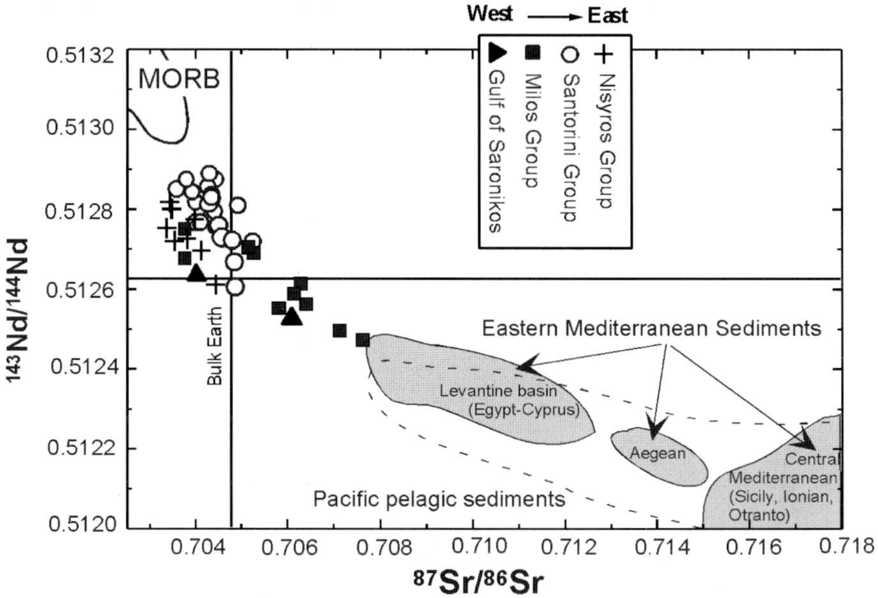

Fig. 19. $^{87}Sr/^{86}Sr$ versus $^{143}Nd/^{144}Nd$ diagram for the most mafic rocks ($SiO_2 < 60$ wt%) of the SAAVA fields. See text for the SAAVA data font. Data font of Eastern Mediterranean and Pacific pelagic sediments from Weldeab et al. (2002) and White et al. (1985), respectively.

be considered to be generated by evolutionary processes starting from more mafic magmas, rather than by crustal anatexis. Indeed, in most of the volcanoes, rhyolites plot at the most acidic extreme of continuous series of evolution (Figs. 3, 21) and their Sr isotope ratios are lower than those of continental crust. Furthermore, isotope ratios show a similar variation from West to East of the arc in the acidic and mafic rocks, thus suggesting a strict link between their lowest and highest values by AFC processes. Some doubts may arise for the Susaki rhyolites, because they are not associated with more mafic magmatism and have quite high Sr isotope ratios (around 0.713) (e.g., Pe, 1975; Pe-Piper and Piper, 1979; Innocenti et al., 1981; Barton and Huijsmans, 1986; Briqueu et al., 1986; Fytikas et al., 1986a,b; Mitropoulos et al., 1987; Huijsmans et al., 1988; Gulen, 1989; Wyers and Barton, 1989; Vougioukalakis, 1993; Francalanci et al., 1994, 1995, 1998, 2003; Vougioukalakis et al., 1995; Pe-Piper and Hatzipanagiotou, 1997; Davis et al., 1998; Aarburg, 1998; Aarburg and Frechen, 1999; Druitt et al., 1999; Zellmer et al., 2000).

SAAVA magmas also underwent to important processes of mixing-mingling between different evolved melts. These processes are particularly evident either when mafic (basaltic to andesitic) magmatic enclaves are included in more evolved (usually dacitic and rhyolitic) lavas and domes or when pyroclastic deposits are characterised by compositionally different juvenile materials. In both cases, they indicate the presence of large zoned magma chamber, which can feed energetic pyroclastic explosions, as

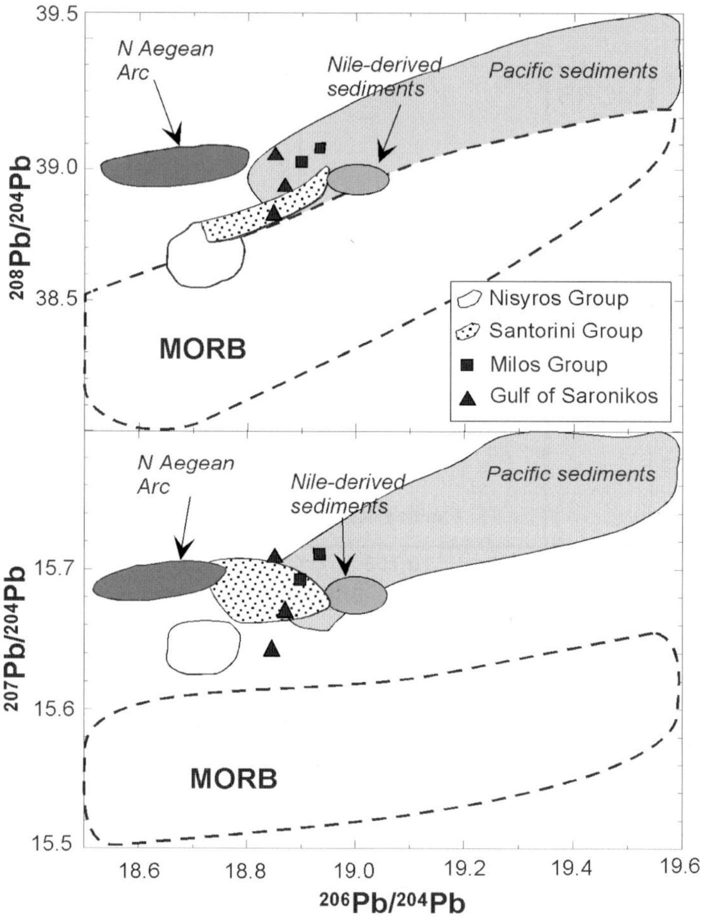

Fig. 20. $^{206}Pb/^{204}Pb$ versus $^{207}Pb/^{204}Pb$ and $^{208}Pb/^{204}Pb$ diagrams for the rocks of the SAAVA fields. See text for the data font.

mainly occurred at Santorini and Nisyros. Re-filling by mafic magmas of these shallow reservoirs may also trigger the eruptions. More complex processes of evolution have been also pointed out for Nisyros magmas, where crystal retention and re-cycling are also associated to AFC. Finally, a polybaric evolution occurring in reservoirs sited at different levels in the crust, has been also suggested for Santorini and Nisyros (e.g., Huijsmans et al., 1988; Wyers and Barton, 1989; Francalanci et al., 1995, 1998; Vougioukalakis et al., 1995; Davis et al., 1998; Aarburg, 1998; Aarburg and Frechen, 1999; Druitt et al., 1999; Zellmer et al., 2000; Mortazavi and Sparks, 2004).

It is noteworthy that the variation from CA to tholeiitic trend in the FeO/MgO versus SiO_2 of Fig. 21, can be also determined by shallow level differentiation processes and, in particular, by the minor role of hornblende and Fe-Ti oxides in Santorini island

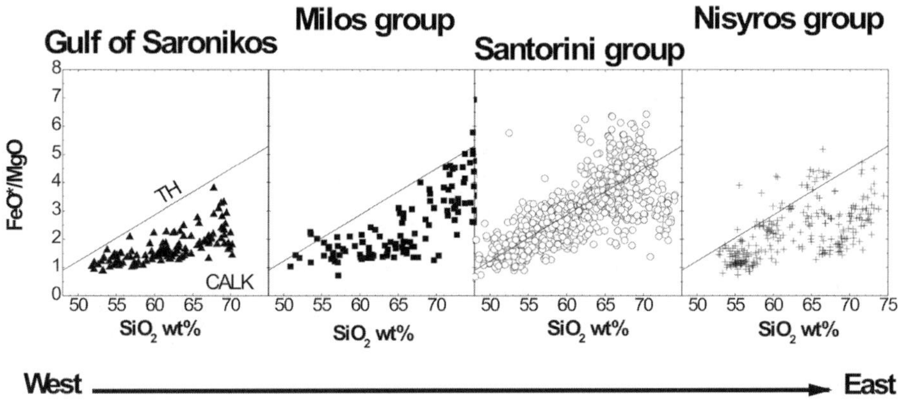

Fig. 21. Variations of FeO/MgO versus silica from West to East of the SAAVA. See text for the data font.

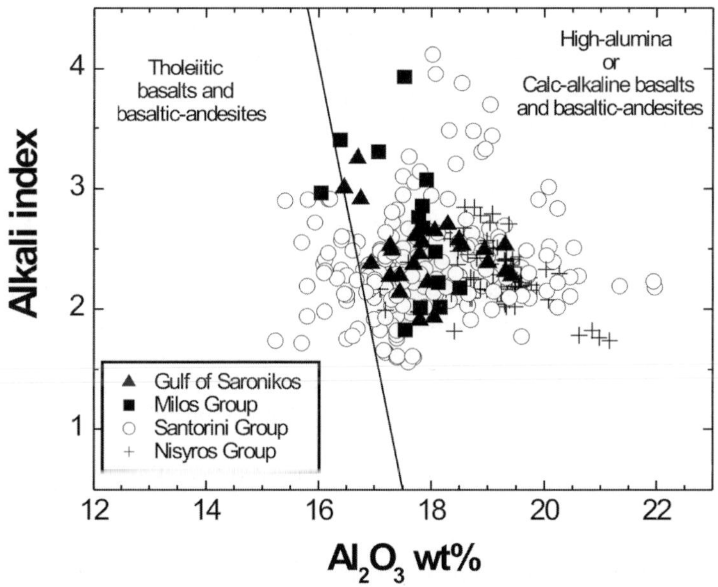

Fig. 22. Alkali index (=(Na_2O+K_2O)/[(SiO_2-43)*0.17] versus alumina for basalts and basaltic-andesites of the SAAVA. See text for the data font.

group, whose crystallisation leads to decrease the FeO/MgO value of the residual magmas. Hornblende is an important crystallising phase in most of the SAAVA fields, including the early volcanic centers of Santorini, at Akrotiri, but it is practically absent in the younger volcanics of Santorini. The Y variation versus silica points out the different role of amphibole in the different sector of the arc (Fig. 24). Indeed, because the solid/liquid partition coefficients of Y for amphibole is always > 1, only when hornblende does not crystallise, Y contents will increase with the degree of magma

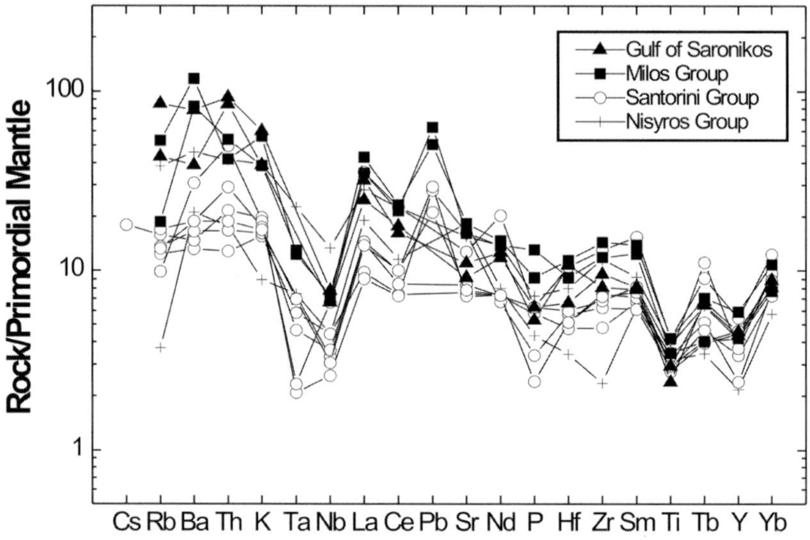

Fig. 23. Patterns of incompatible trace element contents normalised to the Primordial Mantle composition (Sun and McDonough, 1989) for the SAAVA most mafic rocks. See text for the data font.

evolution. The latter case is, in fact, only found for the youngest series of Santorini, where hornblende is not present. Along the same series, moreover, TiO_2 contents increase from basalts to basic andesites for decreasing at higher silica levels. On the contrary, in most of the magmatic series from the other volcanic fields, TiO_2 abundances start to decrease from basalts to more evolved compositions. On the other hand, lower oxygen fugacity has been hypothesised for the Santorini youngest magmas, based on the presence of Ti-rich titanomagnetites (Mitropoulos and Tarney, 1992). A lower oxygen fugacity can prevent the early crystallisation of Fe-Ti oxides, leading to increase FeO in the residual liquids.

It can be also argued that the absence of hornblende crystallisation in the youngest Santorini magmas is probably linked to a higher lithospheric extension in this zone of the arc, which lead magma to rest at shallower levels where amphibole is not stable. Furthermore, a possible lower aqueous fluid content in the youngest Santorini magma could also explain the lack of hornblende in these rocks.

A higher lithosphere extension below Santorini could be also the cause for a greater abundance of basalts occurring at this volcano (Innocenti et al., 1981). Indeed, mafic magmas more easily reach the shallow levels, thus increasing their possibility to be erupted.

4.2. Magma differentiation processes in the mantle source

The genesis of subduction-related magmas can be influenced by many factors, which include the composition of the subducting lithosphere, the type of mass transfer

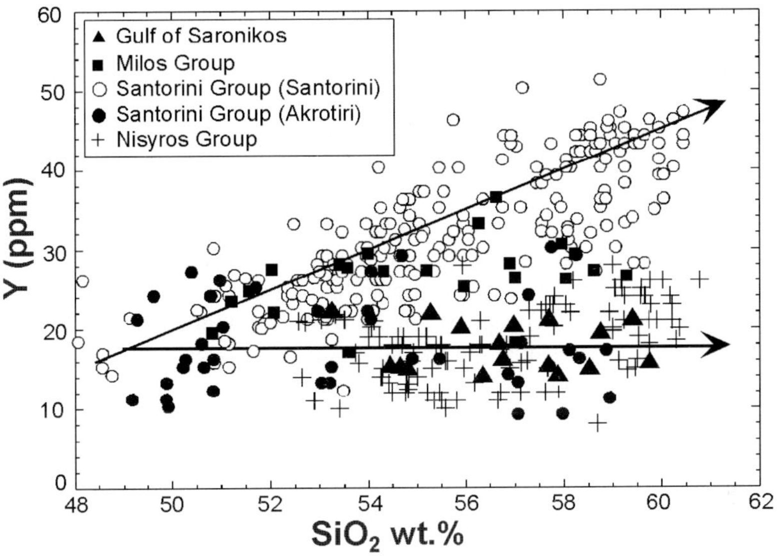

Fig. 24. Y versus SiO_2 diagram for the SAAVA rocks having <60 wt% of silica. See text for the data font.

of crustal material into the mantle wedge (fluids or melts, amount of metasomatising material added to the mantle) and the partial melting process of the metasomatised mantle (e.g., degree of partial melting). Also the nature of the mantle wedge before the event of metasomatism, e.g., lithosphere, asthenosphere, is of primary importance in determining the characteristics of the subduction-related magmas.

Several geochemical and isotopic variations occurring along the SAAVA from West to East seem to be due to differentiation processes in the mantle source during the magma genesis. Indeed, evolutionary processes along the pathway to the surface would have caused less systematic compositional variations. Even the decrease of Sr and Pb isotope ratios towards East is not correlated with the thickness variation of continental crust, which is lower in the central part of the arc, thus ruling out the possibility that this general trend is due to a decrease of crustal contamination.

Furthermore, the general decrease of K_2O with time in the external volcanic fields of the arc seems also to be related to mantle processes, because, in some cases (Milos and Saronikos Gulf), the K_2O decrease occurs among mafic rocks and it is not correlated with the decrease of Sr isotope ratios (Figs. 3, 18). A lower potassium content in orogenic magmas can be generated either: 1) by a lower amount of subducted material added to the mantle wedge, or 2) by higher partial melting degrees of the mantle wedge, or 3) by a major role of astenosphere as pre-metasomatism mantle source, in respect to the continental, possibly re-enriched, lithosphere.

The process (1), however, can be excluded again on the basis of Sr isotope ratios that should be systematically lower in CA magmas, strictly depending on the amount of crustal material brought into the mantle by subduction. Even geodynamic considerations lead to invalidate this process, because evidence of higher degrees of metasomatism are usually expected in younger orogenic magmas, erupted in an evolved phase of subduction, and not in the older magmas, as it occurs in SAAVA. Regarding to the process (3) several authors, based on spatial petrochemical trends, have hypothesised a higher implication of upwelling asthenospheric mantle in the genesis of Santorini magmas, due to a greater lithosphere extension (Mitropoulos et al., 1987; Mitropoulos and Tarney, 1992; Pe-Piper and Piper, 2002, 2004). This implicates the involvement of lithosphere as a pre-metasomatism mantle wedge in most of the SAAVA magma genesis. It is difficult to understand if the mantle involved in the magma genesis above the subducted slab is an asthenospheric or lithospheric mantle. Lithosphere, however, normally constituted by residual peridotite (usually harzburgite), is difficult to be melted. Even if the old continental lithospheric mantle can have suffered several events of incompatible element re-enrichment, the magmas generated by this mantle should have acquired some geochemical characteristics (e.g., high MgO and compatible element contents, low La/Sm values) indicating the residual character of their mantle source. At SAAVA neither the Pliocene HK-CA rocks nor all the rocks from the external parts of the arc show these kind of characteristics. Instead, higher Mg-values and lower La/Sm are found in the central part of the arc. Furthermore, Papazachos et al. (2004) found that SAAVA magmas derive from a "low velocity layer" sited at a depth of 60-90 km along the arc, which is probably the asthenosphere. Thus, different partial melting degrees of a similar and metasomatised mantle source (process 2) seem to be more suitable for explaining the compositional trend with time of the external parts of the arc.

Similar considerations arise in accounting for the lower alkali and incompatible element contents of Santorini magmas in respect with the magmas from the external sectors of the arc (Figs. 22, 23). Indeed, a higher involvement of astenospheric mantle in respect with the continental lithosphere might explain the geochemical characteristics of Santorini rocks. On the other hand, higher partial melting degrees of a similar astenospheric metasomatised mantle can originate primary magmas with a more tholeiitic character, as the mafic rocks of Santorini.

Evidences from the diagram of Fig. 25 could also indicate the involvement of subcontinental lithosphere in the petrogenesis of the magmas from the external sector of the arc, due to the higher Ta/Yb ratios, plotting in the field of active continental margins. In any case, lower Ta/Yb for the Santorini rocks can be explained again by higher degrees of source partial melting, taking in mind that Ta is greatly more incompatible than Yb for the peridotite mineralogical assemblage. This hypothesis can be also corroborated by the lower La/Sm values of Santorini rocks, considering a mantle partial melting where lower abundance of clinopyroxene remains as residue.

Isotope ratios of orogenic magmas are usually considered to indicate the amount of subducted crustal material added to the mantle wedge. This can be represented by both aqueous fluids from a subducted oceanic crust or subducted pelagic sediments.

The U-Th isotope ratio disequilibria suggest some role of slab-derived fluids for the

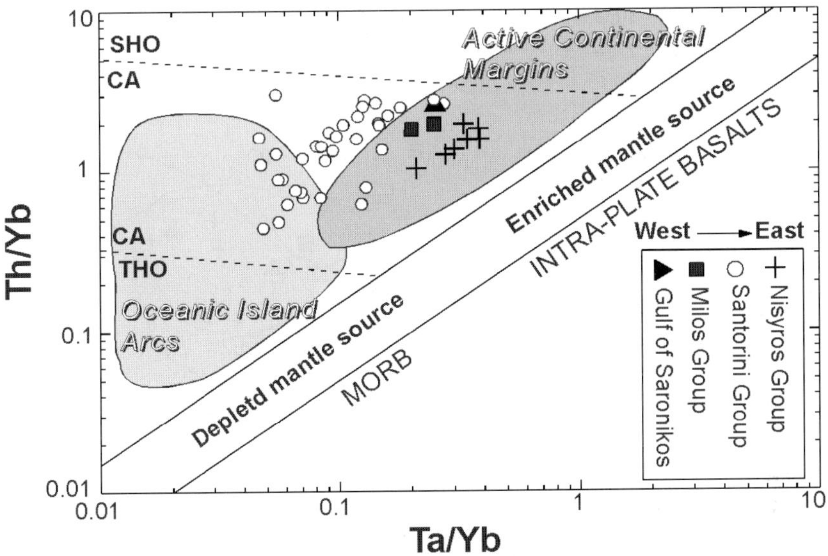

Fig. 25. Th/Yb vs Ta/Yb discrimination diagram for the least evolved SAAVA rocks. Fields for Oceanic Island Arcs and Active Continental Margins from Wilson (1989). See text for the other data font.

genesis of Kameni parental magmas (Zellmer et al., 2000). On the other hand, the diagram of Fig. 26 seems to indicate a low amount of fluids involved in the genesis of SAAVA magmas. Indeed, all the SAAVA mafic rocks plot at the lower Ba/La values of the island arc basalt (IAB) field, with the lowest values found in the Santorini magmas. The latter evidence could also account for the lack of amphibole crystallisation and the lower oxygen fugacity found for the youngest Santorini magmas.

On the other hand, the involvement of melts from slab-derived pelagic sediments is pointed out by the variations of Sr and Nd isotope ratios and B/Be ratios. Indeed, the former plot at the more radiogenic extreme of the SAAVA mafic rock trend (Fig. 19), whereas the high B/Be values measured by Clift and Blusztajn (1999) were considered as an evidence for a flux of sedimentary material from the subducted slab (Plank and Langmuir, 1998). A small amount of sediment melts added to the mantle wedge is also proposed by Zellmer et al. (2000) for the genesis of Santorini parental magmas, on the basis of incompatible trace element budget calculations.

Accordingly, the decrease of Sr and Pb isotope ratios from West to East of the arc can be due to a decreased amount of metasomatising sediments added to the mantle. Alternatively, based on the variations of Sr and Nd isotope ratios of the Eastern Mediterranean sediments from West (Aegean of Fig. 19) to East (Levantine basin of Fig. 19) (Weldeab et al., 2002), a smooth change in the isotopic composition of the subducted sediments can be suggested (Fig. 19).

Nd isotope ratios tend to decrease passing from Santorini to Nisyros island groups, showing a different variation in respect with the other isotope ratios and making

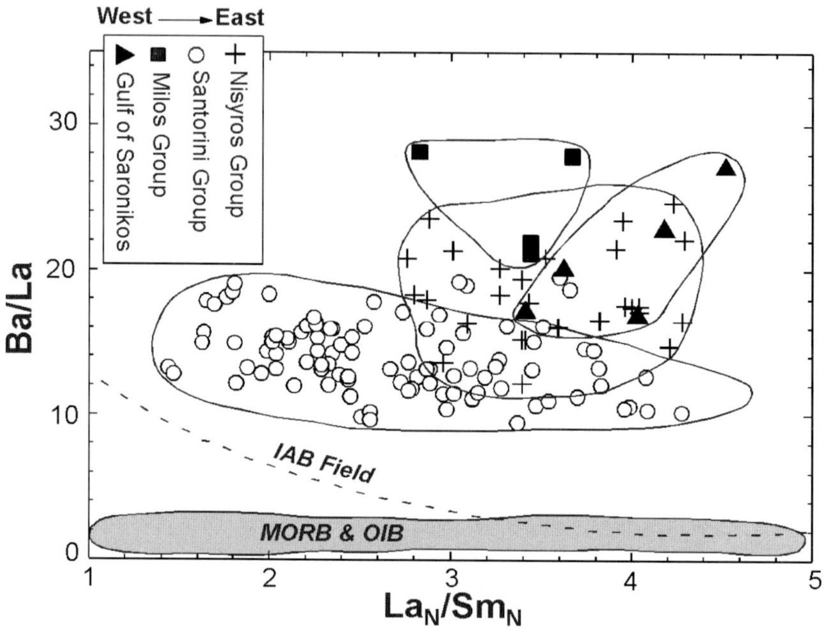

Fig. 26. La_N/Sm_N vs Ba/La ratios plot for the SAAVA rocks. Fields for MORB, OIB and IAB from Wilson (1989). See text for the other data font.

Santorini the volcanic center with the highest $^{143}Nd/^{144}Nd$ (Fig. 19). This characteristic could lead again to hypothesize for the Santorini magma genesis a major role of depleted, MORB-like, asthenospheric mantle, enriched in radiogenic Nd. Nevertheless, an other explanation can be searched in the possible different composition of sediments, which could contain lower Nd abundances in the central part of the arc. Indeed, keeping constant the sediment content added to the mantle, a lower Nd abundance in the slab-derived sediments leads to maintain higher the $^{143}Nd/^{144}Nd$ of the metasomatised mantle. Since now, however, this can be only an inferred hypothesis, due to the lack of available Sr and Nd content data for the Mediterranean sediments (Weldeab et al., 2002). On the other hand, the West-East Sr and Nd isotope variations in the same sediments may suggest a possible bulk compositional variation.

4.3. Magmatism and tectonics

The presence of Pliocene volcanism in all but the central (Santorini) SAAVA area is not easy to understand. Both subducting slab geometry and tectonic regime of the overlying Aegean lithosphere do not have an evident relation with the time onset of the volcanism. On the other hand, as already suggested by Innocenti et al. (1981), the presence of composite volcanoes with central calderas can be related to the thinned lithosphere in the Santorini area and the higher extension in both Nisyros and Santorini areas resulted from the palinspastic models (Le Pichon and Angelier, 1979). Indeed, the tensional tectonics would favour the formation of large shallow magma chambers. Present day extension rates, however, measured by modern geodetic satellite techniques (Reilinger, 1997) do not indicate any important difference along the arc.

In Fig. 27 the main characteristics of the Plio-Quaternary volcanic fields of the SAAVA are schematically resumed. Different Pliocene and Quaternary volcanic areas, trends and polarity of the lineaments controlling the volcanic vents are pointed out (Fig. 27). It is clearly evident that:

- The distribution of the volcanic centers over the zone of the 110-130 km dipping African lithosphere (see Papazachos et al. this volume) is restricted to five very limited areas.
- The trending of these areas is roughly perpendicular to the subduction front: from E-W at Susaki, to ENE-WSW at Milos, NE-SW at Santorini and NW-SE at South Kos-Nisyros.
- The active tectonic lineaments, which control the vent distribution, are not generally concordant with the general, previously mentioned, trending of these areas. In Susaki, Santorini and at some extent in Milos, both areal distribution and vent controlling lineaments are concordant.
- Vent controlling lineaments shift in space and time. In Milos and South Kos-Nisyros fields, where both Pliocene and Quaternary activity coexist, there is a clear change in the direction of the vents controlling lineaments with time: from NE-SW to N-S in Milos, from NW-SE to NE-SW in South Kos-Nisyros. This is not so clear at Susaki and Egina-Methana, but it seems to exist a shifting in the controlling lineaments from E-W to NW-SE.
- Vent migration with time in each field has not an univocal polarity: in Susaki they move from W to E, in Saronikos from ENE to WSW, in Milos from both NE and SW to the center of the field, in South Kos-Nisyros from SW to NE.
- There are large portions over the zone of the 110-130km dipping African lithosphere that are not affected by volcanic activity, even if active faulting, crustal thickness and extension are the same or even more prominent than in the volcanic field areas (see Fig. 7 of Pe-Piper and Piper this volume). Taking into account that most of these areas are submarine, the presence of some small Plio-Quaternary submarine volcanoes is possible, but the marine geology research carried out in the area excludes the presence of extensive volcanic fields.

5. SUMMARY

Several volcanological, geochemical and mineralogical variations occur along the SAAVA volcanic fields.

Many compositional characteristics are determined by differentiation processes during the pathway to the surface, which are dominated by crystal fractionation associated to crustal contamination and mixing-mingling processes.

General systematic variations are generated by differentiation processes occurring at the magma mantle source. SAAVA parental magmas are generated by partial melting of

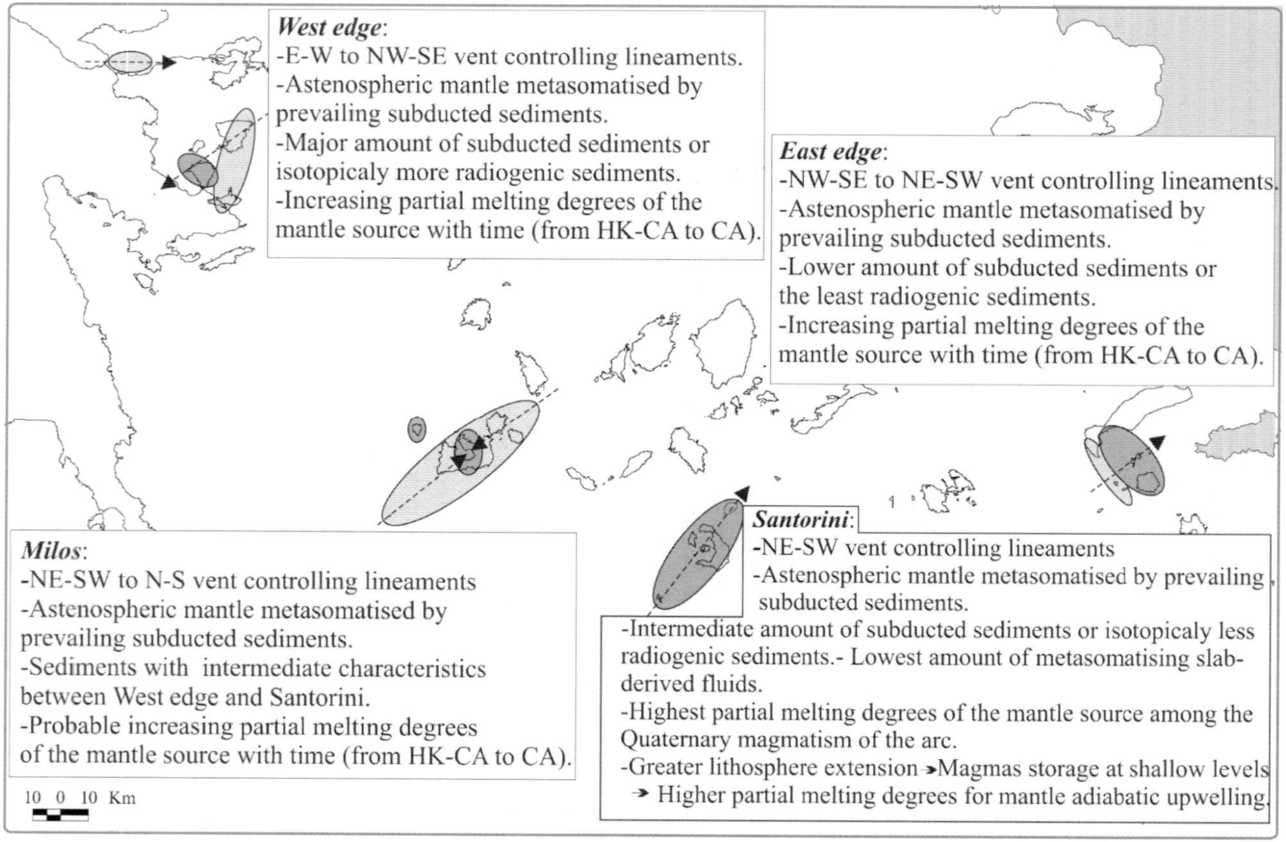

Fig. 27. Main characteristics of the SAAVA. Light gray areas: Pliocene volcanism. Dark gray areas: Quaternary volcanism. Arrows: Time polarity of the volcanism.

a depleted, MORB-like, athenospheric mantle, metasomatised by prevailing subducted sediments and low amount of aqueous fluids (e.g., low Ba/La values, Fig. 26). The latter seems also to be lower in the genesis of Santorini magmas, thus representing a possible explanation for the lower oxygen fugacity and the lack of amphibole crystallisation (probably associated to the higher lithosphere extension, see below) in these magmas.

The abundances of slab-derived sediments added to the mantle probably decrease from West to East (Sr and Nd isotope variations). Alternatively, the West-East $^{87}Sr/^{86}Sr$ and $^{144}Nd/^{143}Nd$ variations of the Mediterranean sediments lead to suggest a similar variation in the subducted sediment composition (Fig. 19). The slightly higher Nd isotope ratios of some Santorini magmas may also lead to hypothesise an inferred lower Nd content of the subducted sediments in the central part of the arc.

The degrees of mantle partial melting are generally high, but variable along the SAAVA or in a same sector of the arc. They seem to increase with time in the magmatism of the external parts (Gulf of Saronikos, Kos-Nisyros and probably Milos) and, among Quaternary volcanism, they are higher during the genesis of Santorini parental magmas.

The greater lithosphere extension below Santorini volcanic field may be responsible of several petrological processes characterising the magmatism in this sector of the arc. Magmas can stop at shallower levels, thus preventing amphibole crystallisation and allowing a higher amount of mafic magmas to reach the surface without undergoing to extensive processes of evolution. The greater lithosphere extension is also able to trigger a higher partial melting degrees observed for the Santorini magma source, due to the adiabatic upwelling of the asthenospheric mantle.

It has been also pointed out that from the beginning of SAAVA magmatism, occurred in Pliocene in the external parts of the arc, the partial melting degrees of the magma mantle source have increased with time and going towards the central sector of the arc.

It can be also concluded that volcanic activity along the South Aegean area is clearly controlled by both the active tectonic lineaments (Papazachos and Panagiotopoulos, 1993) and the subduction kinematics dynamics (Pe-Piper and Piper this volume). Nevertheless, it is difficult to apply a simple evolutionary model, as the whole area is a patchwork of fragmented crustal blocks which move one respect to the other in a very complex relationship, not predictable by any simple general model.

Acknowledgements

Discussion on the SAAVA characteristics with Prof. M. Fytikas, F. Innocenti and G. Pe-Piper greatly enhanced the authors ideas and conclusions.

Research activity was mainly founded by the EU 2nd Framework, under the IGME project "Study of the Hellenic Cenozoic Volcanism".

REFERENCES

Aarburg, S., 1998. Charakterisierung der pyroklastischen Abfolgen der Christiana-Inseln) Sud-Ageis, Griechenland) nach lithologischen, sedimentologischen, petrographischen und geochemischen Gesichtspunkten. Diplomarbeit: 93S.; Koln, Germany.

Aarburg, S. und Frechen, M., 1999. Die pyroklastischen Abfolgen der Christiana-Inseln (Sud-Agais, Griechenland). Terrestrische Quartargeologie: 260-276.

Allen, S.R. and Cas, R.A.F., 1998. Rhyolitic fallout and density current deposits from a phreatoplinian eruption in the eastern Aegean Sea. J. Volcanol. Geotherm. Res., 86: 219-251.

Barberi, F. and Carapezza, M.L., 1994. Helium and CO2 soil gas emissions from Santorini (Greece). Bull. Volcanol., 56: 335-342.

Barberi, F., Navarro, J.M., Rosi, M., Santacroce, R. and Sbrana, A. 1988. Explosive interaction of magma with ground water: insights from xenoliths and geothermal drillings. Rend. Soc. It. Mineral. Petrol., 43: 901-926.

Barton, M., Huijsmans, J.P.P., 1986. Post-caldera dacites from the Santorini volcanic complex, S. Aegean Sea, Greece : an example of the eruption of lavas of near-constant composition over a 2,200 year period. Contrib. Mineral. Petrol., 94: 472-495.

Barton, M., Salters, V.J.M., Huijsmans, J.P.P., 1983. Sr isotope and trace element evidence for the role of continental crust in calc-alkaline volcanism on Santorini and Milos, Aegean Sea, Greece. Earth Planet. Sci. Let., 63: 273-291.

Bellon, H., Jarrige, J.J. and Sorel, D., 1979. Les activites magmatiques egeennes de l´Oligocene a nos jours et leurs cadres geodynamiques. Donnees nouvelles et synthese. Rev. Geol. Dynam. Geogr. Phys., 21: 41-55.

Bohla, M. and Keller, J., 1987. Petrology of plinian eruptions of Nisyros volcano, Hellenic arc. Terra Cognita, 7: 171.

Bond, A., 1976. Multiple sources of pumice in the Aegean. Nature, 259: 194-195.

Briqueu, L., Javoy, M., Lancelot, J.R., Tatsumoto, M., 1986. Isotope geochemistry of recent magmatism in the Aegean arc: Sr, Nd, Hf,and O isotopic ratios in the lavas of Milos and Santorini-geodynamic implications. Earth Planet. Sci. Let., 80: 41-54.

Chiodini, G., Cioni, R., Leonis, C., Marini, L., and Raco, B., 1993. Fluid geochemistry of Nisyros island, Dodecanese, Greece. J. Volcanol. Geotherm. Research, 56: 95-112.

Clift, P., and Blusztajn, J., 1999. The trace-element characteristics of Aegean and Aeolian volcanic arc marine tephra. J. Volcanol. Geotherm. Res., 92: 321-347.

Collier, R.E.L. and Dart, C.J., 1991. Neogene to Quaternary rifting, sedimentation and uplift in the Corinth Basin, Greece. J. Geol. Soc. London, 148: 1049-1065.

Dalabakis, P., 1986. Une des plus puissantes erruptions phreatomagmatique dans la Mediterranee orientale: L´ignimbrite de Kos (Grece). CR Acad. Sci. Paris, ser. II 303: 505-508.

Dalabakis, P. and Vougioukalakis, G. (1993). The Kefalos tuff ring (W.Kos): Depositional mechanisms, vent position and model of the evolution of the

eruptive activity. Bull. Geol. Soc. Greece, XXVIII/2: 259-273.
Davis, E., Gartzos, E., Pavlopoulos, A., Tsagalidis, A., 1993. Petrological and geochemical research of perlites and rhyolites of Kefalos peninsula (S. Kos) and their quality appreciation. In: Special Volume of the Geol. Soc. Greece in honour of A.G.Panagos, Athens, pp. 284-303.
Davis, E., Gartzos, E. and Dietrich V.J., 1998. Magmatic evolution of the Pleistocene Akrotiri volcanoes. In Casale R., Fytikas M., Sigvaldasson G. & Vougioukalakis G.E. (Eds) "The European Laboratory Volcanoes", Proceedings of the 2d Workshop, Santorini, Greece – 2 to 4 May 1996. EUR 18161 EN, European Commission, Luxembourg, pp. 49-67.
Dimitriadis, I.M., Panagiotopoulos, D.G., Papazachos, C.B., Hatzidimitriou, P.M., Karagianni, E.E., and Kane, I., 2004. Recent seismic activity (1994-2002) of the Santorini Volcano using data from a local seismological network. This Volume.
Di Paola, G.M., 1974. Volcanology and petrology of Nisyros island (Dodecanese, Greece). Bull. Volcanol., 38: 944-987.
Doglioni, C., Agostini, S., Crespi, M., Innocenti, F., Manetti, P., Riguzzi, F. and Savasçin, Y., 2002. On the extension in western Anatolia and the Aegean Sea. J. Virtual Expl. 8: 169-184.
Druitt, T.H. and Francaviglia, V., 1992. Caldera formation on Santorini and the physiography of the islands in the late Bronze Age. Bull. Volcanol., 54: 484-493
Druitt, T.H., Mellors, R.A., Pyle, D.M. and Sparks, R.S.J., 1989. Explosive volcanism on Santorini, Greece. Geol. Mag., 126: 95-126.
Druitt, T.H., Edwards, L., Mellors, R.M., Pyle, D.M., Sparks, R.S.J., Lanphere, M., Davies, M. and Barreirio, B., 1999. Santorini Volcano. Geol. Soc. Mem., 19: 165.
Federman, A.N. and Carey, S.N., 1980. Electron microprobe correlation of tephra layers from eastern Mediterranean abyssal sediments and the island of Santorini. Quatern. Res., 13: 160-171.
Ferrara, G., Fytikas, M., Giuliani, O. and Marinelli, G., 1980. Age of the formation of the Aegean active volcanic arc. In: C. Doumas (ed.), Thera and the Aegean world II, vol. 2. London, pp. 37-41.
Fouqué, F., 1879. Santorin et ses eruptions. Masson et cie, Paris.
Fytikas, M. and Vougioukalakis, G., 1993. Volcanic Structure and Evolution of Kimolos and Polyegos (Milos island group). Greek Geol. Soc. Bull., XXVIII/2: 221-237.
Fytikas, M. and Vougioukalakis, G., 1995. Volcanic hazard in the Aegean Islands. In: T. Horlick-Jones, A. Amendola and R. Casale (Eds), Natural Risk and Civil Protection.. E & FN Spon, Brussels, pp. 117-130 .
Fytikas, M., Innocenti, F., Marinelli, G. and Mazzuoli, R., 1976. Geochronological data on recent magmatism in the Aegean Sea. Tectonophysics, 31: T29-T34.
Fytikas, M., Giuliani, O., Innocenti, F., Kolios, N., Manetti, P., Mazzuoli, R., 1986a. The Plio-Quaternary volcanism of Saronikos area (western part of the active Aegean volcanic Arc). Ann. Geol. Pays Hell., 33: 23-45.

Fytikas, M., Innocenti, F., Kolios, N., Manetti, P., Mazzuoli, R., Poli, G., Rita, F., Villari, L., 1986b. Volcanology and petrology of volcanic products from the island of Milos and neighbouring islets. J. Volcanol. Geotherm. Res., 28: 297-317.

Francalanci, L., Fytikas, M. and Vougioukalakis, G.E., 2003. Kimolos and Polyegos volcanoes, South Aegean Arc, Greece: volcanological and magmatological evolution based on stratigraphic and geochemical data. Meeting on "The South Aegean Volcanic Arc: Present knowledge and future perspectives", in the frame of the "Milos Conferences – Magmatism in convergent plate margins", Milos island, Greece. Book of abstracts, pp. 25-26.

Francalanci, L., Fytikas, M. and Vougioukalakis, G.E., 1994. Volcanological and geochemical evolution of Kimolos and Polyegos centers, Milos Island Group, Greece. IAVCEI Congress, Sember 1994, Ankara, Turkey, Book of Abstracts.

Francalanci, L., Varekamp, J.C., Vougioukalakis, G., Defant, M.J., Innocenti, F., Manetti, P., 1995. Crystal retention, fractionation and crustal assimilation in a convecting magma chamber, Nisyros Volcano, Greece. Bull. Volcanol., 56: 601-620.

Francalanci, L., Vougioukalakis, G., Eleftheriadis, G., 1995. The intracaldera post-minoan volcanics of Santorini, Greece: preliminary petrographic and geochemical new data. In: F. Barberi, R. Casale & M. Fratta (edts.), The European Laboratory Volcanoes, Workshop Proceeding, . European Commission, European Science Foundation, Luxembourg, pp. 184-188.

Francalanci, L., Vougioukalakis, G., Eleftheriadis, G., Pinarelli, L., Petrone, C., Manetti, P., Christofides, G., 1998. Petrographic, chemical and isotope variations in the intracaldera post-minoan rocks of the Santorini volcanic field, Greece. In Casale, R., Fytikas, M., Sigvaldasson, G. & Vougioukalakis, G.E. (Eds) "The European Laboratory Volcanoes", Proceedings of the 2d Workshop, Santorini, Greece – 2 to 4 May 1996. EUR 18161 EN, European Commission, Luxembourg, pp. 175-186.

Gansecki, C., 1991. Petrology of the domes and inclusions of Nisyros volcano, Dodecanese islands, Greece. Graduate Thesis, Wesleyan University, USA.

Georgalas, G., 1953. L' éruption du volcan de Santorin en 1950. Bull. Volcanol., 13: 39 -55.

Georgalas, G., Papastamatiou, J., 1953. L' eruption du volcan du Santorin en 1939-41. L' éruption du dome Fouqué. Bull. Volcanol., 13: 3-18.

Geotermica Italiana, 1983. Nisyros 1 geothermal well final report. Pisa.

Giovannetti, M., 1994. L'Eruzione Minoica di Santorini, Grecia: Variazioni Composizonali dei Prodotti e Caratterizzazione dei Frammenti Litici. Graduate Thesis, Università degli Studi di Pisa, Italy, 302 pp.

Gorceix, M.H., 1873. Sur la récente eruption de Nisyros. C R Ac Sc 77, 1039

Gorceix, M.H., 1874. Phenomenes volcaniques de Nisyros. C R Ac Sc 78, 444-446.

Gulen, L., 1989. Isotopic characterization of Aegean magmatism and geodynamic evolution of the Aegean subduction. In: S.R. Hart et al. (eds): Crust/Mantle recycling at convergence zones. NATO Advanced Studies Institute Series, C258, pp. 143-166.

Van Hinsbergen, D.J.J., Snel, E., Garstman, S.A., Marunteanu, M., Langereis, C.G., Wortel, M.J.R. and Meulenkamp, J.E., 2004. Vertical motions in the Aegean volcanic arc: Evidence for rapid subsidence preceding volcanic activity on Milos and Aegina. Submitted to Mar. Geol..

Hardiman, J.C., 1996. Quaternary Volcanism on Nisyros, Greece. Unpublished PhD Thesis. University of Cambridge.

Heiken, G. and McCoy F. Jr., 1984. Caldera development during the Minoan eruption, Thira, Cyclades, Greece. J. Geophys. Res., 89: 8441-8462.

Huijsmans, J., Barton, M., Salters, V., 1988. Geochemistry and evolution of the calc-alkaline volcanic complex of Santorini, Aegean Sea, Greece. J. Volcanol. Geotherm. Res., 34: 283-306.

Innocenti, F., Manetti, P., Peccerillo, A., Poli, G., 1981. South Aegean volcanic arc: geochemical variations and geotectonic implications. Bull. Volcanol., 44: 377-391.

Innocenti, S., 1998. La serie piroclastica di Panagia Kyra: Processi di mescolamento tra magmi a diverso grado di evoluzione a Nisyros (Grecia). Graduate Thesis, University of Florence, Italy, 143pp.

Keller, J., 1971. The major volcanic events in recent Mediterranean volcanism and their bearing on the problem of Santorini ash-layers. Acta 1st Internat. Sci. Congress on the Volcano of Thera, Athens, pp. 152-169.

Keller, J., 1980. Prehistoric pumice tephra on Aegean islands.- C. Doumas (Ed.) "Thera and the Aegean World II", vol 2, London, pp. 49-56.

Keller, J., 1982. Mediterranean Island Arcs. In: Thorpe, R.S. (ed.), Andesites. John Wiley, New York, pp. 307-325.

Keller, J., Gillot, P.Y., Rehren, T. and Stadlbauer, E., 1989. Chronostratigraphic data for the volcanism in the eastern Hellenic Arc: Nisyros and Kos.- TERRA abstracts, 1: 354.

Keller, J., Rehren, T. and Stadlbauer, E., 1990. Explosive volcanism in the Hellenic Arc: A summary and review. In: Hardy, D. A., Keller, J., Galanopoulos, V. P., Flemming, N. C., Druitt, T. H. (Eds), Thera and the Aegean World III. The Thera Foundation, London, vol 2, pp. 13-26

Ktenas, K., 1927. L' eruption du volcan des Kammenis (Santorin) en 1925, II, Bull. Volcanol., 4: 7-46.

Limburg, E.M., 1986. Young pyroclastics on Nisyros, Greece: physical studies. B.A. thesis, Wesleyan University, Middletown CT USA, 104pp.

Limburg, E.M., Lodice, L., Varekamp, J.C., 1986. Volcanology and petrology of Nisyros, Greece. In: Sigurdsson H (ed) Environmental impact of volcanism. Norman Watkins Symposium, University of Rhode Island, Narrangansett RI USA, pp 47-49.

Limburg, E. and Varekamp, J.C., 1991. Young Pumice deposits on Nisyros, Greece. Bull. Volcanol., 54: 68-77.

Lodice, L., 1987. Petrology and geochemistry of Nisyros volcano (Dodecanese, Greece). M.A. Thesis, Wesleyan University, Middletown CT USA, 245pp.

Manetti, P., 1997. Geology and volcanic history of the South Aegean volcanic arc, Greece. In: M. Cortini, B. De Vivo and C. Livadie (Editors), Volcanism and

Archaeology in Mediterranean Area. Research Signpost (Trivandrum, India), 175-191.

Marinos, G., 1960. The Antimilos Volcano in Aegean Sea. Bull. Geol. Soc. Greece, 4 (1): 38-49 (in Greek).

Martelli, A., 1917. Il gruppo eruttivo di Nisyros nel mare Egeo. Mem. Soc. Geol. Ital. Sc., Serie 3a, 20: 258-309.

Matsuda, J., Senoh, K., Maruoka, T., Sato, H. and Mitropoulos, P. 1999. K-Ar ages of the Aegean volcanic rocks and their implications for the arc-trench system. Geochem. J., 33: 369-377.

Mettos, A., Rondogianni, Th. and Bavay, Ph., 1988. Plio-Pleistocene deposits of the Susaki-Ag. Theodori area (Corinth). Stratigraphy and deformation. Bull. Geol. Soc. Greece, XX/2: 91-111.

Mitropoulos, P., Tarney, J., 1992. Significance of mineral composition variations in the Aegean Island Arc. J. Volcanol. Geotherm. Res., 51: 283-303.

Mitropoulos, P., Tarney, J., Saunders, A.D., Marsh, N.G., 1987. Petrogenesis of cenozoic volcanic rocks from the Aegean island arc. J. Volcanol. Geotherm. Res., 32: 177-193.

Mortazavi, M. and Sparks, R.S.J., 2004. Origin of rhyolite and rhyodacite lavas and associated mafic inclusions of Cape Akrotiri, Santorini: the role of wet basalt in generating calc-alkaline silicic magmas. Contrib. Mineral. Petrol., 146: 397-413.

Mountrakis, D., 2004. Tertiary and Quaternary tectonics in Aegean area. This volume.

Mountrakis, D., Pavlides, S., Chatzipetros, A., Meletlidis, S., Tranos, M., Vougioukalakis, G. and Kilias, A., 1998. Active deformation of Santorini. In Casale R., Fytikas M., Sigvaldasson G. & Vougioukalakis G.E. (Eds) "The European Laboratory Volcanoes", Proceedings of the 2d Workshop, Santorini, Greece – 2 to 4 May 1996. EUR 18161 EN, European Commission, Luxembourg. Pp. 13-22.

Muller, P., Kreutzer, H. and Harre, W., 1979. Radiometric dating of two extrusives from a Lower Pliocene marine section on Aegina Island, Greece. Newslett. Stratigr., 8: 70-78.

Papanikolaou, D., Chronis, G., Lykousis, V., Pavlakis, P., Roussakis, G. and Syskakis, D., 1988. Submarine neotectonic map of Saronikos gulf. Earthquake, Planning and Protection Organization.

Papanikolaou, D.J., Lekkas, EL, Sakelariou, D., 1991. Volcanic stratigraphy and evolution of the Nisyros volcano. Bull. Geol. Soc. Greece, XXV: 405-419.

Papazachos, B.C. and Panagiotopoulos, D.G., 1993. Normal faults associated with volcanic activity and deep rupture zones in the southern Aegean volcanic arc. Tectonophysics, 220: 301-308.

Papazachos, B.C., Dimitriadis, S.T., Panagiotopoulos, D.G., Papazachos, C.B. and Papadimitriou, E.E., 2004. Deep structure and active tectonics of the Southern Aegean Volcanic Arc. This volume.

Pavlides, S., 1993. Active faulting in multi-fractured seismogenic areas; examples from Greece. Z. Geomorph. N.E., Suppl.-Bd. 94: 57-72.

Peccerillo, A. and Taylor, S.R., 1976. Geochemistry of Eocene calc-alkaline volcanic rocks from Kastamonu area, Northern Turkey. Contrib. Mineral. Petrol., 58: 63-81.

Pe, G.G., 1975. Strontium isotope ratios in volcanic rocks from the Northwestern part of the Hellenic Arc. Chem. Geol., 15: 53-60.

Pe-Piper, G. and Piper, D.J.W., 2002. The igneous rocks of Greece, The anatomy of an orogen. Beitrage regionalen geologie der erde. Gebruder borntraeger. 573 pp.

Pe-Piper, G. and Piper, D.J.W., 2004. The South Aegean active volcanic arc: relationships between magmatism and tectonics. This volume.

Pe-Piper, G., Hatzipanagiotou, K., 1997. The Pliocene volcanic rocks of Crommyonia, western Greece and their implications for the early evolution of the South Aegean arc. Geol. Mag., 134: 55-66.

Pe-Piper, G., Piper, D.J.W., 1979. Plio-Pleistocene Age of High-Potassium Volcanism in the Northwestern Part of the Hellenic Arc. Tschermarks Min. Petr. Mitt., 26: 163-165.

Pe-Piper, G., Piper, D.J.W. and Reynolds, P.H., 1983. Paleomagnetic stratigraphy and radiometric dating of the Pliocene volcanic rocks of Aegina, Greece. Bull. Volcanol., 46:1-7

Petrone, C.M., 1995. Petrologia delle vulcaniti post-calderiche del vulcano di Santorini (Grecia). Graduate Thesis, University of Florence, Italy, 198pp.

Plank, T., Langmuir, C.H., 1998. The chemical composition of subducting sediment and its consequences for the crust and mantle. Chem. Geol., 145: 325-394.

Principe, C., Arias, A. and Zoppi, U., 2003. Hydrothermal explosions on Milos: from debris avalanches to debris flows deposits. In "The South Aegean Active Volcanic Arc: Present Knowledge and Future Perspectives" international conference, Milos 2003. Book of abstracts 95.

Puchelt, H., Murad, E. And Hubberten, H-W, 1977. Geochemical and petrological studies of lavas, pyroclastics and associated xenoliths from the Christiana Islands, Aegean Sea. N. Jb. Miner. Abh. 131, 2, pp. 140-155.

Puchelt, H., Hubberten, H.-W., Stellrecht, R., 1990. The Geochemisty of the Radial Dykes of the Santorini Caldera and its Implications. In: Hardy D A, Keller J, Galanopoulos V P, Flemming N C, Druitt T H. (Eds), Thera and the Aegean World III. The Thera Foundation, London, 2, pp. 229-236.

Pyle, D.M., 1990. New estimates for the volume of the Minoan eruption. In: Hardy D A, Keller J, Galanopoulos V P, Flemming N C, Druitt T H. (Eds), Thera and the Aegean World III. The Thera Foundation, London, 2, pp. 113-121.

Reck, H., 1936. Santorini. - Der Werdergang eines Inselvulcans und sein Ausbruch 1925 - 1928. Dietrich Reimer, Berlin, 3 vols.

Rehren, T.H., 1988. Geochemie und Petrologie von Nisyros (Östliche Ägäis). Ph.D. Thesis, University of Freiburg, 167pp.

Seymour, K.ST., and Vlassopoulos, D., 1989. The potential for future explosive volcanism associated with dome growth at Nisyros, Aegean volcanic arc, Greece. J. Volcanol. Geotherm. Res., 37: 351-364

Seymour, K.ST., and Vlassopoulos, D., 1992. Magma mixing at Nisyros volcano, as inferred from incompatible trace-element systematics. J, Volcanol, Geotherm. Res., 50: 273-299

Schroder, B., 1976. Volcanism, neotectonics and post-volcanic phenomena east of Corinth (Greece). Proceedings of the International Congress on Thermal Waters, Geothermal Energy and Volcanism of the Mediterranean area. Athens, pp. 240-248.
Seidenkrantz, M. and Friedrich, W.L., 1993. Santorini, part of the Hellenic arc: age of the earliest volcanism documented by foraminifera. Bull. Geol. Soc. Greece, XXVIII/3: 99-115.
Seward, D., Wagner, G.A. and Pichler, H., 1980. Fission track ages of Santorini volcanics. In: C. Doumas (ed.), Thera and the Aegean world II. London, vol. 2, pp. 101-108.
Smith, P.E., York, D., Chen, Y., Evensen, N.M., 1996. Single crystal 40Ar/39Ar dating of a Late Quaternary paroxysm on Kos, Greece: Concordance of terrestrial and marine ages. Geophys. Res. Letters, 23: 3047-3050.
Stewart A.L. and McPhie J., 2003. Facies architecture of the submarine-to-subaerial volcanic succession on Milos, Greece. In "The South Aegean Active Volcanic Arc: Present Knowledge and Future Perspectives" international conference, Milos 2003. Book of abstracts 24.
Stiros, S.C., 1995. The 1953 seismic surface fault: implications for the modelling of the Susaki (Corinth area, Greece) geothermal field. J. Geodynamics, 20, No2: 167-180.
Strabo. Goegraphica, Lib. I, 3, 59.
Sun, S.S. and McDonough, W.F., 1989. Chemical and isotopic systematics of oceanic basalts: implications of mantle composition and processes. In: Saunders A.D., Norry, M.J. (Eds), Magmatism in the Ocean Basins. Geol. Soc., Spec. Publ., 42, pp. 313-346.
Traineau, H. and Dalabakis, P., 1989. Mise en evidence d´une éruption phreatique historique sur l´ile de Milos (Grece). C.R. Acad. Sci. Paris, t. 308, Serie II:247-252.
Triantaphylis, M., 1994. Geological map of South Kos island, 1:50000. IGME, Athens.
Varekamp, J.C., 1992. Some remarks on volcanic vent evolution during plinian eruptions. J. Volc. Geotherm. Res., 54: 309-318.
Vitaliano, C.J., Taylor, S.R., Norman, M.D., McCulloch, M.T., Nicholis, I. A., 1990. Ash Layers of the Thera Volcanic Series: Stratigraphy, Petrology and Geochemistry. In: Hardy, D. A., Keller, J., Galanopoulos, V. P., Flemming, N.C., Druitt, T.H. (Eds), Thera and the Aegean World III. The Thera Foundation, London, vol. 2, pp. 53-78.
Volentik, A., Vanderkluysen, L. and Principe, C., 2002. Stratigraphy of the caldera wals of Nisyros volcano, Greece. Eclogae geol. Helv., 95: 223-235.
Vougioukalakis, G., 1984. Studio vulcanologico e geochimico -petrografico dell'isola di Nisyros (Dodecaneso, Grecia). M.Sc. Thesis, University of Pise, Italy, 235pp
Vougioukalakis, G., 1993. Volcanic stratigraphy and evolution of Nisyros island. Bull. Geol. Soc. Greece, XXVIII: 239-258.
Vougioukalakis, G., 1989. Geological map of Nisyros island, 1:25.000, IGME, Athens.
Vougioukalakis, G., Mitropoulos, D., Perissoratis, C., Andrinopoulos, A. and Fytikas, M., 1994. The submarine volcanic centre of Kolumbo, Santorini, Greece. Bull.

Geol. Soc. Greece, XXX/3: 351-360.

Vougioukalakis, G., Francalanci, L., Sbrana, A., Mitropoulos, D. 1995. The 1649-1650 Kolumbo submarine volcano activity, Santorini, Greece. In: F. Barberi, R. Casale & M. Fratta (edts)., "The European Laboratory Volcanoes, Workshop Proceeding", European Commission, European Science Fondation, Luxembourg, p. 189-192.

Vougioukalakis, G., Sachpazi, M., Perissoratis, C. and Lyberopoulou, Th., 1998. The 1995-1997 seismic crisis and ground deformation on Nisyros volcano, Greece: a volcanic unrest?. 6th Int. Meeting on Colima volcano, Abstracts Vol.

Wagner, G.A., Storzel, C. and Keller, J., 1976. Spaltspurendatierung quartärer Gesteinsgläser aus dem Mittelmeerraum. N. Jb. Miner. Mh., 2: 84-94.

Washington, H. S., 1926. Santorini eruption of 1925. Bull. Geol. Soc. Am., 37: 349 - 384.

Weldeab, S., Emeis, K.C., Hemleben, C. and Siebel, W., 2002. Provenance of lithogenic surface sediments and pathways of riverine suspended matter in the Eastern Mediterranean Sea: evidence from 143Nd/144Nd and 87Sr/86Sr ratios. Chem. Geol., 186: 139-149.

White, W.M., Dupre, B. and Vidal, P., 1985. Isotope and trace element geochemistry of sediments from the Barbados Ridge-Demerara Plain region, Atlantic ocean. Geochim. Cosmochim. Acta, 49: 1875-1886.

Wilson, M., 1989. Igneous Petrogenesis. Chapman & Hall, London.

Wyers, P.G., Barton, M., 1989. Polybaric evolution of calc-alkaline magmas from Nisyros, Southeastern Hellenic Arc, Greece. J. Petrol., 30: 1-37.

Zellmer, G., Turner, S., Hawkesworth, C., 2000. Timescales of destructive plate margin magmatism: new insights from Santorini, Aegean volcanic arc. Earth Planet. Sci. Lett., 174: 265-281.

The South Aegean active volcanic arc: relationships between magmatism and tectonics

G. Pe-Piper[1,*] and D.J.W. Piper[2]

[1] Department of Geology, Saint Mary's University, Halifax, Nova Scotia, B3H 3C3, Canada

[2] Geological Survey of Canada (Atlantic), Bedford Institute of Oceanography, P.O. Box 1006, Dartmouth, Nova Scotia, B2Y 4A2, Canada

ABSTRACT

Two principal volcanic associations, together with a third minor association, occur in the South Aegean active volcanic arc, differing in magma type, age, spatial distribution, relationship to faulting, and petrogenesis, even though geophysical data indicate a continuous subducted slab at 130 - 150 km beneath the volcanic centres. Variation in magmatism is related to changes in tectonics during the evolution of the arc, as a result of collision of African continental crust with the Aegean-Anatolian microplate, that set up changing patterns of strike-slip faulting in the arc. A synthesis is presented of the age and the variation in major and trace elements and radiogenic isotopes of the rocks of the arc. The western part of the arc (including Aegina, Methana, and the older rocks of Milos and Santorini) has typical arc-related andesite - dacite volcanism, predominantly of Pliocene age, associated with E-W listric faulting with slow slip rates. Nd and Sr isotopes and trace elements show that magmas resulting from volatile-induced melting in the asthenospheric mantle wedge subsequently either underwent assimilation with fractional crystallisation (AFC) or mixed with local partial melts within the lithospheric mantle. Viscous felsic magmas were likely trapped in the lower crust. In contrast, the mid to late Quaternary of the central and eastern part of the arc (Milos, Santorini, Nisyros) consists of lavas and voluminous pyroclastics, with lava compositions including both tholeiitic and calc-alkaline minor basalt, andesite, dacite and minor rhyolite. These younger magmas resulted from melting both of hydrated mantle (calc-alkaline magmas) and depleted asthenospheric mantle (tholeiitic magmas), influenced by regional extension, which is greatest in the central part of the arc. This

* Corresponding author: e-mail: gpiper@smu.ca

young volcanism began at the same time as the ENE-trending strike-slip faulting that resulted from collision with an indentor of thinned continental crust of the African plate. Stepover faulting and extension on early Quaternary NE-trending strike slip faults as a result of this mid- to late Quaternary ENE-trending sinistral strike slip motion provided pathways for magmas (including mantle-derived felsic magmas) to rise and fractionate and also pathways for water to enter shallow magma chambers. The third minor group of rocks comprises felsic lavas at Crommyonia and Kos, principally of Pliocene age, derived from mid-crustal anatexis. The ultimate cause of magmatism in the South Aegean active volcanic arc is subduction-related release of hydrous volatiles, but there are important differences between the petrogenesis of the older western and younger eastern parts of the arc.

Keywords: tectonics, strike-slip, volcanism, petrogenesis, Greece, Pliocene, Quaternary.

1. INTRODUCTION

The South Aegean active volcanic arc (Fig. 1) has long been regarded as a single entity, stretching from Crommyonia (Sousaki) in the northwest to Nisyros in the southeast (Paraskevopoulos, 1957; Pe and Piper, 1972). The evidence for this includes the general calc-alkaline geochemical character of all the major centres, their Pliocene-Quaternary age, and the occurrence of active volcanism in historic times all along the arc: at Methana, Milos, Santorini, Kolumbo Bank and Nisyros. A single Benioff zone, defined by earthquake hypocentres, occurs beneath the arc (Papazachos and Comninakis, 1971; Makropoulos and Burton, 1984) and the subducting slab can be mapped by seismic tomography (Spakman et al., 1993; Papazachos and Nolet, 1997). The geochemical character of the arc volcanic rocks is quite different from the trachyte and adakite found in minor volcanic centres in the back-arc region (Pe-Piper and Piper, 1989, 2002).

In the last decades, more geochronological data has been acquired, high quality geochemical data including radiogenic isotopes have become available, and there is more understanding of tectonic variation through time along the arc. These data suggest that the evolution of the arc is more complex than previously thought. In this paper, we propose that three distinct groups of volcanic rocks occur within the arc. The evolution of the arc can be better understood if each of these three groups is considered separately.

The South Aegean active volcanic arc is remarkable for the very rapid extension and crustal thinning of the Aegean microplate on which it has been erupted, manifested by shallow depth to Moho (Fig. 1), deep basins (Mascle and Martin, 1989) and seismically active faults (Hatzfeld, 1999). Furthermore, in the span of the active volcanism from early Pliocene to the present, there were major changes in regional faulting in the South Aegean active volcanic arc as a result of the collision of continental African crust with the Aegean-Anatolian plate (Piper and Perissoratis, 2003).

In this paper, we first synthesise the geochemistry and petrology of the volcanic rocks of the arc, in order to infer how magmas were produced. We then relate the

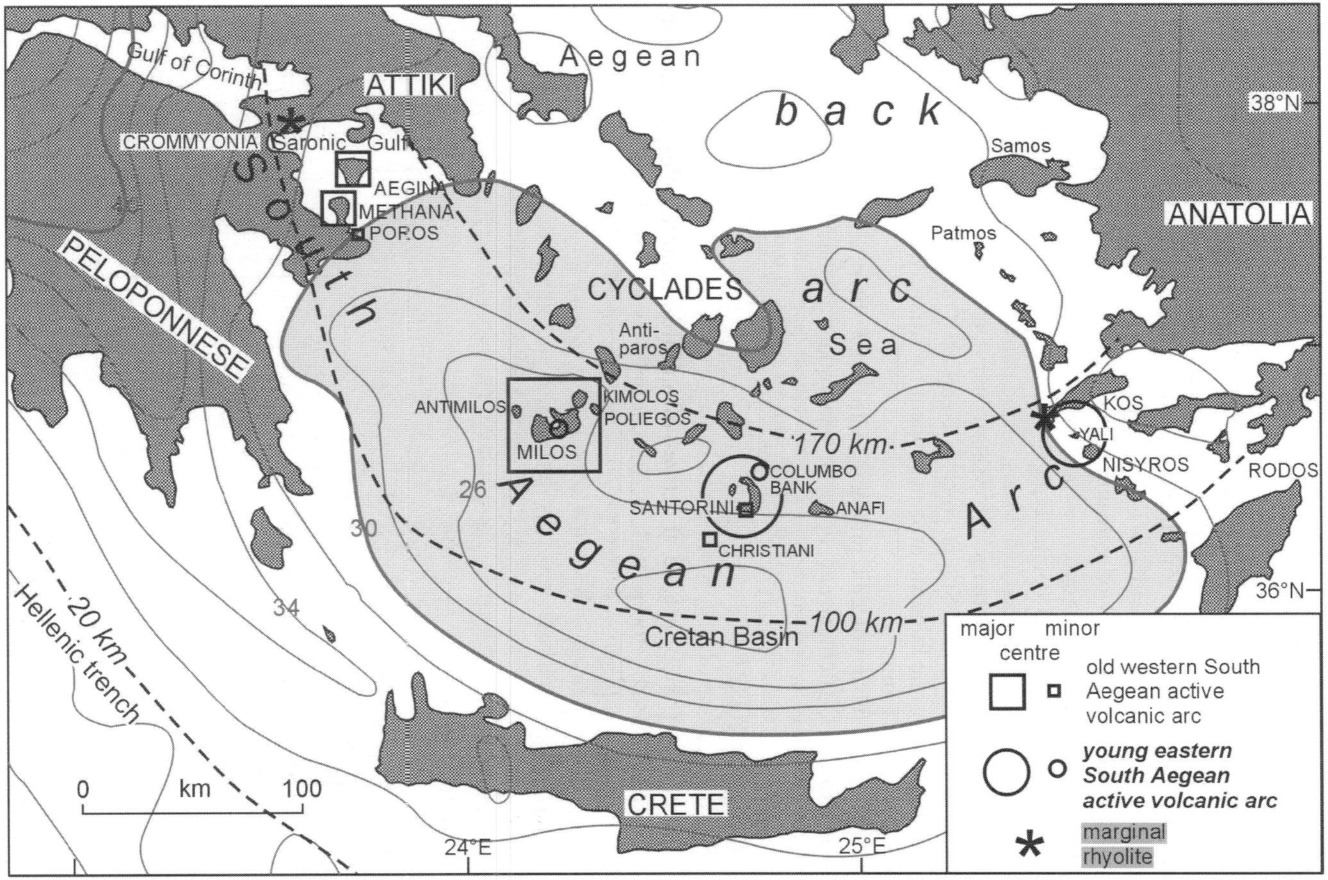

Fig. 1. Volcanic centres of the South Aegean active volcanic arc showing the three groups recognised in this paper. Depth to Moho (in km) from Makris (1977), >30 km shaded. Depth to Benioff zone (20, 100 and 170 km contours) from Papazachos et al. (2000).

temporal and spatial changes in magma petrogenesis to the changing tectonics, which created new pathways for magma to pass through the crust.

2. VARIATION IN VOLCANIC PRODUCTS IN SPACE AND TIME

The variation in abundance of different rock types in the South Aegean active volcanic arc is shown schematically in Fig. 2, together with available radiometric dates. This shows three main groups of volcanic rocks on the basis of relative volumes of different magma types and their age:

1. Voluminous andesite and dacite, typical calc-alkaline arc-related volcanic rocks, occur at Aegina, Methana and Milos, principally of Pliocene age on Aegina, Pliocene to early Quaternary on Milos, and late Pliocene and mid to late Quaternary on Methana. Similar rocks of earliest Quaternary age are found on Santorini. Minor late Pliocene tuffs may occur on Anafi. We refer to these rocks as the old western group.
2. Predominantly felsic volcanic products, with lesser intermediate and mafic lavas, of mid- to late Quaternary age, occur in small quantities on Milos and much more

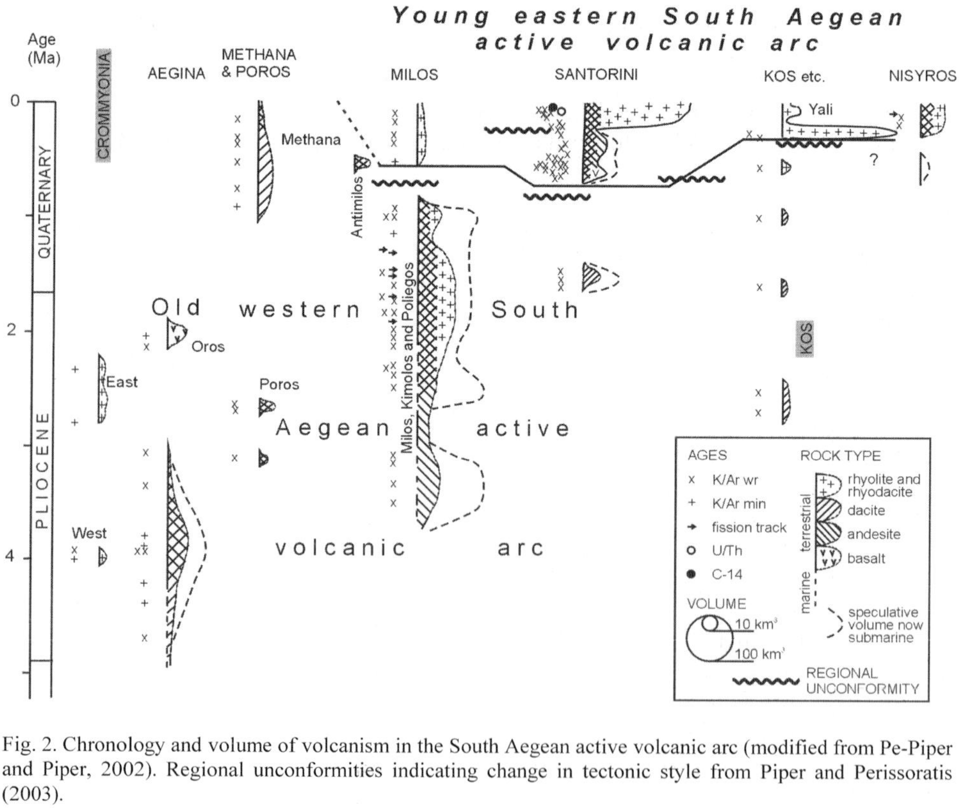

Fig. 2. Chronology and volume of volcanism in the South Aegean active volcanic arc (modified from Pe-Piper and Piper, 2002). Regional unconformities indicating change in tectonic style from Piper and Perissoratis (2003).

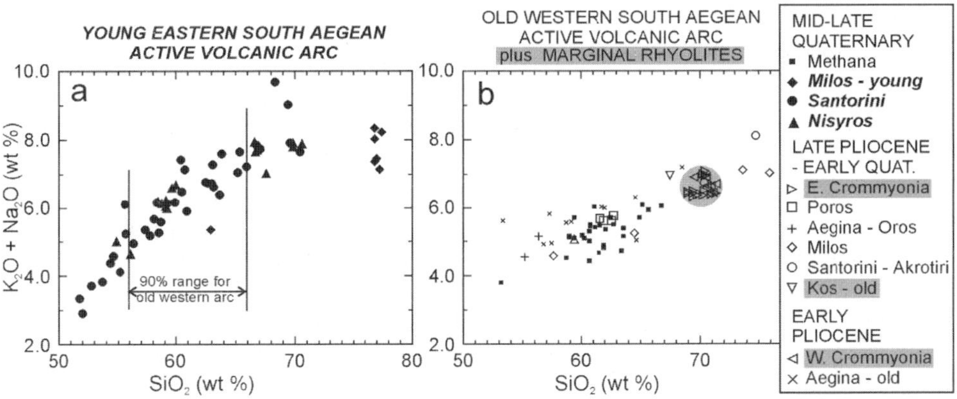

Fig. 3. Variation in total alkalies with silica by age and volcanic type in the South Aegean active volcanic arc. (Modified from Fig. 275 of Pe-Piper and Piper 2002; data from numerous sources listed in Appendix 3 of Pe-Piper and Piper, 2002). In this and subsequent figures, the old western group is indicated by normal font, the young eastern group by bold italic font, and the marginal rhyolites by a grey background.

voluminously at Santorini and the Nisyros-Yali volcanic centre, where the major explosive eruption formed the Kos Plateau Tuff. We refer to these rocks as the young eastern group.

3. Minor, small volume volcanic centres of rhyodacite or rhyolite composition occur at Crommyonia (Sousaki) (Pliocene) and the Kefalos peninsula of Kos (Pliocene to mid Quaternary). We refer to the rocks of Crommyonia and Kos as the marginal rhyolites.

3. GEOCHEMICAL AND ISOTOPIC VARIABILITY

The three groups of rocks differ significantly in geochemistry and radiogenic isotopes. The old western group of rocks, the Pliocene - Quaternary andesite-dacite centres of the western and central arc, range principally from 56 - 66% SiO_2 and over this range, total alkalies increase from 4.5% to 6% (Fig. 3b). High field strength elements show little variation: for example Zr (80-160 ppm), Nb (5-11 ppm), and Y (15-28 ppm) (Fig. 4a, c). Initial ratios of $^{87}Sr/^{86}Sr$ are generally higher than bulk Earth, from 0.705 to 0.708 and ε_{Nd} ranges from +2 to -5 (Fig. 5). Ce/Sr ratio increases only slowly with increasing $^{87}Sr/^{86}Sr_i$ (Fig. 6).

The young eastern group of rocks, the mid- to late Quaternary rhyolitic centres of the central and eastern arc, show a much wider range of SiO_2 contents. These rocks include tholeiitic basalt at Santorini and basaltic andesite at Nisyros (Fig. 3a) and voluminous pyroclastics of dacite and rhyolite composition (Figs. 2, 3a). Total alkalies increase more rapidly with increasing SiO_2 than in the western arc, with values of 7.5% at 66% SiO_2. High-field-strength elements show a much greater range than in the western arc (even for the same SiO_2 range, as indicated by the fields in Figs. 4b and 4d). Individual volcanic series show distinct Y/Zr and Nb/Zr ratios (Tarney et al., 1998).

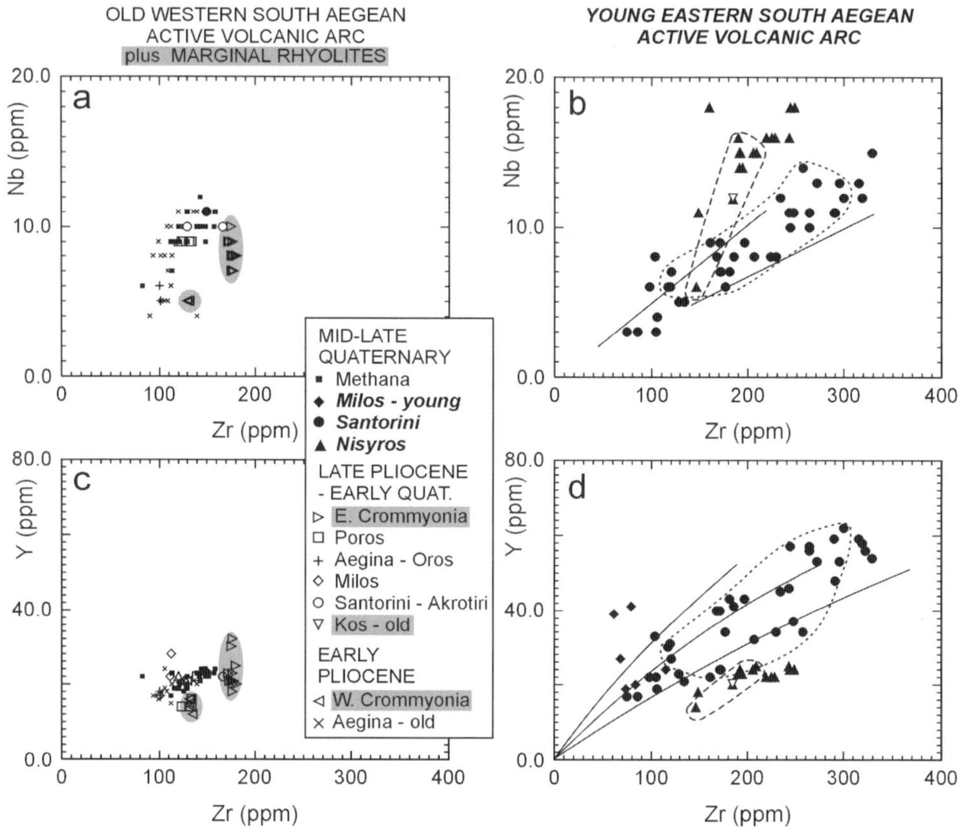

Fig. 4. Variation in selected trace elements with age and volcanic type in the South Aegean active volcanic arc. (a, b) Nb vs. Zr, dashed lines in (b) show range for Santorini and Nisyros for 56% < SiO_2 < 66%; (c, d) Y vs. Zr, dashed lines in (d) show range for Santorini and Nisyros for 56% < SiO_2 < 66%. Solid lines in b and d are trends for different rock associations in Santorini, from Tarney et al. (1998). Data sources as in Fig. 3.

Initial ratios of $^{87}Sr/^{86}Sr$ are generally close to bulk Earth, from 0.7035 to 0.705 and E_{Nd} ranges from +5 to +1 (Fig. 5). Sr isotope composition of dacite and rhyolite at both Santorini (0.7047-0.7053; Druitt et al., 1998, 1999) and Nisyros (0.7040-0.7045; Wyers and Barton, 1989) suggests little crustal involvement. Ce/Sr ratio increases rapidly with increasing $^{87}Sr/^{86}Sr_i$ (Fig. 6).

There is less geochemical data available for the minor dacite to rhyolite Pliocene centres of Crommyonia and Kos. Although the total alkalies continue the trend with increasing SiO_2 for other rocks of the western arc (Fig. 3), the K/Na ratio is much higher for Crommyonia. Sr isotope values from Crommyonia are unusually high (Fig. 5) and Nd isotopes unusually low (Fig. 6), implying a source from melting of continental crust, probably at mid-crustal depths. More mafic rocks are present only as small enclaves and as clasts in fluvial conglomerate. In Kos, only Sr isotope data is available for a mid-Pliocene dacite flow in the Kefalos peninsula (Mitropoulos et al., 1998), which has low

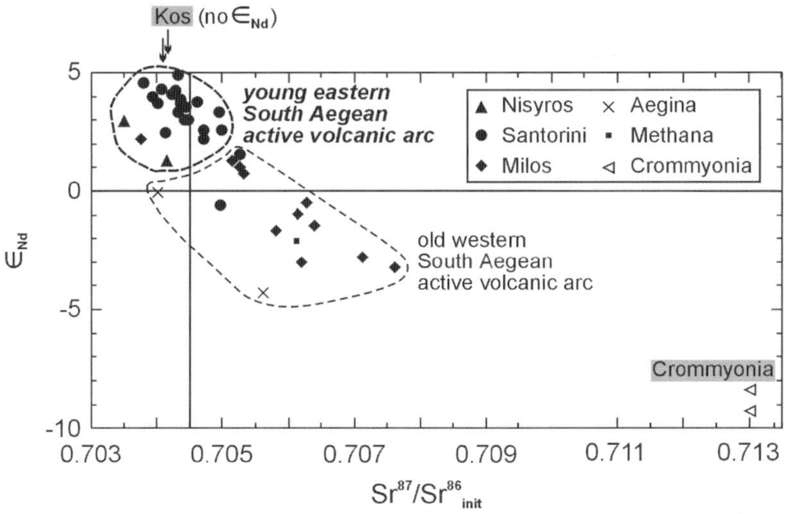

Fig. 5. Nd and Sr isotope ratios of the South Aegean active volcanic arc. (Modified from Fig. 283 of Pe-Piper and Piper, 2002; data from numerous sources listed in Appendix 3 of Pe-Piper and Piper, 2002).

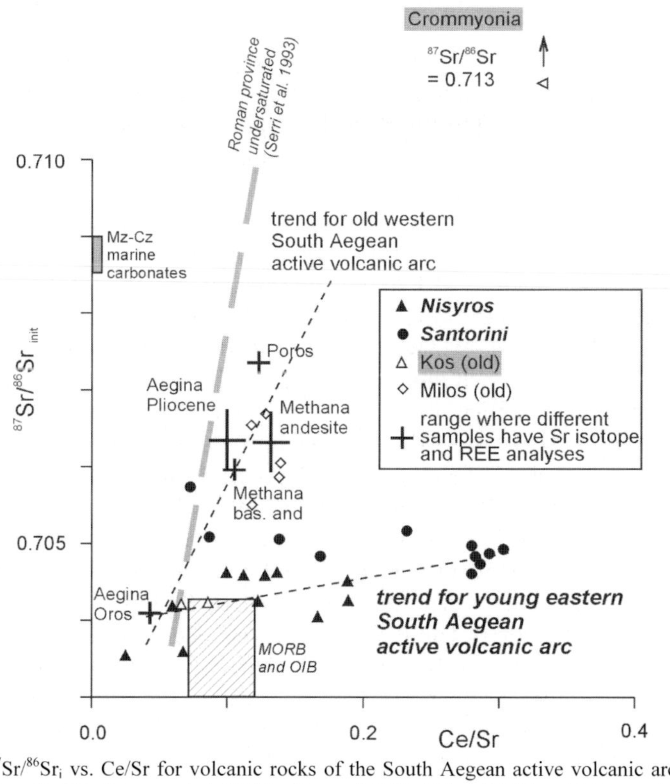

Fig. 6. Plot of $^{87}Sr/^{86}Sr_i$ vs. Ce/Sr for volcanic rocks of the South Aegean active volcanic arc. Data from Pe (1975), Barton et al. (1983), Briqueu et al. (1986), Wyers (1987), Gülen (1989) and Mitropoulos et al. (1998).

values of 0.7042. It is thus likely that the petrogenesis of the Kos volcanic rocks is different from those at Crommyonia. Rocks from both areas differ geochemically from alkaline back-arc rhyolites of late Miocene - early Pliocene age at Antiparos, Patmos and Samos (Pe-Piper and Piper, 2002).

4. TECTONIC SETTING

Both seismic tomography (Papazachos and Nolet, 1997) and the distribution of earthquake epicentres (Papazachos et al., 2000) indicate that there is a dipping subducted slab, with a steepening of the dip at 80-100 km. Both the older western and younger eastern volcanic centres are located about 150 km above the Benioff zone (Fig. 1). In the east, the shallower part of the Benioff zone is poorly defined and tomography suggests a gap between 75 and 95 km depth and the deeper part of the slab may be freely sinking (Papazachos et al., 2000). Papazachos and Panagiotopoulos (1993) suggested that each of the major volcanic centres (Crommyonia, Methana-Aegina, Milos, Santorini, Nisyros-Kos) was characterized by a set of shallow and intermediate depth earthquakes trending about 60E (ENE), but the new earthquake epicentre compilation of Papazachos et al. (2000) provides no support for this interpretation.

The old western group of rocks, the Pliocene - Quaternary andesite-dacite centres of the western and central arc, are all located in areas where Pliocene basins were defined by principally E-W trending faults, terminating against N-S trending faults (Fig. 7b), as mapped by Collier and Dart (1991) for Crommyonia and the Pliocene Gulf of Corinth; Papanikolaou et al. (1988) for the Saronic Gulf; and Piper and Perissoratis (2003) for Milos. In the Pliocene, there is no evidence for rapid basin subsidence: associated sediments are generally of shallow water origin, and where deep-water is at present close to the volcanoes, such as at Crommyonia and Milos, there is evidence that the principal subsidence took place in the mid to late Quaternary (Hsu et al., 1978; Piper and Perissoratis, 2003). The E-W faults are likely listric faults (Collier and Dart, 1991; Jackson, 1994) with N-S wrench faults acting as transfer faults. In Crete, Pliocene E-W trending faults had an important sinistral strike-slip component (ten Veen and Meijer, 1998) and similar faults of late Miocene age in Kos experienced strike-slip motion during emplacement of the Dikeos monzonite. At Methana, mid Quaternary volcanic centres show E-W trends (Gaitanakis and Dietrich, 1995), and E-W faults continue to be seismically active in the Gulf of Corinth, Attiki, and western Anatolia (Hatzfeld, 1999).

In the eastern part of the arc, prominent NE-SW trending strike-slip faults developed in the early Quaternary (ten Veen and Meijer, 1998; Pe-Piper et al., submitted). In Crete, faults on this trend are of late Pliocene age (ten Veen and Kleinspehn, 2003) and in eastern Crete they continue to be seismically active today (Fig. 7c). Fault scarps on this trend are also recognised near Santorini and Milos (Piper and Perissoratis, 2003) and control the location of hot springs on the north coast of Kimolos (Fytikas and Kolios, 1979). In Milos, Fytikas and Vougioukalakis (1993) identified the first evidence for NE-trending lineaments influencing volcanism in rocks dated at 1.6 Ma. At

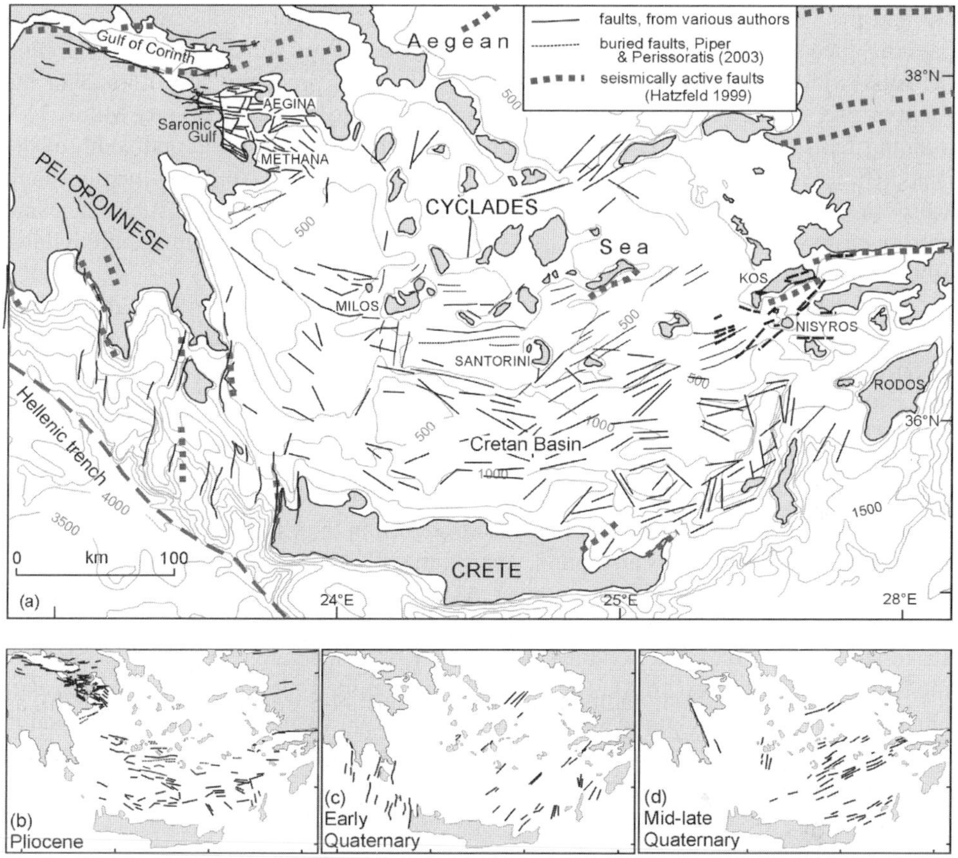

Fig. 7. (a) Bathymetry and major faults of the South Aegean active volcanic arc. Faults within the arc from Piper and Perissoratis (2003); in Cretan basin from Mascle and Martin (1989); near Kos from Pe-Piper et al. (submitted); in and around the Peloponnese from Papanikolaou et al. (1988), Collier and Dart (1991) and Armijo et al. (1992). Seismically active faults from Hatzfeld (1999). Distribution of mapped faults is limited by available data and map is incomplete for faults north of 37EN. Inset maps: (b) faults inferred to have been active in the Pliocene. (c) faults initiated in the early Quaternary. (d) faults initiated in the mid- to late Quaternary.

Santorini, linear trends of volcanic vents have been termed the Kameni and Columbos lines (Heiken and McCoy, 1984) and are also on the NE-SW trend, but appear today to be major normal faults. Dykes on the Santorini caldera wall principally trend NE or NNE (Mountrakis et al., 1998). This NE-trending fault set probably represents the first consequence of collision of African continental crust with the Aegean - Anatolian plate (Lyon-Caen et al., 1988; Bornhoff et al., 2001; Piper and Perissoratis, 2003).

In the mid- to late Quaternary, new active faults developed in the South Aegean arc and adjacent areas (Piper and Perissoratis, 2003) (Fig. 7d) and are spatially associated with the young, eastern group of rocks. This faulting parallels that defined by seismically active faults (Hatzfeld, 1999): N or NNW trending in the western part of the arc and the Peloponnese and ENE-trending in the eastern part of the arc. These fault patterns appear to have been initiated during the Quaternary in response to continuing collision with African continental crust. The faulting is predominantly normal in the western part of the arc and in the Peloponnese (Armijo et al., 1992). At Milos, N-S trending fault scarps are up to 100 m high at the sea floor both north and south of the island (Piper and Perissoratis, 2003). In the eastern part of the arc, ENE-trending fault scarps are widespread in the area from Santorini to Kos and show rapid subsidence and narrow basin inversion characteristic of strike-slip faulting (Papadopoulos and Pavlides, 1992; Piper and Perissoratis, 2003).

The change in dominant fault direction is marked by regional deep-water unconformities on the flanks of basins, the age of which was estimated (with substantial uncertainty) by Piper and Perissoratis (2003) from extrapolation of sedimentation rates and correlation with coastal progradation units controlled by eustatic sea level change (Fig. 3). Around Milos, the major unconformity is at 0.7 ± 0.2 Ma, corresponding approximately to the onset of late rhyolitic magmatism. Near Santorini, two regional unconformities are developed, one at about 0.8 ± 0.2 Ma, and the other at about 0.2 Ma, the latter marking the onset of rapid basin subsidence in the Christiani basin to the west. These two unconformities correspond, approximately, to the onset of mid Quaternary basalt-andesitic volcanism of the younger Akrotiri series and the onset of voluminous pyroclastic eruptions of the Thera Pyroclastic Formation (Fig. 3). The onset of major strike-slip faulting on ENE trending faults near Kos immediately preceded eruption of the Kos Plateau Tuff at 0.16 Ma (Pe-Piper et al., submitted). Regionally, the mid to late Quaternary is marked by rapid subsidence in the Cretan basin (Hsu et al., 1978) and in much of the South Aegean arc. Rates of subsidence of 2.5 mm/a are documented near Santorini (Piper and Perissoratis, 2003) and at Kos (Pe-Piper et al., submitted), an order of magnitude greater than likely basin subsidence rates in the Pliocene.

5. THE INFLUENCE OF FAULTING ON VOLCANISM

The onset of the young, eastern group of volcanism, the mid-to late Quaternary explosive pyroclastic centres of Milos, Santorini, and Yali-Nisyros, is correlated chronologically with the onset of ENE-trending strike-slip faults and the associated deep-water unconformities (Piper and Perissoratis, 2003). The very large Kos Plateau Tuff eruption at 161 ka (Allen, 2001) took place at a time when regional unconformities were developed around Yali - Nisyros and when there was a marked reduction in slip on many older faults (Pe-Piper et al. submitted). Detailed study of the fault patterns in the area of Kos and Nisyros led Pe-Piper et al. (submitted) to hypothesize how this ENE-trending faulting promoted explosive volcanism. Under conditions of extensional sinistral shear along the late Quaternary ENE-trending strike-slip faults, the early Quaternary NE-trending strike-slip faults accommodated extensional deformation by normal faulting (NF in Fig. 8D) and essentially functioned as stepover zones between master strike-slip faults. This tendency for extension on what were previously crustal scale strike-slip faults would have facilitated the upward percolation of magma. As

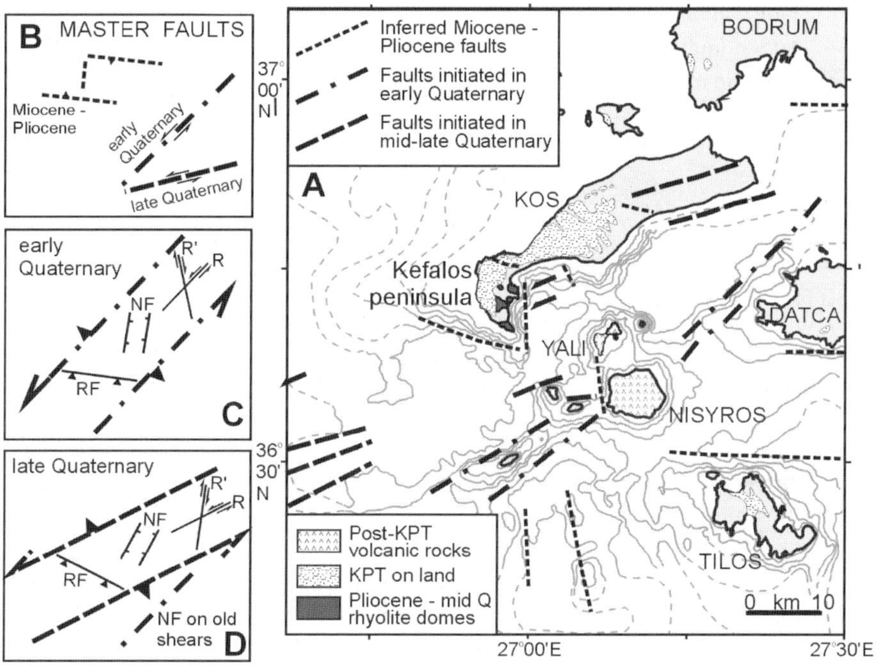

Fig. 8. Tectonic evolution of the Yali-Nisyros volcanic centre. (A) Geological map showing faults (based on Pe-Piper et al. submitted). (B) Cartoon showing dominant fault patterns through time. (C) Secondary fault patterns under early Quaternary extensional sinistral shear (from Woodcock and Schubert, 1994). R, R' are Riedel shears, NF = normal faulting, RF = reverse faulting. (D) Secondary fault patterns under late Quaternary extensional sinistral shear.

argued by Tarney et al. (1998), the downward percolation of water may have been important in some phreatomagmatic eruptions, including the Kos Plateau Tuff (Allen, 2001).

We suggest a similar tectonic history for the volcanism at Santorini (Fig. 9). The early Quaternary volcanism at Akrotiri on Santorini and at Christiani was along a NE-trend and faults of this trend appear to define some basin margins around Santorini. To the east of Santorini, prominent ENE-trending strike-slip faults developed at 0.3 to 0.2 Ma, leading to rapid basin subsidence and some basin inversion in the area from Anafi to Astypalea, and the initiation of voluminous explosive volcanism at Santorini at 0.25 to 0.2 Ma (Piper and Perissoratis, 2003). We suggest a process similar to that at Yali-Nisyros, in which regional motion on the ENE sinistral shears led to extension on the older NW-trending lineament now marked by the Kameni line, which acted as a stepover zone.

The influence of faulting on the old western group of volcanoes and the marginal rhyolites is less clear. In Methana, the trend of mid to late Quaternary volcanic centres mapped by Gaitanakis and Dietrich (1995) is prominently E-W, parallel to one of the dominant basin-bounding fault directions in the Saronic Gulf (Papanikolaou et al.,

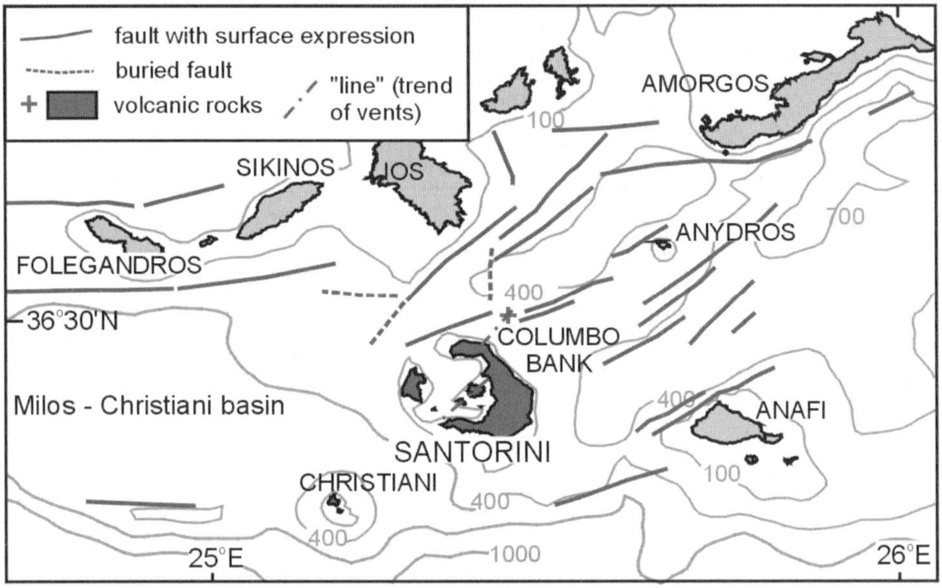

Fig. 9. Tectonic evolution of the Santorini volcanic centre, based on Piper and Perissoratis (2003). Only faults whose trends can be determined from closely spaced seismic-reflection profiles are shown.

1988). Fault influence on other volcanic centres in the Saronic Gulf is uncertain. In the older volcanism of Kos, the domes of the Kefalos peninsula form a general N-S trend, parallel to transfer faults between the East Kos and Kos-Astypalea basins.

6. PETROGENESIS

The abundance of hydrous phenocrysts (hornblende, mica) throughout the arc is strong circumstantial evidence for an important role for slab-derived fluids. On the other hand, the substantial crustal thinning as a result of late Cenozoic extension throughout the South Aegean (Fig. 1) is likely to have played a role in creating decompression melting. Evidence for crustal assimilation and at least some fractional crystallisation is compelling from detailed studies of magma series throughout the arc, particularly on Santorini. AFC modelling of both element and isotopic abundances suggests assimilation to fractionation ratios of 0.1 to 0.2. What is not clear is the relative importance of upper crustal processes (e.g. Santorini: Druitt et al., 1998, 1999), base of crust processes (e.g. Nisyros: Wyers and Barton, 1989), and contamination in the sub-lithospheric mantle by "crustal" material (Pe-Piper and Piper, 2001). Direct geochemical contribution from the subducting slab beneath the arc volcanoes is perhaps indicated by high B/Be ratios (40-60, with outlying values near 90) in tephra derived from the Aegean arc, that Clift and Blusztajn (1999) argued resulted from flux of sedimentary material from the subducted slab. If so, then the high $^{87}Sr/^{86}Sr_i$ for relatively low Ce/Sr

alkaline magmas). Extension on early Quaternary NE-trending strike slip faults as a result of mid- to late Quaternary ENE-trending sinistral strike slip motion provided pathways for magmas (including mantle-derived felsic magmas) to rise and fractionate and also pathways for water to enter shallow magma chambers.

3. Minor felsic volcanic rocks at Crommyonia and Kos are principally of Pliocene age. The lack of intermediate and mafic rocks indicates that the magmas were dominated by mid-crustal anatexis, confirmed by Nd and Sr isotopes and trace element abundances at Crommyonia. A role for arc-related magmas in producing this melting is indicated by rare small mafic enclaves.

Although the ultimate cause of the South Aegean active volcanic arc is subduction-related release of hydrous volatiles, it cannot otherwise be regarded as a single geological phenomenon. The older western and younger eastern parts have quite different origins and should be interpreted separately.

Acknowledgements

Work supported in part by an NSERC Discovery Grant to GPP. Discussions with several participants at the SAAVA 2003 Conference and the critique of an anonymous referee sharpened the ideas in this paper.

REFERENCES

Aldanmaz, E., Pearce, J.A., Thirlwall, M.F. and Mitchell, J.G., 2000. Petrogenetic evolution of late Cenozoic post-collisional volcanism in western Anatolia, Turkey. J. volcan. geotherm. Res., 102: 67-95.

Allen, S.R., 2001. Reconstruction of a major caldera-forming eruption from pyroclastic deposit characteristics: Kos Plateau Tuff, eastern Aegean Sea. J. volcan. geotherm. Res., 105: 141-162.

Altherr, R., Kreuzer, H., Wendt, I., Lenz, H., Wagner, G.A., Keller, J., Harre, W. and Hohndorf, A., 1982. A late Oligocene/early Miocene high temperature belt in the Attic-Cycladic crystalline Complex (S.E. Pelagonian, Greece). Geol. Jb., E23: 97-164.

Armijo, R., Lyon-Caen, H. and Papanastassiou, D., 1992. East-west extension and Holocene normal-fault scarps in the Hellenic arc. Geology, 20: 491-494.

Barton, M., Salters, V.J.M. and Huijsmans, J.P.P., 1983. Sr-isotope and trace element evidence for the role of continental crust in calc-alkaline volcanism on Santorini and Milos, Aegean Sea, Greece. Earth Plan. Sci. Letts., 63: 273-291.

Bornhoff, M., Makris, J., Papanikolaou, D. and Stavrakakis, G., 2001. Crustal investigation of the Hellenic subduction zone using wide-aperture seismic data. Tectonophysics, 343: 239-262.

Briqueu, L., Javoy, M., Lancelot, J.R. and Tatsumoto, M., 1986. Isotope geochemistry of recent magmatism in the Aegean arc: Sr, Nd, Hf and O isotopic ratios in the lavas of Milos and Santorini - geodynamic implications. Earth Plan. Sci. Letts., 80: 41-54.

Brown, M. and Solar, G.S., 1998. Granite ascent and emplacement during contractional deformation in convergent orogens. J. struct. Geol., 20: 1365-1395.

Clift, P. and Blusztajn, J., 1999. The trace-element characteristics of Aegean and Aeolian volcanic arc marine tephra. J. volcan. geotherm. Res., 92: 321-347.

Collier, R.E.L. and Dart, C.J., 1991. Neogene to Quaternary rifting, sedimentation and uplift in the Corinth Basin, Greece. J. geol. Soc. Lond., 148: 1049-1065.

Dietrich, V.J., Mercolli, I. and Oberhänsli, R., 1988. Dazite, High-Alumina-Basalte und Andesite als Produkte amphiboldominierter Differentation (Aegina und Methan, Ägäischer Inselbogen). Schweiz. Min. Pet. Mitt., 68: 21-39.

Druitt, T.H., Edwards, L., Lanphere, M., Sparks, R.S.J. and Davis, M., 1998. Volcanic development of Santorini revealed by field, radiometric, chemical and isotopic studies. In: R. Casale, M. Fytikas, G. Sigvaldasson and G. Vougioukalakis (Editors), The European laboratory volcanoes. European Commission, EUR 18161, pp. 37-48.

Druitt, T.H., Edwards, L., Mellors, R.M., Pyle, D.M., Sparks, R.S.J., Lanphere, M., Davis, M. and Barriero, B., 1999. Santorini Volcano. Geological Society Memoir No. 19, Geological Society, London.

Fytikas, M.D. and Kolios, N.P., 1979. Preliminary heat flow map of Greece. In: V. Cermak and L. Rybach (Editors), Terrestrial Heat Flow in Europe, Springer, New York, pp. 197-205.

Fytikas, M. and Vougioukalakis, G., 1993. Volcanic structure and evolution of Kimolos and Polyegos (Milos island group). [In Greek, English abstract]. Bull. geol. Soc. Greece, 28: 221-237.

Gaitanakis, P. and Dietrich, V., 1995. Geological map of Methana peninsula 1:25 000. ETH, Zurich.

Gautier, P., Brun, J.P and Jolivet, J., 1993. Structure and kinematics of Upper Cenozoic extensional detachment on Naxos and Paros (Cyclades islands, Greece). Tectonics, 12: 1180-1194.

Gill, J., 1971. Orogenic andesites. Springer, New York.

Güleç, N., 1991. Crust-mantle interaction in western Turkey; implications from Sr and Nd isotope geochemistry of Tertiary and Quaternary volcanics. Geol. Mag. 128: 417-435.

Gülen, L., 1989. Isotopic characterization of Aegean magmatism and geodynamic evolution of Aegean subduction. In: S.R. Hart (Editor), Crust/Mantle recycling at convergence zones, NATO Advanced Studies Institute Series. C258: 143-166.

Hatzfeld, D., 1999. The present-day tectonics of the Aegean as deduced from seismicity. In: Durand, B., Jolivet, L., Horváth, F. and Seranne, M. (Editors), The Mediterranean basins: Tertiary extension within the Alpine orogen, Geological Society, London, Special Publication 156: 416-426.

Heiken, G. and McCoy, F. Jr., 1984. Caldera development during the Minoan eruption, Thira, Cyclades, Greece. J. geophys. Res., 89: 8441-8462.

Hsu, K.J., Montadert, L., Bernouilli, D., Bizon, G., Cita, M., Ericson, A., Fabricius, F., Garrison, R.E., Kidd, R.B., Mellières, F., Müller, C. and Wright, R.C., 1978. Site 378, Cretan Basin. Initial Reports of the Deep Sea Drilling Project 42(1): 321-357.

Huijsmans, J.P.P., Barton, M. and Salters, V.J.M., 1988. Geochemistry and evolution of the calc-alkaline volcanic complex of Santorini, Aegean Sea, Greece. J. volcan. geotherm. Res., 34: 283-396.
Jackson, J., 1994. Active tectonics of the Aegean region. Ann. Rev. Earth Plan. Sci., 22: 239-271.
Keller, J., 1982. Mediterranean Island Arcs. In: Thorpe, R.S. (Editor), Andesites. Wiley, New York, pp. 307-326.
Lyon-Caen, H., Armijo, R., Drakopoulos, J., Baskoutas, J., Delibassis, N., Gaulon, R., Kouskouna, V., Latoussakis, J., Makropoulos, K., Papadimitriou, P., Papanastassiou, D., and Pedotti, G., 1988. The 1986 Kalamata (South Peloponnese) earthquake: Detailed study of a normal fault, evidences for east-west extension in the Hellenic arc. J. geophys. Res., 93: 14,967-15,000.
Makris, J., 1977. Geophysical investigations of the Hellenides. Hamburger geophysischer Einzelschriften, 34: 1-124.
Makropoulos, K.C. and Burton, P.W., 1984. Greek tectonics and seismicity. Tectonophysics, 106: 275-304.
Mascle, J. and Martin, L., 1989. Shallow structure and recent evolution of the Aegean Sea: a synthesis based on continuous reflection profiles. Mar. Geol., 94: 271-299.
Mitropoulos, P. and Tarney, J., 1992. Significance of mineral composition variations in the Aegean Island Arc. J. Volcan. Geotherm. Res., 51: 283-303.
Mitropoulos, P., Tarney, J., Stouraiti, C., Notsu, K. and Arakawa, Y., 1998. Sr isotopic variation along the Aegean Arc: constraints on magma genesis on the basis of new Sr isotope data. Bull. geol. Soc. Greece, 32 (3): 215-224.
Mountrakis, D., Pavlides, S., Chatzipetros, A., Meletlidis, S., Tranos, M., Vougioukalakis, G. and Kilias, A., 1998. Active deformation of Santorini. In: R. Casale, M. Fytikas, G. Sigvaldarsson and G. Vougioukalakis (Editors), The European laboratory volcanoes. European Commission, EUR 18161, pp. 13-22.
Papadopoulos, G.A. and Pavlides, S.B., 1992. The large 1956 earthquake in the South Aegean: macroseismic field configuration, faulting, and neotectonics of Amorgos Island. Earth Plan. Sci. Letts., 113: 383-396.
Papanikolaou, D., Lykousis, V., Chronis, G. and Pavlakis, P., 1988. A comparative study of neotectonic basins across the Hellenic arc: the Messiniakos, Argolikos, Saronikos and southern Evoikos gulfs. Basin Res., 1: 167-176.
Papazachos, B.C. and Comninakis, P.E., 1971. Geophysical and tectonic features of the Aegean Arc. J. geophys. Res., 76: 8517-8533.
Papazachos, B.C. and Panagiotopoulos, D.G., 1993. Normal faults associated with volcanic activity and deep rupture zones in the southern Aegean volcanic arc. Tectonophysics, 220: 301-308.
Papazachos, B.C., Karakostas,V.G., Papazachos, C.B. and Scordilis, E. M., 2000. The geometry of the Wadati-Benioff zone and lithospheric kinematics in the Hellenic arc. Tectonophysics, 319: 275-300.
Papazachos, C.B. and Nolet, G., 1997. P and S deep velocity structure of the Hellenic area obtained by robust nonlinear inversion of travel times. J. geophys. Res., 102: 8349-8367.

Paraskevopoulos, G.M., 1957. Über die Chemismus und die provinzialen Verhältnisse der tertiären und quatären Ergussgesteine des ägäischen Raumes und der benachbarten Gebiete. Tsch. Min. Petr. Mitt., series 3, 6: 1-72.

Pe, G.G., 1975. Strontium isotope ratios in volcanic rocks from the northwestern part of the Hellenic Arc. Chem. Geol. 15: 53-60.

Pe, G.G. and Piper, D.J.W., 1972. Vulcanism at subduction zones; The Aegean area. Bull. geol. Soc. Greece, 9: 113-144.

Pe-Piper, G., 1998. The nature of Triassic extension-related magmatism in Greece: evidence from Nd and Pb isotope geochemistry. Geol. Mag., 135: 331-348.

Pe-Piper, G. and Hatzipanagiotou, K., 1997. The Pliocene volcanic rocks of Crommyonia, western Greece and their implications for the early evolution of the South Aegean Arc. Geol. Mag., 134: 55-66.

Pe-Piper, G. and Piper, D.J.W., 1989. Spatial and temporal variation in Late Cenozoic back-arc volcanic rocks, Aegean Sea region. Tectonophysics, 169: 113-134.

Pe-Piper, G. and Piper, D.J.W., 2001. Late Cenozoic, post-collisional Aegean igneous rocks: Nd, Pb and Sr isotopic constraints on petrogenetic and tectonic models. Geol. Mag., 138: 653-668.

Pe-Piper, G. and Piper, D.J.W., 2002. The Igneous Rocks of Greece. Borntraeger, Stuttgart, 645 pp.

Pe-Piper, G., Piper, D.J.W. and Perissoratis, C., submitted. The relationship of tectonism to volcanism: the Kos Plateau Tuff eruption of 161 ka. J. Volcan. Geotherm. Res. *(revised version returned to journal; may be accepted by the time proofs are checked)*

Piper, D.J.W. and Perissoratis, C., 2003. Quaternary neotectonics of the South Aegean arc. Mar. Geol. 198: 259-288.

Puchelt, H., Murad, E. and Hubberten, H.W., 1977. Geochemical and petrological studies of lavas, pyroclastics and associated xenoliths from the Christiana Islands, Aegean Sea. N. Jb. Min. Abh., 131: 140-155.

Robertson, A.H.F., Clift, P.D., Degnan, P. and Jones, G., 1991. Paleoceanography of the Eastern Mediterranean Neotethys. Paleogeog., Paleoclim., Paleoecol., 87: 289-343.

Skarpelis, N.S., Kyriakopoulos, K. and Villa, I., 1992. Occurrence and $^{40}Ar/^{39}Ar$ dating of a granite in Thera (Santorini, Greece). Geol. Rund., 81: 729-735.

Spakman, W., van der Lee, S. and van der Hilst, R.D., 1993. Travel-time tomography of the European-Mediterranean mantle down to 1400 km. Phys. Earth Plan. Int., 79: 3-74.

Tarney, J., Barr, S.R., Mitropoulos, P., Sideris, K., Katerinopoulos, A. and Stouraiti, C., 1998. Santorini: geochemical constraints on magma sources and eruption mechanisms. In: R. Casale, M. Fytikas, G. Sigvaldarsson and G. Vougioukalakis (Editors), The European laboratory volcanoes. European Commission, EUR 18161, pp. 89-111.

ten Veen, J.H. and Meijer, P.Th., 1998. Late Miocene to Recent tectonic evolution of Crete (Greece): geological observations and model analysis. Tectonophysics, 298: 191-208.

ten Veen, J.H. and Kleinspehn, K.L., 2003. Incipient continental collision and plate-

boundary curvature: Late Pliocene – Holocene transtensional forearc, Crete, Greece. J. geol. Soc. Lond., 160: 161-181.
Wyers, G.P., 1987. Petrogenesis of calc-alkaline and alkaline magmas from the southern and eastern Aegean Sea, Greece. Ph.D. thesis, Ohio State University.
Wyers, G.P. and Barton, M., 1987. Geochemistry of a transitional ne-trachybasalt - Q-trachyte lava series from Patmos (Dodecanesos), Greece: further evidence for fractionation, mixing and assimilation. Contr. Min. Pet., 97: 279-291.
Wyers, G.P. and Barton, M., 1989. Polybaric evolution of calc-alkaline magmas from Nisyros, southeastern Hellenic Arc, Greece. J. Petrol., 30: 1-37.
Woodcock, N.H. and Schubert, C. 1994. Continental strike-slip tectonics. In: P.L. Hancock (Editor), Continental deformation. Pergamon Press, pp. 251-263.

Our bubbling Earth

R.D. Schuiling

ABSTRACT

In several places on earth large volumes of gas are seen to escape. These gases are usually dominated by CO_2. The emissions are associated with volcanic activity, and are attributed to magma degassing. It will be shown that in the case of Milos this explanation is unacceptable for quantitative reasons. An alternative explanation will be given, and its consequences for the thermal evolution of certain sections of the crust will be discussed.

1. OUR EXAMPLE

Milos is a volcanic island that forms part of the active volcanic arc in the Aegean. The geothermal gradient near its surface is extreme, of the order of 8 to 10°C per 10 meter, i.e. about 30 times the average geothermal gradient on earth (Fytikas, 1989). In an area of about 35 km^2 on and around Milos 2.2×10^6 tons of CO_2 are released annually (Dando et al., 1995, 2000). This gas release has probably continued for a period of 3 million years (Fytikas et al., 1976), leading to a total gas production so far of 6.6×10^{12} tons. Let us first calculate if magma degassing can be made responsible. Acid and intermediate magmas like those found on Milos can dissolve 1 wt% or less of CO_2 (Holloway and Blank, 1994) so we need 6.6×10^{14} tons of magma to dissolve the CO_2 escaped so far, or approximately 3×10^{14} m^3 = 3×10^5 km^3. A magma chamber with a cross section of 35 km^2 (the area over which degassing has been observed) should then have a vertical extension of 8.600 km! It is evident that magma degassing alone cannot offer a satisfactory explanation for the amount of gas observed.

Carbon isotopic compositions of the emissions around Milos are between +1.1 and –1.0 ‰ (Botz et al., 1996), typical of marine carbonates (Keith and Weber, 1964). Milos is lying over a subducting slab which originates in the Hellenic Trench, South of Crete. In an earlier episode of subduction, material was subducted in a more northerly zone, and part of this subducted material has been exhumed again, and is visible as high-pressure/low-temperature metamorphic rocks among others on Syros and Sifnos. From the mineral parageneses it is clear that the material must have been subducted to 90 km depth or more (Schliestedt et al., 1987). Apart from eclogites, jadeite rocks, glaucophane schists etc., the subducted formations are rich in carbonate sediments. The subducted slab dips steeply under the Aegean and even penetrates into the lower mantle (Wortel and Spakman, 2000).

If the composition of the subducted material is similar to the exhumed part on Syros/Sifnos, this means that huge volumes of carbonate rocks has been dragged down to great depth in the mantle, and are now in an almost vertical position. This permits the generation of very large volumes of CO_2 under a limited area (the vertical extent of a carbonate formation in vertical position becomes very large). The temperature of the subducted slab lags behind the temperature of the surrounding mantle at any depth, but both temperatures continuously rise. As a result, carbonate formations will start to dissociate due to mineral reactions with the surrounding silicate rocks. The enthalpy of these reactions is dominated by the enthalpy change of $CaCO_3 \rightarrow CaO + CO_2$, and is rather insensitive to the type of Ca-silicate formed. The amount of heat required to provide for the CO_2-flux is between 1 and 2 orders of magnitude more than can be provided by normal heat flow in the mantle, but by drawing the required heat from a much wider zone in the mantle surrounding the subducting slab a sufficient amount of heat can be provided. From the seismic tomographic sections as shown by Wortel and Spakman *(op.cit.)*, it seems that the colder part of the mantle is indeed much wider than the slab itself.

The generated fluids, including water that is produced by dehydration reactions of hydrous minerals in the slab, are rapidly transported upward They leave their point of origin at a lower temperature than the surrounding mantle, but they will reach a point at which the temperatures of the fluid and the mantle are equal. Farther up, they will start to exert a heating effect on the surrounding rocks, which can be considerable, provided they move upward in a somewhat focused manner, possibly along the slab itself.

2. EFFECTS OF FLUIDS ON THE MANTLE

At a certain point they have to break through the mantle wedge. Wherever such mixed CO_2-H_2O fluids meet dry mantle on their way up, reactions will take place like:

$$2\,CaMg(SiO_3)_2 + 5\,MgSiO_3 + H_2O \rightarrow Ca_2Mg_5Si_8O_{22}(OH)_2 + Mg_2SiO_4$$

(in the reaction above, the tremolite proxies for hornblende).

The net result is that water is preferentially retained, whereas the gas phase will become enriched in CO_2. A likely consequence of this change from a dry mantle to a water-rich mantle, and the temperature rise by the introduction of hot gases is the formation of acid to intermediate magmas.

3. THERMAL EFFECTS OF FLUIDS IN THE CRUST

The thermal state of the crust is also affected by the passage of such massive, hot, focused flows of fluids.

Below Milos, and below the Cyclades islands in general, the crust is relatively thin, about 25 km, and the top of the mantle is at high temperatures, around 1300 K (Makris, 1977). So the gases enter the base of the crust at 1300 K, and bubble out lukewarm or cold at the earth's surface, at 298 K. The enthalpy difference between 1 g of CO_2 at 1300 K and 298 is about 1.13 kJ (Robie et al., 1978). The total mass of gas passing through the system in 3 million years is 6.6 x 10^{15} kg, so these gases add 74 x 10^{17} kJ to a crustal section with a horizontal extension of 35 km^2. This exceeds considerably the contribution from a normal heat flow over the same surface area and during the same time span, which is only 2 x 10^{17} kJ. Gas fluxes from the mantle can lead to significant heating of the crustal sections through which they pass. This heating will result in the formation of relatively high-temperature/low-pressure metamorphic assemblages, expressed as thermal domes in metamorphic terrains, like on Naxos, where fluid inclusions are seen to consist mostly of CO_2 (Kreulen, 1977). It has been calculated that 7.5 x 10^{15} kg of fluid have passed through the metamorphic system at Naxos (Schuiling and Kreulen, 1979), similar to the amount escaping around Milos. Synmetamorphic quartz veins and pods mark the channels through which the fluids passed. Naxos is not associated with known volcanic activity, although there is a strong suspicion that the Stylida peninsula hosts an altered rhyolite dome capped by silica sinters.

When the fluids finally meet the boiling line of water at a depth of 1 to 2km, a major part of the heat transport is taken over by permanently boiling geothermal systems or intermittent phreatic explosions. Such steam eruptions have been numerous and sometimes spectacular on Milos, and the resulting depressions have diameters up to 1000m (Fytikas et al., 1986).

REFERENCES

Botz, R., Stüben, D., Winckler, G., Bayer, R., Schmitt, M. and Faber, E., 1996. Hydrothermal gases from offshore Milos Island, Greece. Chemical Geology, 130: 161-173.

Dando, P.R. et al., 2000. Hydrothermal studies in the Aegean Sea. Phys.Chem.Earth (B), 25: 1-8.

Dando, P.R., et al., 1995. Gas venting rates from submarine hydrothermal areas around the island of Milos, Hellenic Volcanic Arc. Continental Shelf Research, 15: No.8, 913-929.

Fytikas, M., Guiliani, O., Innocenti, F., Marinelli, G. and Mazzuoli, R., 1976. Geochronological data on recent magmatism of the Aegean Sea. Tectonophysics, 31: 29-34.

Fytikas, M., et al., 1986. Volcanology and petrology of volcanic products from the island of Milos and neighbouring islets. J.Volc.Geoth.Res. 28: 297-317.

Fytikas, M., 1989. Updating of the geological and geothermal research on Milos Island. Geothermics, 18: No 4, 485-496.

Holloway, J.R. and Blank, J.G., 1994. Application of experimental results to C-O-H speciation in natural melts. In: Volatiles in magmas, Rev.Mineral., 30: 187-230.

Keith, M.L. and Weber, J.N., 1964. Isotopic composition and environmental classification of selected limestones and fossils. GCA, 28: 1787-1806.

Kreulen, R., 1977. CO_2-rich fluids during regional metamorphism on Naxos, a study on fluid inclusions and stable isotopes. PhD-thesis, Utrecht.

Makris, J., 1977. Geophysical investigations of the Hellenides. Hamb.Geophys. Einzelschr., 34: 1-124.

Robie, R.A., Hemingway, B.S. and Fisher, J.R., 1978. Thermodynamic properties of minerals and related substances at 298.15 K and 1 bar (10^5 Pascals) pressure and at higher temperatures. Geol.Surv.Bull., 1452, 456 p.

Schliestedt, M., Altherr, R. and Matthews, A., 1987. Evolution of the Cycladic crystalline complex: petrology, isotope geochemistry and geochronology. In: Chemical transport in metasomatic processes. NATO ASI Ser.C, 218: 389-428.

Schuiling, R.D. and Kreulen, R., 1979. Are thermal domes heated by CO_2 –rich fluids from the mantle? Earth Planet. Sci. Lett., 43: 298-302.

Wortel, M.J.R. and Spakman, W., 2000. Subduction and slab detachment in the Mediterranean-Carpathian Region. Science, 290: 1910-1917.

Magmatic evolution processes as recorded in plagioclase phenocrysts of Nea Kameni rocks (Santorini Volcano, Greece)

A.P. Santo

Dipartimento di Scienze della Terra, Università di Firenze, Italy, e-mail: asanto@geo.unifi.it

ABSTRACT

Compositional zoning in plagioclase (plg) mineral phase plays a relevant role in petrogenetic studies due to the wide possible compositional variation and high sensitivity to changes of chemical and physical conditions. In this work, major and some trace element abundances by electron microprobe (EMP) and secondary ion mass spectrometry (SIMS) analyses together with textural features by Nomarski differential interference contrast (NDIC) microscopy were obtained from plagioclase crystals of Nea Kameni dacitic rocks (Santorini Volcano, Greece).

In the studied rocks, plagioclase crystals display a variety of textures such as fine-scale oscillatory zoning, resorption surfaces, sieved and patchy zones, and a wide variability of Anorthite (An) and trace element (TE) content. Based on some textural and compositional characteristics, an overall subdivision of plagioclase phenocrysts into two main groups was done: (1) clear crystals displaying a core composition in the range An 40-65%, high Sr, Ba, and LREE abundance; (2) crystals, generally sieved-core, showing a very Ca-rich core (An >75%) and low Sr, Ba and LREE content.

The trace element composition of the melt that precipitated plagioclase was calculated; changes in melt composition generally correspond to changes in TE abundance in plagioclase. However, some high Ba_{melt} or Sr_{melt} values are not correspondingly present in the crystals or vice versa.

The obtained data indicate a complex growth history of the studied plagioclase crystals. The broad range of An and TE abundance suggests that the plg under investigation experienced a broad spectrum of melt composition and/or chemical-physical condition of crystallisation. Group-2 plg cores are too anorthitic, Sr-rich and Ba-poor to have crystallised from an evolved melt, thus indicating disequilibrium with the host rocks; therefore, they were considered of xenocrystic origin. Evolutionary processes such as mixing and crystal-liquid mixing were proposed to have played an important role in determining many of the Nea Kameni plagioclase characteristics.

Keywords: Plagioclase; NDIC texture; Trace elements; Melt composition; Magmatic evolution processes.

1. INTRODUCTION

Compositional zoning is a common characteristic of magmatic minerals and reflects changes of chemical and physical conditions during mineral growth. In particular, the study of plagioclase feldspar zoning is found to be extremely useful in the comprehension of crystallization history; indeed, this mineral is capable to record even subtle changes in the magmatic system, due to the possibility of wide compositional variation and high sensitivity to variations of chemical and physical conditions. The slow interdiffusion in plagioclase guarantees that its chemical and textural zoning reflects primary growth. The composition of plagioclase in equilibrium with a certain melt composition depends on intensive and kinetic variables such as melt composition and water fugacity, temperature and pressure (e.g. Tsuchiyama, 1985), decompression and growth rate (Nelson and Montana, 1992; Lofgren, 1980). Unfortunately, it is usually difficult to distinguish the role exerted by each single variable in determining the final textural and compositional characteristics of the crystals.

Several methods are generally used to investigate mineral zoning (e.g. optical microscopy, Nomarski differential interference contrast, back-scattered electron imaging, cathodoluminescence, electron microprobe, laser interferometry). In the studies of mineral zoning the most detailed representation of the compositional profile is desirable. The traditional transmitted light polarizing microscope is not always appropriate to this purpose. Instead, the Nomarski differential interference contrast (NDIC) imaging technique is able to reveal very fine textural details (e.g. Fig. 1) otherwise almost invisible in transmitted light (Pearce et al, 1987; Pearce and Kolisnik, 1990); however, it does not readily indicate composition. NDIC microscopy has a greater space resolution than the electron microprobe but it gives only qualitative information. Nevertheless, the availability of the textural details of zonation is of great help to appropriately choose the areas to analyse. This fact is of relevant importance keeping into account the limits of space resolution of the electron microprobe which is inadequate when studying very fine zoning. Obtaining a compositional continuous profile is beyond the purpose of the present work. However, once the high and low topographic zones in the NDIC images have been "calibrated" by means of EMP analyses one can have an idea (at least to a semi-quantitative level) of the Ca content variation throughout the entire crystal. A careful selection of appropriate zones for the analysis even without a continuous profile makes it possible to obtain relevant information on the plagioclase growth and to compare data obtained by techniques having intrinsic different space resolution.

Furthermore, electron-microprobe, while providing accurate measurements of the major element content, does not permit to decide between different rival magmatic

is by far the most abundant phenocryst phase (typical mode 9 vol. %) and occurs as euhedral or subhedral and variously twinned crystals. In addition to the primary mineral phases, plg, olivine (ol), cpx and opx of clear xenocrystic nature are sometimes present: olivine displays rounded and resorbed crystals, clinopyroxene and orthopyroxene xenocrysts show a major element composition similar to that of primary crystals but different trace element abundances (Conticelli et al., 1998; Santo, unpubl. data). Groundmasses consist of abundant plg microliths, small amounts of pyroxene and oxides and brown glass. Plg compositions vary from about An_{30} to An_{90} in the rocks as a whole and occasionally from An_{45} to An_{90} within a single crystal. Core phenocrysts display a wide compositional variability (An 40-90 %), even though they exhibit a bimodal distribution with a gap existing between An_{62} and An_{72}. The composition of microphenocryst cores (An 50-60 %) as well as that of plagioclase in groundmass (An 32-50 %) lies in a more restricted range (Santo and Bomparola, 1998). Clinopyroxene is diopsidic or augitic in composition, orthopyroxene is enstatitic (En 56-65 %), olivine is forsteritic (Fo 78-80 %).

4. ANALYTICAL TECHNIQUES

The Nomarski differential interference contrast imaging technique (see Anderson, 1983; Clark et al., 1986; Pearce and Clark, 1989 for further details) was used to enhance zonal and textural features in selected plagioclase crystals. The selection of crystals for the present study was made with great care in order to avoid that the observed phenocryst characteristics might be due to the section orientation (Pearce and Kolinsnik, 1990; Pearce, 1984). Crystals oriented in order to provide as much of information as possible were selected. The NDIC technique involved etching of polished thin sections with concentrated fluoboric acid (HBF_4) for 40 seconds. Differential interference contrast images were obtained by using a Nikon Labophot-2 microscope equipped with Nomarski DIC objectives.

Major and minor-element composition of plagioclase was obtained by a wavelenght-dispersive system using a JEOL JXA-8600 electron microprobe equipped with a Series II Tracor Northern system, at 15 kV accelerating voltage and variable counting time and beam current (10-20 nA).

The etched samples were repolished prior to analysis in order to eliminate any possible contribution due to the etching (Anderson, 1983; Pearce et al., 1987); the points to be analysed were located precisely by using photomicrographs of etched crystals taking the zoning and surface characteristics - as revealed by the NDIC observations - into account. During analyses, particular care was taken to the zone width. Zones displaying very narrow width (< 10 μm) have not been analysed. The obtained data were corrected for matrix effects according to the method of Bence and Albee (1968).

Trace element analyses of the same selected plagioclase crystals were carried out with a Cameca IMS 4f ion-microprobe at the "Centro Studio per la Cristallochimica e Cristallografia", University of Pavia. The techniques used are reported in Shimizu and

Hart (1982), Shimizu and Richardson (1987), Bottazzi et al. (1990). Analytical uncertainties are <10% for REE and <5% for the other elements.

SIMS analyses produce a hole on the sample surface, which is clearly visible through the optical microscopy observation. After the SIMS analyses, the samples were etched again so that it was possible to distinctly and simultaneously see the NDIC zoning and the SIMS point analysis. Representative major and trace element analyses of Nea Kameni plg crystals are given in Table 2. The complete set of data is available on request from the author.

Table 2. Major (%) and trace (ppm) element composition of representative plagioclase crystals.

Eruptive cycle	1570												
Sample	TH 8					TH 8			TH 8				TH 8
	rim	inner rim	core	inner rim	rim	rim	core	rim	inner rim	outer core	outer core	rim	inner rim
SiO_2	46.4	47.6	47.4	47.7	46.8	46.7	48.5	47.5	57.0	56.5	56.8	59.3	58.4
Al_2O_3	34.3	33.7	33.4	33.5	34.1	34.2	33.5	34.3	27.9	27.4	28.1	26.2	26.5
FeO	0.69	0.67	0.68	0.63	0.76	0.76	0.66	0.70	0.51	0.48	0.49	0.52	0.61
CaO	17.51	16.75	16.68	16.49	17.42	17.22	16.10	16.99	9.48	9.23	9.47	7.41	8.07
Na_2O	1.53	1.87	1.84	1.90	1.50	1.45	2.03	1.71	5.57	6.02	5.98	7.02	6.56
K_2O	0.04	0.04	0.05	0.02	0.03	0.04	0.03	0.04	0.25	0.20	0.18	0.29	0.29
Total	100.5	100.6	100.0	100.2	100.6	100.4	100.8	101.2	100.7	99.9	101.1	100.7	100.4
An %	86.1	83.0	83.1	82.6	86.4	86.6	81.3	84.4	47.7	45.3	46.2	36.2	39.8
Ab %	13.6	16.8	16.6	17.2	13.5	13.2	18.5	15.4	50.8	53.5	52.8	62.1	58.5
Or %	0.23	0.24	0.30	0.12	0.18	0.24	0.18	0.24	1.50	1.17	1.05	1.69	1.70
Ti	115	155	244	277	126	104	122	120	206	374	249	210	208
Sr	316	346	351	336	330	280	324	324	438	437	477	515	510
Y	0.19	0.38	0.66	1.14	0.28	0.19	0.19	0.19	0.46	1.47	0.45	0.47	0.47
Zr	0.09	0.38	1.33	n.d.	0.09	b.d.	0.19	b.d.	0.11	n.d.	0.11	0.12	b.d.
Ba	10.64	19.75	19.12	16.27	15.62	8.22	13.86	12.17	108	107	109	118	93.5
La	0.31	0.74	0.84	1.00	0.71	0.36	0.58	0.29	3.15	3.40	2.86	2.79	2.69
Ce	1.03	1.41	1.77	2.03	1.55	0.60	1.04	1.07	5.47	6.08	5.67	4.99	4.79
Nd	0.27	0.36	0.45	0.78	0.44	0.18	0.59	0.40	1.15	1.89	1.87	1.98	1.59
Sm	0.07	0.09	0.16	0.24	0.09	0.05	0.16	0.14	0.28	0.28	0.33	0.27	0.42
Eu	0.20	0.27	0.26	0.35	0.19	0.16	0.21	0.16	1.40	1.33	1.52	1.19	1.20

5. PLAGIOCLASE FROM NEA KAMENI

The NDIC observations of plagioclase revealed a great variety of textural characteristics even coexisting in the same rock sample or crystal: sieved, fritted and patchy textures, clear unzoned cores, fine-scale zoned rims, and resorption surfaces (e.g. Fig. 1). These features are displayed by plg from all rock samples, independently of eruptive cycles. Despite the variety of features observed, which sometimes coexist even within the same crystal, it is possible to broadly define two major textural groups of plagioclase: a) group-1 plg, generally characterised by a clear poorly zoned core and by a variously zoned mantle and rim; in a few cases concentric fine-scale zones occupy the entire crystal; resorption surfaces observable in the crystals are generally associated with a non-significant An variation; b) group-2 plg that generally displays sieved, fritted or patchy cores and irregular zones in the entire crystal or exclusively in the rim. Interestingly, these two textural groups correspond to an overall compositional subdivision: cores of group-1 plg show An content in the range An 40-55 %; group-2 plg shows always a very Ca-rich core that commonly has An content > 75%. In terms of An content, rim composition is rather variable. An-zoning, as obtained by EMP analyses, is generally normal with rims more sodic with respect to the cores; however, reverse-zoned crystals are also present even though zoning is light. In addition, a third "intermediate" group of crystals ("group-3" plg) could be also identified by considering those phenocrysts showing a Ca-rich core and a sudden decrease of An content in the rim associated with an abrupt resorption surface (e.g. KAM 74-1; Fig. 5). Interestingly, the An % of these rims is similar to the rim composition of the group-1 from the same eruptive cycle.

Trace element abundances, as obtained by SIMS analyses, confirm the existence of two main different groups of crystals. Group-1 plg generally exhibits high Ti, Sr, Ba and LREE abundance whereas group-2 crystals are characterised by a lower content of Ti, Sr, Ba and LREE. According to the main subdivision, rims in crystals of "intermediate" group show high Ti, Sr, Ba and LREE abundance whereas cores display low content of Ti, Sr, Ba and LREE. Y and Zr are low, scattered and often their content is below the detection limit of the analytical method.

Some of the analysed grains are shown in detail in Figs. 3-6; for each crystal, the composition of the analysed points in terms of An, Sr and Ba content, is reported graphically. Eu (not showed) often displays behaviour similar to Sr. An important feature of these figures is the general correspondence between An and trace element zoning, although large changes in An content may not correspond to large changes in Ba or Sr content. In detail, in group-1 the behaviour of Sr and Ba is more irregular than in group-2 where a negative correlation between An and Sr and Ba is generally observed. This feature should indicate a relationship between the magma composition and the trace element behaviour during plg crystallisation. Anyway, in both groups, in a few cases, Sr and Ba contents seem to be completely independent from the An content. This observation implies an important conclusion i.e. the trace element variation is not always linked to An zonation thus confirming what suggested by Blundy and Shimizu (1991).

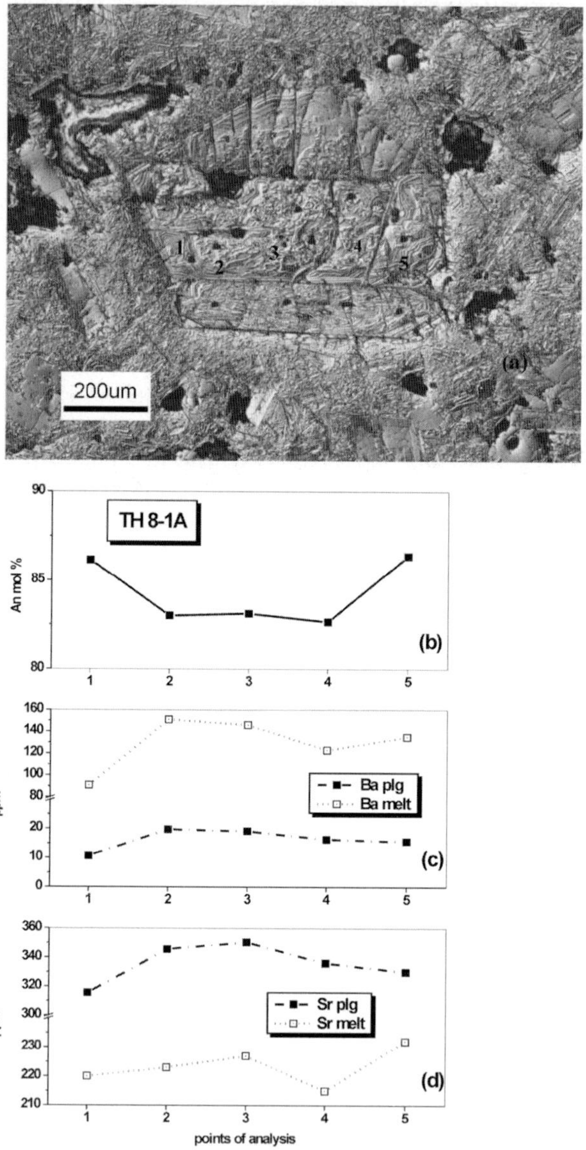

Fig. 3. (a) Nomarski DIC image of a plagioclase phenocryst (group-2) from 1570 lavas showing SIMS point analysis; (b) Anorthite content in the above crystal; (c) and (d) Sr and Ba content in the plagioclase crystal and in the correspondent melt.

Some representative REE patterns are reported in Fig. 7. They are those typically found in plagioclase (Drake and Weill, 1975) and are characterised by decreasing abundance of LREE from La to Sm and positive spikes of Eu. Crystals are generally

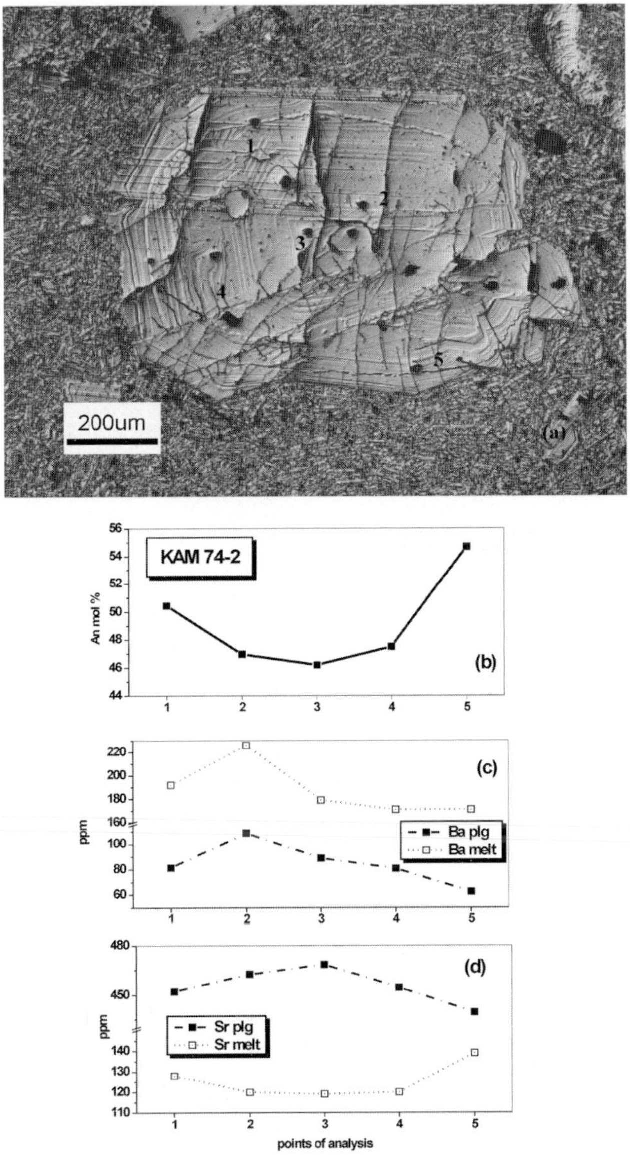

Fig. 4. (a) Nomarski DIC image of a plagioclase phenocryst (group-1) from 1866 lavas showing SIMS point analysis; (b) Anorthite content in the above crystal; (c) and (d) Sr and Ba content in the same plagioclase crystal and in the correspondent melt.

homogeneous with no evidence of zoning (e.g. KAM 74-2 and KAM 1-4) may be indicating the low REE sensitivity to the variation of crystallisation conditions. The different abundance of REE between rim and core in the "group-3" plg is evident in

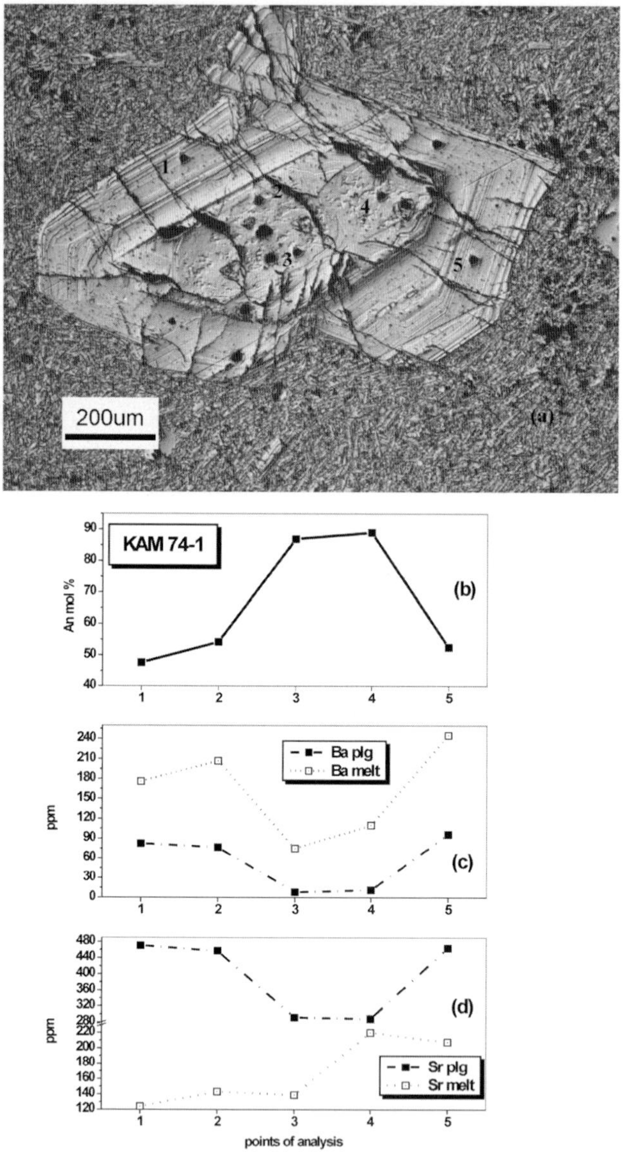

Fig. 5. (a) Nomarski DIC image of a plagioclase phenocryst (group-3) from 1866 lavas showing SIMS point analysis; (b) Anorthite content in the above crystal; (c) and (d) Sr and Ba content in the same plagioclase crystal and in the correspondent melt.

Fig. 7d and suggests the crystallisation of the different two groups of plagioclase (respectively, rim and core of group-3) from chemically different liquids.

In the variation diagrams of Fig. 8, crystals of the different groups plot in quite

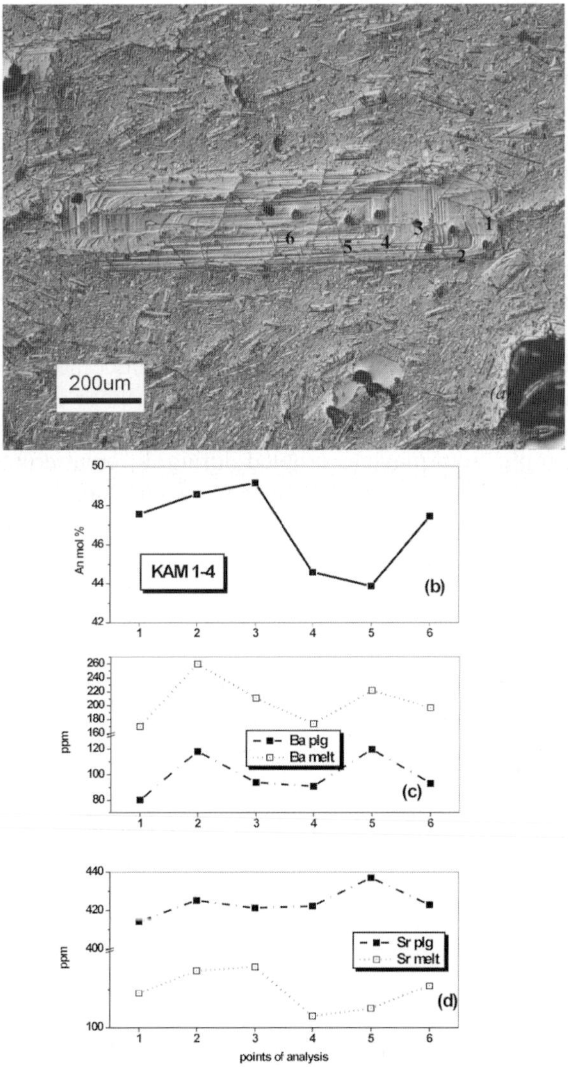

Fig. 6. (a) Nomarski DIC image of a plagioclase phenocryst (group-1) from 1950 lavas showing SIMS point analysis; (b) Anorthite content in the above crystal; (c) and (d) Sr and Ba content in the same plagioclase crystal and in the correspondent melt.

distinct areas; an overall negative correlation of Sr and Ba versus An is in any case observable; however, in each group, crystals from different eruptive cycles display distinctive abundances of Sr. Ba abundance displays a great dispersion of values in group-1 of plagioclase where the variability is wide among crystals from different eruptive cycles but also among plagioclase crystals present in the same rock sample; on the contrary, Ba content is almost constant (with low values) within group-2. Core crystals from "group-3" plot in the group-2 area whereas rims plot in the group-1 area, both for Sr and Ba values, and show composition similar to rims of group-1 plg (not showed).

Thus, even though with a few exceptions, plg in magmas erupted at different time and displaying similar dacitic composition shows different abundances of trace elements. Ba and Sr abundances seem to indicate the existence of different magmas. Partition coefficients for Sr and Ba strongly depend on temperature. However, partition coefficients for Sr are greater than unity at all geological temperatures whereas for Ba they are less than unity above 1060°C (Drake and Weill, 1975). The negative correlations existing in each eruptive cycle between Sr and Ba with An are clearly due to the fact that both Sr and Ba enter albitic plagioclase more readily than calcic plagioclase (Blundy and Wood, 1991). However, it is more difficult to explain the variability of trace element content at a constant value of An. Whereas different content of trace elements in plg from magmas erupted during different eruptive cycles can be imputable to modifications occurred in the magmas with time, the variation during the same eruptive cycle seems to indicate a more complex crystallisation history.

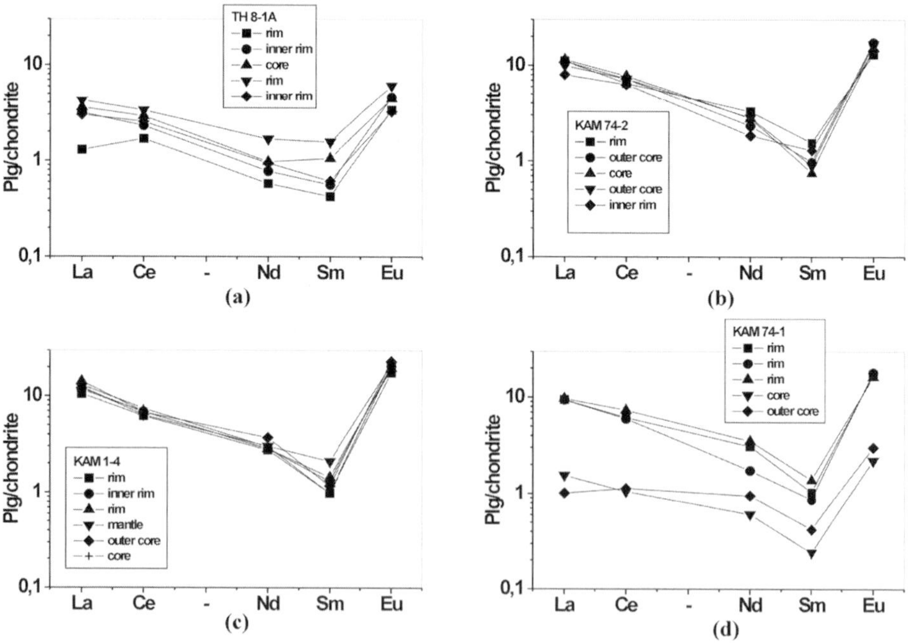

Fig. 7. Chondrite-normalised REE patterns of representative Nea Kameni plagioclase crystals. (a) plg crystal of Fig. 3; (b) plg crystal of Fig. 4; (c) plg crystal of Fig. 5; (d) plg crystal of Fig. 6. Normalisation values from Sun and McDonough (1989).

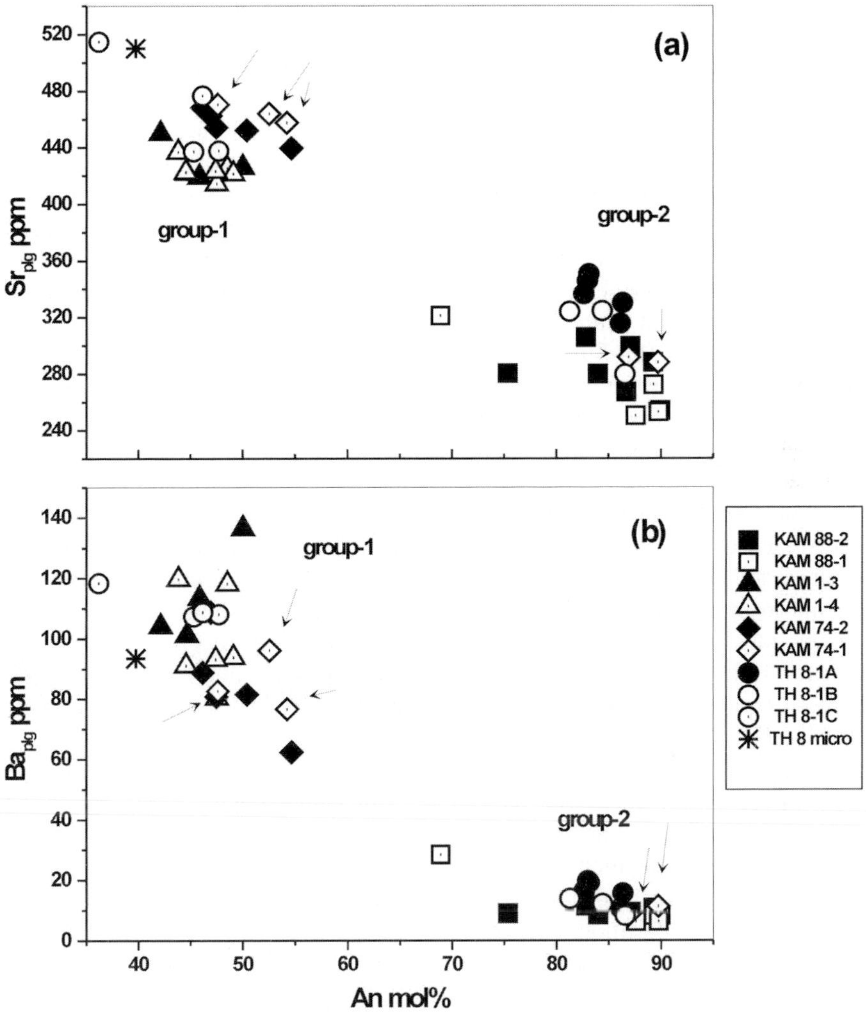

Fig. 8. Variation diagram of (a) Sr and (b) Ba content in the plagioclase versus An. Arrows indicate group-3 crystals.

6. SR AND BA MELT COMPOSITION

Blundy and Wood (1991) have shown that the partitioning of Sr and Ba between plagioclase and silicate melts is strongly dependent on the An content of the plagioclase; variation of parameters such as temperature, pressure and melt composition has a subordinate effect producing only small variation of distribution coefficients at a given plagioclase An content. However, subsequently Blundy and Shimizu (1991) verified that trace element concentrations were not necessarily controlled by the An content and Morse (1992) suggested a very important role of the liquid composition in the partitioning of Sr in plg.

By using the semi-empirical equations derived by Blundy and Wood (1991), the trace element content of each plagioclase zone of known An content can be related to the trace element concentration in the melt from which it has precipitated. In order to use these expressions, it is necessary to estimate the equilibration temperature. Calculated crystallisation temperatures of Nea Kameni lavas on the basis of different geothermometers gave values in the range 970-1020°C (Conticelli et al., 1998). These values are in agreement with those calculated by Barton and Hujsmans (1986) and directly measured by Washington (1926). Higher values (1245-1275°C) have been found trough the homogenisation of glass inclusions in plagioclase (Vaggelli and Francalanci, 1998). However, as the temperature varies during the crystallisation, a definite estimate of temperature is impossible to obtain; in this work an equilibration temperature of 1000°C has been considered to be a reasonable estimate. In addition, Blundy and Shimizu (1991) demonstrated the small effect of temperature on the calculated melt composition. Calculation of element concentrations in the melt assumes that each crystal zone grows in equilibrium with the melt. This assumption is considered to be valid in plutons (Hart and Allegre, 1980; Blundy and Shimizu, 1991) where the plg growth rates are low. However, at large undercooling, crystal growth may take place in non-equilibrium conditions. As a consequence, the composition of melts obtained by using these equations must be used with caution particularly in the case of volcanic rocks (Singer et al., 1995).

With all due reservations, however, trace element concentration of the melt in correspondence of the measured spots has been calculated by using the equations reported by Blundy and Shimizu (1991). The results are showed in Figs. 3-6 where Sr and Ba abundances in the melt are compared to the An and trace element abundances in plg crystals. Group-1 melts display lower Sr and higher Ba content in respect to group-2 melts. The different partition coefficients of Sr and Ba for plagioclase during the crystallisation of plg yields residual liquids impoverished in Sr relative to Ba (Drake and Weill, 1975). Clearly, An-rich liquids will contain higher Sr and lower Ba content in opposition to An-poor liquids. Changes in melt trace element composition generally are associated with changes in plagioclase trace element composition. However, in Figs. 3-6, some differences can be observed: they can be partly explained by the dependence of partition coefficients from the An content (see equations 1 and 2 of Blundy and Shimizu, 1991). However, they may represent real geochemical feature of the melt or/and may also bear witness to diffusion effects across the core-rim boundary (Blundy and Shimizu, 1991).

In terms of the Sr_{melt} and Ba_{melt} variation with An content (Fig. 9), it is possible to observe a positive correlation for both groups of crystals between Sr_{melt} and An. Melt abundances display a certain variability in both groups among crystals from different eruptive cycles and/or within the same crystal even at a constant An value.

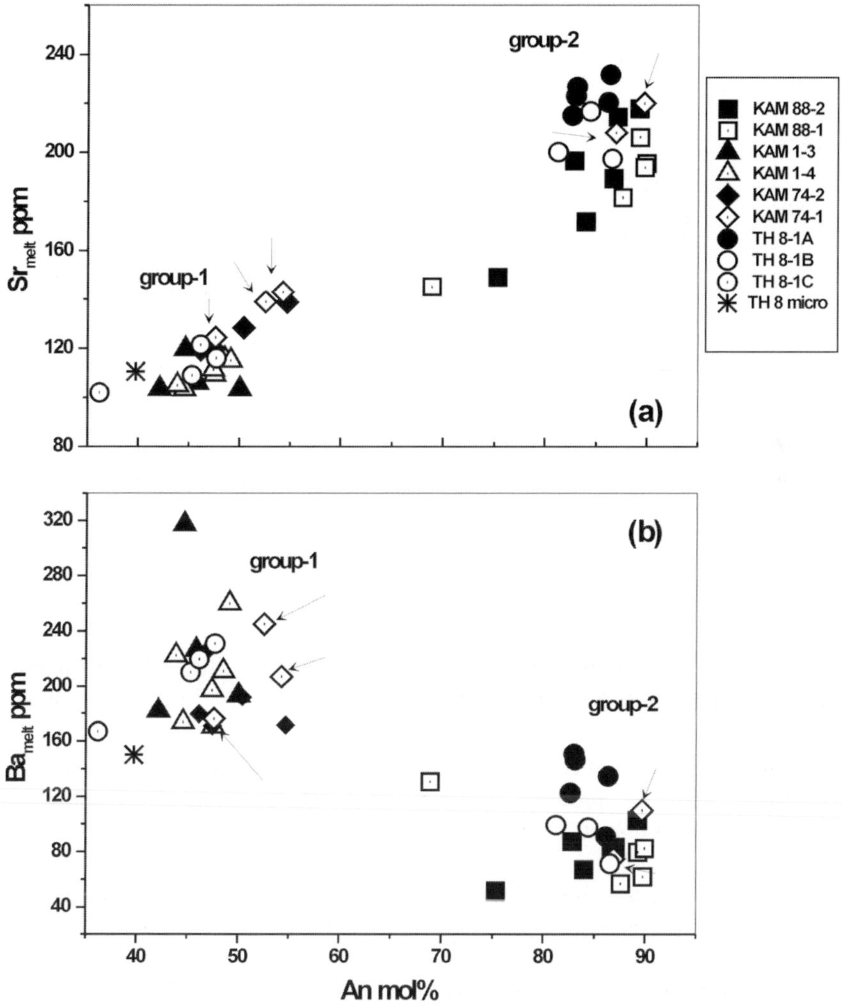

Fig. 9. Variation diagram of (a) Sr and (b) Ba content in the calculated melt versus An. Arrows indicate group-3 crystals.

From the Ba_{melt} vs. Sr_{melt} plot (Fig. 10) one can observe that: i) "group-1" exhibits higher Ba content and lower Sr content with respect to "group-2"; ii) "group-1" shows a large inter- and intra-samples variation, particularly evident for Ba; iii) "group-2" shows a variable Sr content and a positive correlation with Ba_{melt}.

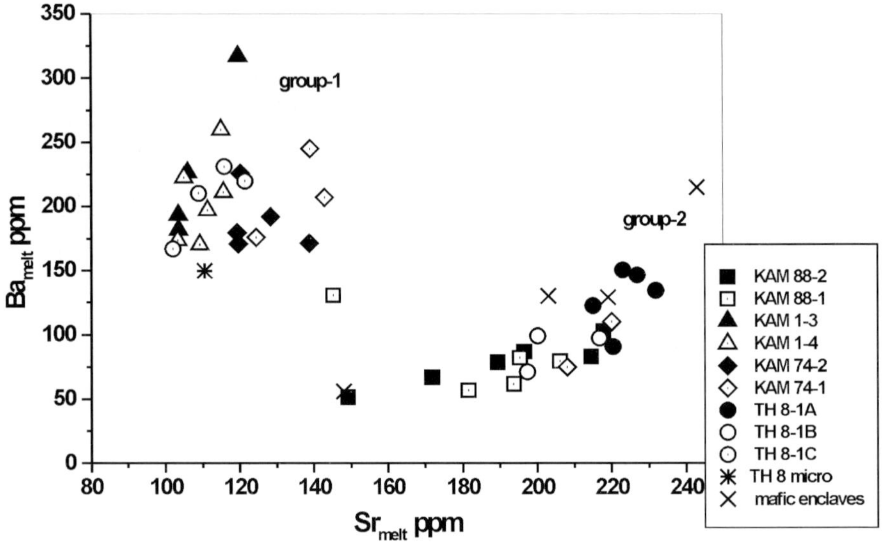

Fig. 10. Variation diagram of Ba versus Sr composition of the calculated melts.

7. DISCUSSION AND CONCLUSION

The presence of different types of plagioclase crystals in the Nea Kameni rocks was already recognised in previous works (e.g. Barton and Huijsmans, 1986; Stamatelopoulou-Seymour et al., 1990; Conticelli et al., 1998; Santo and Bomparola, 1998).

Barton and Huijsmans (1986) divided the large (up to 2 mm) plagioclase crystals in "phenocrysts" and "xenocrysts". They interpreted these latter, showing a calcic core (An > 70 %), as the product of disaggregation of the cognate xenoliths. In addition to the zoning pattern, Stamatelopoulou-Seymour et al. (1990) studied also the internal growth surfaces of the plagioclase megacrysts and they found disequilibrium textures, which led to the conclusion that magmas were formed by mixing. They also called "xenocrysts" the plagioclase crystals showing very calcic core and considered them as inherited from an old mixing event.

As reported in the previous paragraphs, textural and compositional (major and trace elements) characteristics of plagioclase studied in the present work, confirm the existence of two main different groups of plagioclase crystals. Clearly, plagioclase of group-2, having very high An content of cores, is of xenocrystic origin. Its texture and composition indicate chemical disequilibrium with the host magma. Additional evidence of disequilibrium is represented by the presence of other xenocrystic minerals such as olivine, orthopyroxene and clinopyroxene. In the following discussion crystals of group-1 will be considered as "phenocrysts" whereas plg cores of group-2 will be termed "xenocrysts". In addition, a third group of crystals has been recognised, showing intermediate characteristics, i.e. high-An core (typical of group-2 plg) and low-An rim (typical of group-1 plg).

A process of magma mixing between dacitic and mafic magmas could in part adequately explain the plagioclase characteristics and it is also testified by several evidence (petrography, trace element and isotope variation of Nea Kameni lavas; e.g. Conticelli et al., 1998). However, the small shift in Anorthite content corresponding to many dissolution surfaces observed in the plagioclase crystals does not represent an evidence of extensive magma mixing processes. The presence of An, Ba, and Sr zoning and the lack of REE zoning in the group-1 and group-2 plagioclase are in agreement with small compositional variations of the liquid; otherwise it could be imputed to variation of crystallisation conditions for which REE have low sensitivity.

Alternatively and/or in addition to magma mixing, the mechanism of *crystal-liquid mixing* proposed by Blundy and Shimizu (1991) is considered capable of producing the chemical and textural discontinuities occurring across most Nea Kameni plg crystals. According to this mechanism, plagioclase crystals are periodically *recycled* during magma evolution, as widely documented (e.g. Nixon and Pearce, 1987; Cox and Mitchell, 1988) and *retained* during magma fractionation or *re-entrained* by disruption of early cumulates from the margin of the chamber. *Retained* crystals generally display continuous zoning pattern or small discontinuities whereas *re-entrained* crystals show discontinuous textures. Whereas small-scale zoning and minor resorption surfaces can be attributed to local effects of small variation in the kinetic parameters, major resorption surfaces and zoning must be due to more important changes. The presence of Ca-rich plagioclase (group-2) in felsic lavas record influxes of hotter mafic magma, but not necessarily through mixing. This hypothesis is consistent with the presence in the Nea Kameni lavas of several mafic enclaves probably crystallised by more mafic melts than those found as lavas (Conticelli et al., 1998).

In Fig.10 some representative aphyric mafic enclaves compositions are also reported (Petrone, 1995). The Ba and Sr composition of the calculated melts from which plagioclase core of group-2 plg crystallised is close to that of aphyric mafic enclaves collected in the Kameni lavas. This fact represents an important result which demonstrates also the validity of Blundy and Wood (1991) equations, in calculating the trace element concentration in the melt from which plagioclase has precipitated. As previously mentioned, Kameni lavas contain numerous mafic aphyric enclaves displaying different texture, shape and size. They are considered to derive from different layers of a stratified magma chamber (Conticelli et al., 1998). In particular, the aphyric mafic enclave compositions may represent the less evolved magmas underplating the dacitic melts (from which group-1 plg crystallised) and therefore may correspond to the melt, which originated the group-2 plagioclase. According to the mechanism of "crystal-liquid mixing" the rims of group-3 plg, which display similar composition to rims of group-1 plg, might have been crystallised more or less simultaneously after the re-entrainment of calcic cores in the host melt. The variability of enclave composition allows also the hypothesis that the Kameni magmas experienced different and cyclic arrivals of mafic magmas and that during the crystallisation of plagioclase the system conditions were changing continuously. Furthermore, the calculated Ba-Sr composition of the melts, from which the plagioclase rims crystallised vary considerably among samples, perhaps indicating different physical conditions or the existence in the magma chamber of magmas with different composition.

Acknowledgements

The authoress would like to thank A. Peccerillo, University of Perugia, for critical reading of the manuscript and helpful suggestions. Tsegaye Abebe is also thanked for his kind help. This research has been financially supported by the Italian Council of Research (CNR).

REFERENCES

Anderson, A.T. Jr., 1983. Oscillatory zoning of plagioclase: Nomarski interference contrast microscopy of etched polished sections. Am. Mineral., 68: 125-129.

Barton, M. and Huijsmans, J.P.P., 1986. Post-caldera dacites from the Santorini volcanic complex, Aegean Sea, Greece: an example of the eruption of lavas of near-constant composition over a 2200 year period. Contrib. Mineral. Petrol., 1986, 94: 472-495.

Bence, A.E. and Albee, A.L., 1968. Empirical correction factors for the electron microanalyses of silicates and oxides. J. Geol., 76: 382-483.

Blundy, J.D. and Shimizu, N., 1981. Trace element evidence for plagioclase recycling in calc-alkaline magmas. Earth and Planetary Sci. Lett., 102: 178-197.

Blundy, J.D. and Wood, B.J., 1991. Crystal-chemical controls on the partitioning of Ba and Sr between plagioclase feldspar, silicate melts and hydrothermal solutions. Geochim. Cosmochim. Acta, 55: 193-209.

Bottazzi, P., Ottolini, L. and Vannucci, R., 1990. Quantitative SIMS analysis of rare earth elements in mafic-ultramafic rock samples. In: Benninghoven A., Evans C.A., McKeegan K.D., Storms H.A., Werner HV (eds). Secondary ion mass spectrometry SIMS VII. Wiley, Chichester, England, pp. 413-416.

Clark, A.H., Pearce, T.H., Roeder, P.L. and Wolfson, I., 1986. Oscillatory zoning and other microstructures in magmatic olivine and augite: Nomarski interference contrast observations on etched polished surfaces. Am. Mineral., 7: 734-741.

Conticelli, S., Francalanci, L., Santo, A.P. and Petrone, C., 1998. Mineral chemistry data as a contribution to the understanding of the post-Minoan magmatic system of Santorini, Greece. In: "The European Laboratory Volcano, Proceedings of the 2nd Workshop" R. Casale, M. Fytikas, G. Sigvaldasson, G. Vougioukalakis (eds), European Commission, Volcanic Risk, vol. EUR 18161 EN, Luxembourg, pp. 157-174.

Cox, K.G. and Mitchell, C., 1988. Importance of crystal settling in the differentiation of Deccan Trap basaltic magmas. Nature, 333: 447-449.

Drake, M.J. and Weill, D.F., 1975. Partition of Sr, Ba; Ca, Y, Eu^{2+}, Eu^{3+}, and other REE between plagioclase feldspar and magmatic liquid: an experimental study. Geochim. Cosmochim. Acta, 39: 689-712.

Ferrara, G., Fytikas, M., Giuliani, O. and Marinelli G., 1980. Age of formation of the Aegean active volcanic arc. In: Doumas C. (Ed.) Thera and the Aegean World II. Thera Foundation, London, pp. 37-41.

Francalanci, G., Vougioukalakis, G., Eleftheriadis, L., Pinarelli, C., Petrone, P., Manetti, P. and Christofides, G., 1998. Petrographic, chemical and isotope variations in the intracaldera post-Minoan rocks of the Santorini volcanic field, Greece. In: "The European Laboratory Volcano, Proceedings of the 2nd Workshop" R. Casale, M. Fytikas, G. Sigvaldasson, G. Vougioukalakis (eds), European Commission, Volcanic Risk, Luxembourg, vol. EUR 18161 EN, pp. 176-186.

Fytikas, M., Innocenti, F., Manetti, P., Mazzuoli, R., Peccerillo, A. and Villari, L., 1984. Tertiary to Quaternary evolution of volcanism in the Aegean region. Geol. Soc. London, spec publ., 17: 687-699.

Fytikas, M., Kolios, N. and Vougioukalakis, G.E., 1990. Post-Minoan volcanic activity of the Santorini volcano. Volcanic hazard and risk. Forecasting possibilities. In Hardy D A, Keller J, Galanopoulos V P, Flemming N C, Druitt T H. (Eds), Thera and the Aegean World III. The Thera Foundation, London,vol 2, 183-198.

Hart, S.R. and Allègre, C.J., 1980. Trace element constraints on magma genesis. In: Hargraves, R.B. (Ed.) Physics of Magmatic Processes, Princeton University Press, Princeton, pp. 121-160.

Heiken, G. and McCoy, F. Jr., 1984. Caldera development during the Minoan eruption Thira, Cyclades, Greece. J. Geophys. Res., 89: 8441-8462.

Higgins, M.D., 1996. Magma dynamics beneath Kameni volcano, Thera, Greece, as revealed by crystal size and shape measurements. J. Volcanol. Geotherm. Res., 70: 37-48.

Lofgren, G.E., 1980. Experimental studies on the dynamic crystallization of silicate melts. In: Hargraves, R.B. (Ed.) Physics of Magmatic Processes, Princeton University Press, New Jersey, pp. 487-551.

Morse, S.A., 1992. Partitioning of strontium between plagioclase and melt: a comment. Geochim. Cosmochim. Acta, 56: 1735-1737.

Nelson, S.T. and Montana, A., 1992. Sieve-textured plagioclase in volcanic rocks produced by rapid decompression. Am. Mineral., 77: 1242-1249.

Nicholls, I.A., 1971. Petrology of Santorini volcano, Cyclades, Greece. J. Petrol., 12: 67-119.

Ninkovitch, D. and Hays, J.D., 1972. Mediterranean island arcs and origin of the high potash volcanoes. Earth Planet. Sci. Lett., 16: 331-345.

Nixon, G.T. and Pearce, T.H., 1987. Laser-interferometry study of oscillatory zoning in plagioclase: the record of magma mixing and phenocryst recycling in calc-alkaline magma chambers, Iztacchiuatl volcano, Mexico. Am. Mineral., 72: 1144-1162.

Pearce, T.H., 1984. The analysis of zoning in magmatic crystals with emphasis on olivine. Contrib. Mineral. Petrol., 86: 149-154.

Pearce, T.H., Russell, J.K. and Wolfson, I., 1987. Laser interference and Nomarski interference imaging of zoning profiles in plagioclase phenocrysts from the May 18, 1986 eruption of Mt. St. Helens, Washington. Am. Mineral., 72: 1131-1143.

Pearce, T.H. and Clark, A.H., 1989. Nomarski interference contrast observations of textural details in volcanic rocks. Geology, 17: 757-759.

Pearce, T.H. and Kolisnik, A.M., 1990. Observations of plagioclase zoning using interference imaging. Earth-Science Reviews, 29: 9-26.

Petrone, M.C., 1995. Petrologia delle vulcaniti post-calderiche del vulcano Santorini (Grecia). Thesis, University of Florence, Dipartimento di Scienze della Terra.

Pichler, H. and Kussmaul, S., 1980. Comments on the geological map of the Santorini islands. In: Doumas C. (Ed.) Thera and the Aegean World II. London, pp. 413-426.

Santo, A.P. and Bomparola, R.M., 1998. NDIC and EMP Study of Mineral Zoning: An Example from Nea Kameni lavas. Microchim. Acta, 15: 219-225.

Seward, D., Wagner, G.A., and Pichler, H., 1980. Fission track ages of Santorini volcanics. In: Doumas C. (Ed.) Thera and the Aegean World II. The Thera Foundation, London, pp. 101-108.

Shimizu, N. and Hart, S.R., 1982. Applications of the ion-microprobe to geochemistry and cosmochemistry. Ann. Rev. Earth Planet. Sci., 10: 483-526.

Shimizu, N. and Richardson, S.H., 1987. Trace element abundance patterns of garnet inclusions in peridotite-suite diamonds. Geochim. Cosmochim. Acta 51: 755-758.

Singer, B.S., Dungan, M.A. and Layne, G.D., 1995. Textures and Sr, Ba, Mg, Fe, K and Ti compositional profiles in volcanic plagioclasi: Clues to the dynamics of calc-alkaline magma chambers. American Mineral., 80: 776-798.

Stamatelopoulou-Seymour, K., Vlassopoulos, D., Pearce, T.H. and Rice, C., 1990. The record of magma chamber processes in plagioclase phenocrysts at Thera Volcano, Aegean Volcanic Arc, Greece. Contrib. Mineral. Petrol., 104: 73-84.

Sun, S.S. and McDonough, W.F., 1989. Chemical and isotopic systematics of oceanic basalts: implications for mantle composition and processes. In: Saunders A.D. and Norry M.J. (eds.), Magmatism in ocean basins. Geol. Soc. London. Spec. Pub., 42: 313-345.

Tsuchiyama, A., 1985. Dissolution kinetics of plagioclase in the melt of the system diopside-albite-anorthite, and origin of dusty plagioclase in andesites. Contrib. Mineral. Petrol., 89: 1-16.

Vaggelli, G. and Francalanci, L., 1998. The composition of post-Minoan parental magmas of Santorini inferred from the study of the silicate-melt inclusions in mafic enclaves of dacitic lavas. In: "The European Laboratory Volcano, Proceedings of the 2nd Workshop" R. Casale, M. Fytikas, G. Sigvaldasson, G. Vougioukalakis (eds), European Commission, Volcanic Risk, vol. EUR 18161 EN, Luxembourg, pp. 187-192.

Washington, H.S., 1926. Santorini eruption of 1925. Bull. Geol. Soc. of America, 37: 349-384.

Volcanic hazards in the Aegean area, relative risk evaluation, monitoring and present state of the active volcanic centers

G.E. Vougioukalakis[1,*] and M. Fytikas[2]

[1] Institute for Geology and Mineral Exploration. Athens, Greece

[2] Dept of Geology, Aristotle University of Thessaloniki, Greece

ABSTRACT

Along the SAAVA, two subaereal volcanic fields host known historically active volcanic centers: 1. Methana, with an historic eruption registered in ~230 BC, and 2. Santorini, with 10 registered historic eruptions, the last one in 1950. Two other volcanic fields are considered as potentially active volcanoes: 1. Nisyros, with historic hydrothermal explosions, the last one in 1887, and 2. Milos, with probable historic hydrothermal explosions during the 1-2nd century AD.

In Methana peninsula, volcanic activity was mainly extrusive and effusive. Dangerous pyroclastic products (block and ash flows and related surges) are restricted nearby to the extrusive volcanic edifices. Volcanic hazard and risk in the area is low.

Santorini is the most active volcanic field of the South Aegean volcanic arc, one of the world's most violent caldera volcanoes. It comprises two of the three active volcanic Aegean centres, these of Kameni and Kolumbo.

The *maximum expected event* on Santorini is a catastrophic paroxysmal eruption like the Minoan event. The probability of occurrence of such an event in the near future is very low. The *most probable event to occur* on Santorini is one of post-Minoan type, like those that built up Nea Kameni and Kolumbo.

The data available on the post-Minoan activity do not permit an accurate *long-term forecasting* of the next volcanic unrest. A *short-term forecasting* for the onset of such a type of activity is considered possible with an integral and efficiently operating monitoring network.

In Milos volcanic field, the reactivating of the volcano with hazardous *explosive hydrovolcanic* eruptions (the maximum expected event) cannot be excluded. The *most probable to occur* event at Milos is that of a *hydrothermal explosion*, thus consisting a considerable natural hazard for the area. Volcanic risk in Milos island is relatively very low.

---

* Corresponding author: e-mail: gevagel@otenet.gr

On Nisyros, a very young stratovolcano (subaerial cone <160 ka), the hazard and risk evaluation on the maximum expected event, a *phreatomagmatic eruption*, cannot be estimated accurately. Such an event pose in grave danger not only Nisyros inhabitants but even the nearby area of Kos and the neighbouring coasts of Turkey. The hazard and risk of a hydrothermal explosion is relatively high, as numerous recent hydrothermal craters are present in the Nisyros caldera floor, area that is being visited by thousands of tourists during the summer time.

Yali islet is an Upper Quaternary rhyolitic volcanic edifice. The very young age (probably Neolithic) of the last *explosive eruption* of Yali, lead us to consider even this centre as a *potentially active volcano*.

Regarding the monitoring state of the active and potentially active volcanic centers, Santorini is the only area in which has been set up an integral monitoring system and a Volcano Observatory. Nisyros volcano is not efficiently monitored up to now. The same is valid for Milos and Methana volcanic fields

1. INTRODUCTION

Hellenic Quaternary volcanic activity focused in a narrow arcade belt, extending from Methana peninsula westward to Nisyros island eastward, the so-called ***South Aegean Active Volcanic Arc*** (SAAVA) (Francalanci et al., this volume, Pe-Piper and Piper this volume and therein references). Plio-Quaternary volcanic activity in this area is considered to be related to the subduction of the African plate beneath the Aegean

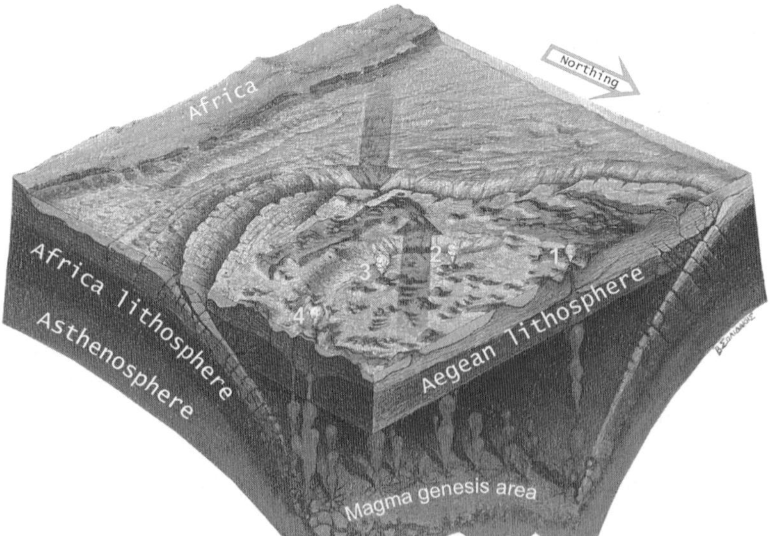

Fig. 1. A schematic cartoon indicating the present geotectonic situation in the South Aegean area and the position of the active or potentially active volcanic centers. 1: Methana, 2: Milos, 3: Santorini, 4: Nisyros.

region (Fig. 1, Papazachos et al., this volume and therein references).

Along the SAAVA, two subaereal volcanic fields host known active volcanic centers: 1. Methana, with an historic eruption registered in ~230 BC, and 2. Santorini, with 10 registered historic eruptions, the last one in 1950. Two other volcanic fields are considered as potentially active volcanoes: 1. Nisyros, with historic hydrothermal explosions, the last one in 1887, and 2. Milos, with probable historic hydrothermal explosions during the 1-2nd century AD.

1.1. Volcanic activity and hazard estimation of the different active centers

1.1.1. Methana

The main part of the peninsula consists of calc-alkaline volcanic products of andesitic - dacitic composition. Volcanic activity started at about 0.9 Ma (Fytikas et al., 1986a; Pe-Piper and Piper, 2002). The last eruption was registered at 230 BC (Strabo, Geographica): relatively calm, effusive and extrusive activity build up the lava domes and flows of Kameno Vouno, at the NW edge of the peninsula (Fig. 2).

During all the eruptive periods, volcanic activity at Methana was mainly extrusive and effusive, producing lava domes and flows with subordinate explosive activity. A few outcrops of block and ash flows as well as related surge deposits, indicate that hazardous events have been manifested in the area. This imposes the necessity of a more integral study of this volcanic field to be carried out, aiming at the formulation of

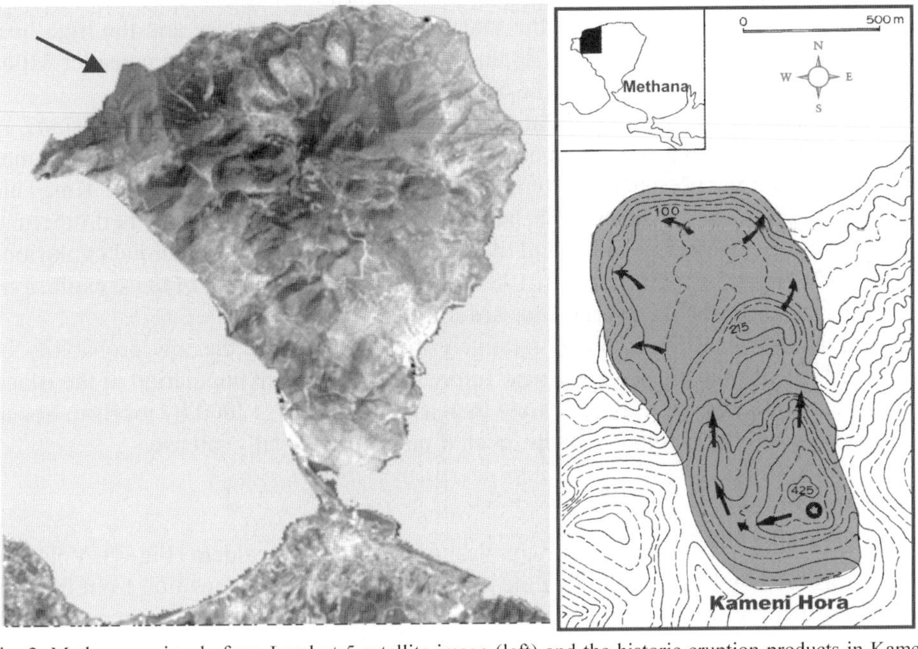

Fig. 2. Methana peninsula from Landsat 5 satellite image (left) and the historic eruption products in Kameni Hora area (right). The star indicate the vent and the arrows the flow directions.

a more accurate model for the evolution of the volcanic activity and the evaluation of the volcanic hazard in this area.

Our present estimation, based in the existing knowledge of the volcanic activity in the area, is that volcanic hazard in the area is low. Volcanic risk is also low as the inhabitants of the peninsula are few (about 2000), social and economic activity is also relatively low. The NW coast, where historic volcanic activity was manifested, is quite uninhabited. From June to October the population increases considerably because of the presence of well-known Spas in the SE shore of the peninsula where the relative important touristic center is found.

1.1.2. Milos

Since 3.5 Ma volcanic activity was manifested in the area of Milos volcanic field, fed by typical calc-alkaline magmas (Fytikas et al., 1986b; Fytikas and Vougioukalakis, 1993; Francalanci et al., 1994, 2003; Stewart and McPhie, 2003).

The most recent magmatic events consist of two big explosive phreatomagmatic eruptions that built up Trahilas and Fyriplaka tuff rings (Fig. 3). These events have been dated at 380 and 90 ka respectively (Fytikas et al., 1986b). New data (Principe et al., 2003) indicate a much younger age for the Fyriplaka eruption (19 ka). Historic magmatic eruptions have not been registered.

Considering that the future behaviour of the volcano will be conditioned by the last volcanic activity cycle, and the repose time periods will continue to have the same duration, the reactivating of the volcano with hazardous ***explosive hydromagmatic eruptions*** is not improbable. Both, the intensive shallow seismicity and the high heat flow registered in Milos area argue for an active region and lead us to consider Milos volcano as a potentially active volcanic centre.

Another serious volcanic hazard at Milos is that of the ***hydrothermal explosions***. A high enthalpy geothermal field has been explored at central Milos (Zefiria and Adamas areas) confirming the presence of overheated steam (320°C) at a depth of < 1 Km. This argument, besides the presence of a large number of relatively young hydrothermal craters nearby the geothermal area, and the presence of historic hydrothermal explosions (80-200 AD, Traineau and Dalabakis, 1989) indicates that triggering of these explosions is probable, thus consisting a serious hazard for this area.

Volcanic risk in Milos island is relatively low, considering the low probability of occurrence of volcanic events in the near future. The permanent population of the island is 5000 persons, and during summer time increases to 10000 – 12000 by tourism. Social and economic activity is relatively low, even if quarrying is really intense.

1.1.3. Nisyros

Nisyros island is a very young stratovolcano with a central caldera (Fig. 4). Volcanic activity was feed by typical calc-alkaline magmas, ranging in composition from basaltic andesite to rhyolite. The oldest sub aerial volcanic products have an age less than 160 Ka (Keller, 1971, 1980; Di Paola, 1974; Vougioukalakis, 1989, 1993; Seymour and

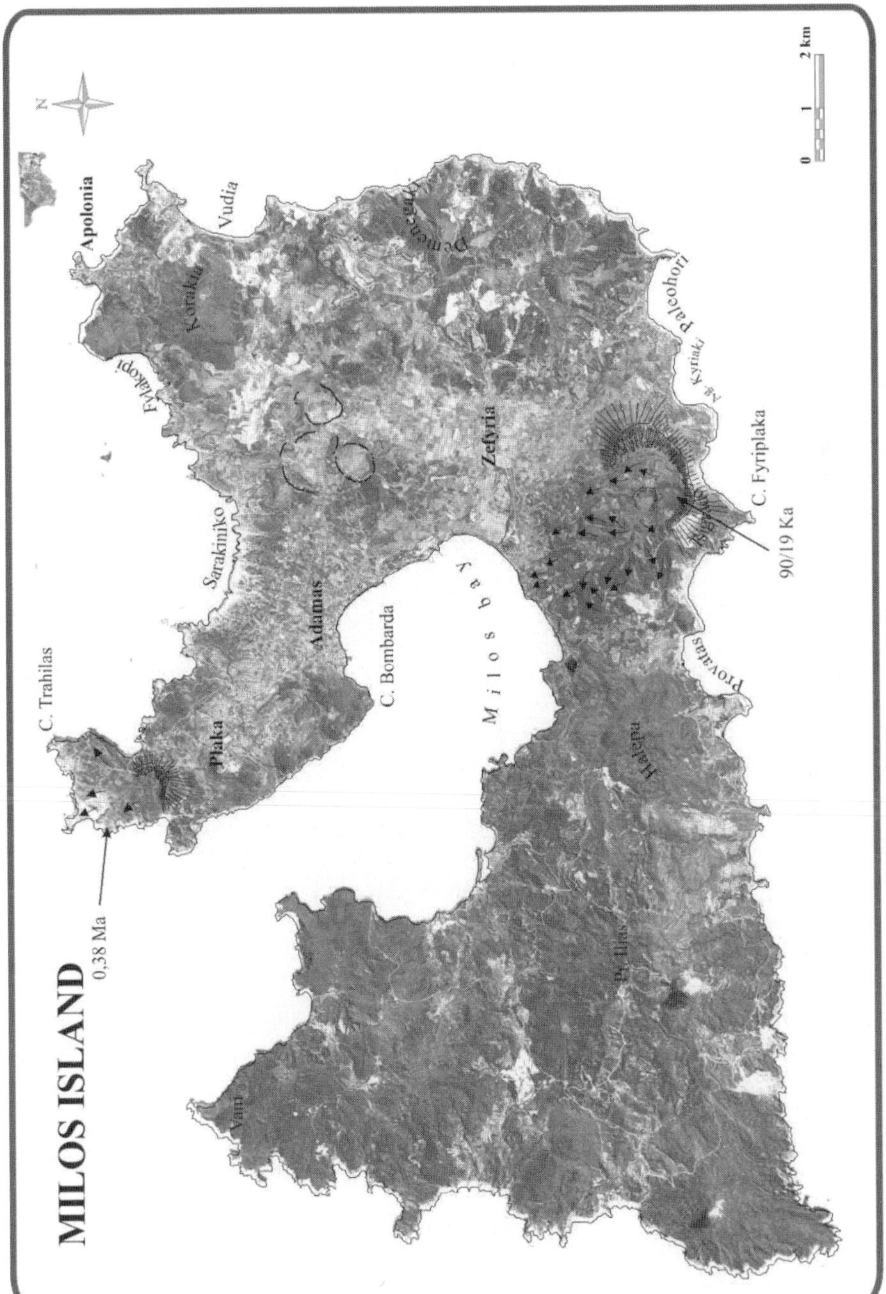

Fig. 3. Milos island from Landsat 5 satellite image. Trahilas (380 ka) and Fyriplaka (90 or 19 ka) centers, as well as the large hydrothermal craters rim are indicated.

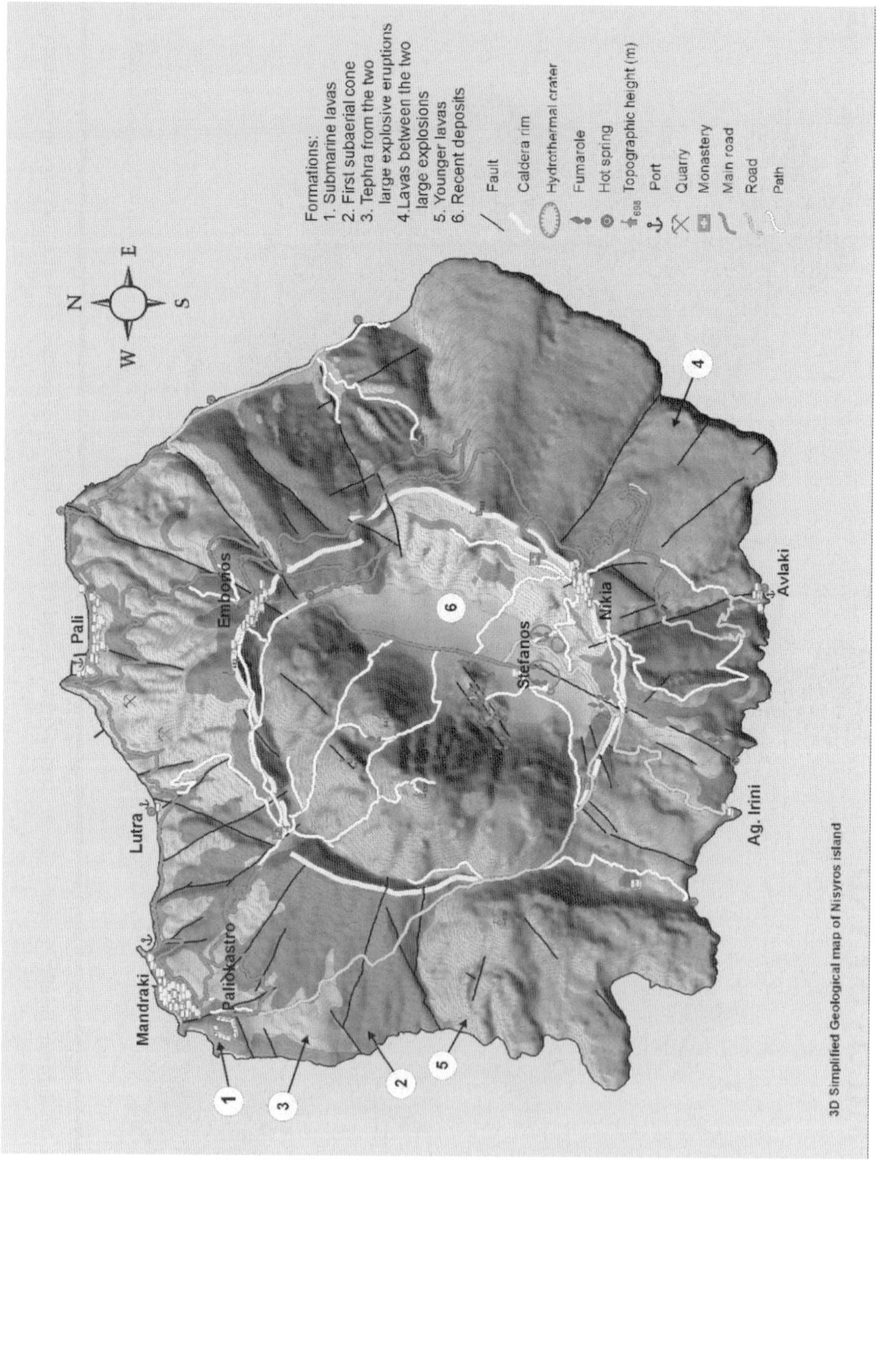

Fig. 4. Simplified 3D geological map of Nisyros volcano.

Vlassopoulos, 1989; Keller et al., 1989, 1990; Limburg and Varekamp, 1991; Francalanci et al., 1995a; Smith et al., 1996).

Two main eruptive cycles are distinguished in the evolutionary model of Nisyros volcanic activity: the first cycle includes the cone-building eruptive activity and the second one the caldera forming eruptive activity. Second cycle consists of two different phases. Each phase commenced with a low intensity - low magnitude phreatomagmatic eruption fed by rhyolitic magmas. This triggered a central calderic collapse that was followed by extrusion of rhyolitic-dacitic domes and lava flows.

The existing radiometric ages (K-Ar and ^{14}C) are contrasting but they show that the major part of Nisyros was built up during the last 160.000 years, and the two ***catastrophic explosive events*** have an age less than 40.000 years, with a considerable time gap between them. This imposes that volcanic hazard is relevant at Nisyros and leads us to consider Nisyros as a ***potentially active volcano***. The high shallow seismicity and the high geothermal anomaly argue for this estimation also (Sachpazi et al., 2002).

A serious volcanic hazard present at Nisyros is that of the ***hydrothermal explosions***. In the 19th century two periods of reactivation with hydrothermal explosions have been registered into the Nisyros caldera depression (1871-1873 and 1887) (Gorceix, 1873, 1874; Martelli, 1917). More than 20 older hydrothermal explosion craters in the caldera floor, ten of them well preserved with a maximum diameter of 300 m, indicate that this type of activity was frequent in historic and prehistoric time (Marini et al., 1993; Vougioukalakis, 1998) (Fig. 5). This registered activity and the presence of a high

Fig. 5. The hydrothermal craters on Nisyros caldera floor, seen from the top of Pr. Ilias lava dome. In ascending age order, 1: Mikros Polyvotis (1887), 2: Alexandros or Flegethron (1873), 3: Polyvotis (1873), 4: Ahileas, 5: Megalos Polyvotis, 6: Logothetis, 7: Mikros Stefanos, 8: Stefanos, 9,10: Kaminakia,

enthalpy geothermal field at Nisyros (fluids with >450°C at 1800 m depth) (Geotermica Italiana, 1983; Karydakis Gr., pers. comm.) makes probable the manifestation of hydrothermal explosions in the near future.

Regarding the risk evaluation, we cannot estimate the volcanic risk in the area regarding a phreatomagmatic eruption as we cannot estimate with the existing data, the probability of occurrence of such an eruption. In the case of the manifestation of a phreatomagmatic eruption not only Nisyros inhabitants (about 1.000 people) but even the nearby area of Kos and the neighbouring coasts of Turkey could be in grave danger.

The risk during a hydrothermal explosion is relatively high. Nisyros caldera is being visited by thousands of tourists during the summer time (about 60.000 people per season nowadays). In the case of an outburst in this period, the human losses could be too heavy.

Yali islet is the youngest volcanic centre of Nisyros island group. This is an Upper Quaternary rhyolitic volcanic edifice. Two volcanic cycles are distinguished, each with an initial explosive eruption that deposited rhyolitic pumice, followed by an extrusion of obsidian-perlitic lava domes and flows (Vougioukalakis, 1989). The very young age (probably Neolithic) of the last *explosive eruption* of Yali (Yali3-4 of Keller 1980 stratigraphy), lead us to consider even this centre as a ***potentially active volcano***.

1.1.4 Santorini

Santorini volcanic field is the most active of the South Aegean volcanic arc. It comprises two of the three active volcanic Aegean centres, these of Kameni and Kolumbo (Fig. 6) (Fouqué, 1879; Washington, 1926; Ktenas, 1927; Reck, 1936; Georgalas, 1953; Georgalas and Papastamatiou 1953; Druitt et al., 1989, 1999;).

Santorini, a multi-centre volcanic field dissected by a flooded caldera, is one of the

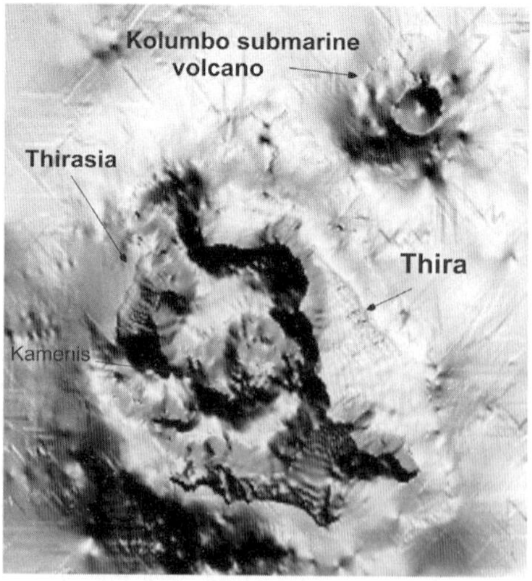

Fig. 6. 3D topographic representation of the Santorini and Kolumbo centers.

world's most violent caldera volcanoes. The last 400.000 years over a hundred explosive eruptions were manifested. Twelve of these discharged volumes of magma exceeding a few cubic kilometres, and triggered at least four caldera collapses (Druitt et al., 1989; Druitt and Francaviglia, 1992). Last of them was the Minoan eruption of the late Bronze Age (~3600 BP) (Bond and Sparks, 1976; Doumas, 1983; Heiken and McCoy, 1984, 1990; Sparks and Wilson, 1990; Pyle, 1990; Cioni et al., 2000). It discharged about 30 km^3 of rhyodacitic magma, destroying the advanced civilisation of the island, and distributed ash over a large area of eastern Mediterranean and Turkey. The effects of tsunamis and ash fallout were hardly felt on Crete, 120 km to the south.

The position of the vents during the last 600 ka was largely controlled by two NE-SW volcanotectonic lines, the *Kameni and Kolumbo lines*, which acted as path for the magma. At least five major explosive eruptions was focused on **Kameni line** as well as Aspronisi tuff ring. **Kolumbo line** controlled the vent position of two major explosive events; as well as dyke swarm, cinder cones and Cape Kolumbo tuff ring on north Thira.

After the Minoan eruption, volcanic activity continued, localizing mainly in the intracaldera area. Extrusive, effusive and slightly explosive activity produced the lava domes, flows and pyroclasts that built up Palea and Nea Kameni islets, between 197 BC and 1950 AD. All the volcanic centres are distributed along a zone 600 m wide of N65°E direction, the so-called Kameni line (Fytikas et al., 1990). Even if intracaldera activity was mild and did not threaten human lives, an explosive event occurred in 726 AD causing considerable destruction in Santorini.

Outside the caldera depression volcanic activity was manifested only once, during 1649-1650 AD (Fytikas and Vougioukalakis, 1995; Vougioukalakis et al., 1994, 1995). Initial extrusion of about 2 km^3 dacitic-andesitic magma, built up Kolumbo submarine volcano; an edifice with ellipsoid shape, 300 m high and 8 km max. axis long, on the extension of Kolumbo line. Consequent hydromagmatic and magmatic explosive activity discharged about 1 km^3 DRE of rhyodacite to andesite pumice and ash, triggering a caldera collapse (3 km diameter and 500 m deep) and a large tsunami that devastated Santorini's coasts and caused damages within a radius of 150 km.

Magmatic gases and ash fallout caused serious problems on Santorini inhabitants during this volcanic unrest. More than 70 people lost their lives and more than 1000 animals were killed, most probably by the H_2S produced (Fouque, 1879; P. Baxter, pers. comm.).

The ring island of Santorini is heavily populated (14000 inhabitants on ca 90 km^2). A continuously increasing amount of people (about 1000000 persons per year nowadays) visit Santorini yearly. More than 1/3 of them visit Nea Kameni islet and walk on the historic craters. Santorini contributes to the national income millions of dollars yearly (Vougioukalakis, 2002).

That imposes an accurate evaluation of the volcanic hazard and risk.

The *maximum expected event* on Santorini is a catastrophic paroxysmal eruption like the Minoan event. The probability of occurrence of such an event is very low, as geological and radio dating indicate that major explosive events on Santorini are separated by more than 20.000 years time gap. The second magmatic cycle of Santorini lasted about 200 ka and comprises 7 major events. The repose time between the two late major explosive events, (Cape Riva and Minoan eruptions) is 17,4 ka. So, is improbable that a major event occurs after only 3,6 ka. Furthermore, recent seismic data argue against the presence of a big magma chamber in Santorini area.

In the case of occurrence of the maximum expected eruption, the most probable location of the vent is along Kameni or Kolumbo line. The caldera walls are of great importance in the distribution of the pyroclastic flow and surge deposits, which are controlled by the topography. Anyway, the pyroclastic fall and surge deposits would cover all the area and devastate the Santorini island. In this case there is no reason for plotting a volcanic hazard zonation map. Should be interesting an ash fallout distribution modelling and the evaluation of the expected tsunami effects; as both interest a vast area of Eastern Mediterranean.

The *most probable expected event* on Santorini is one of post-Minoan type, like those that built up Nea Kameni and Kolumbo.

In favour of the hypothesis that Kameni and Kolumbo are still active now-days are:

- Kamenis magmatology, which argue for a continuous refilling of the Kameni magma chamber with injection of new mafic magma (Francalanci et al., 1995a, 1995b, 1995c, 1998).
- Seismological data, that indicate seismic activity concentrated mainly on NE and E trending lineamants beneath the Kameni and mainly Kolumbo active centre (Dimitriadis et al., this volume).
- Ground deformation measurements, which indicate a spreading of the area between Thirasia and Nea Kameni (Stiros et al., this volume).
- The soil gas survey carried out on Thira (Barberi and Carapezza, 1994) that detected the main anomalies on Helium and CO_2 concentrations corresponding to Kolumbo and Kameni lines, proving that both lineaments continue to be active systems.

In the case of the repetition of a post-Minoan type volcanic activity, the radius of the hazardous effects can be traced with relative precision, based on the already known effects of this type of activity.

Based on this and in some other probable scenarios, a volcanic hazard zonation map was plotted (Fig. 7) reporting:

- The probable position of the future eruptive centres,
- the phreatic explosions hazard zones
- the ballistic ejecta hazard zones,
- the tsunamis hazard zones,
- the toxic gases and ash fallout exposed zones,
- the hydroclastic ejecta hazard zones,
- the lava flows and scoriae cones hazard zones,
- the landslides hazard zones.

It is needed to point out that a large number of sub-plinian events are registered in the Santorini stratigraphy. Very few is known about the periodicity and the magnitude of these potentially distructive events. This is a subject that needs more accurate study, aiming at a complete hazard evaluation of the area.

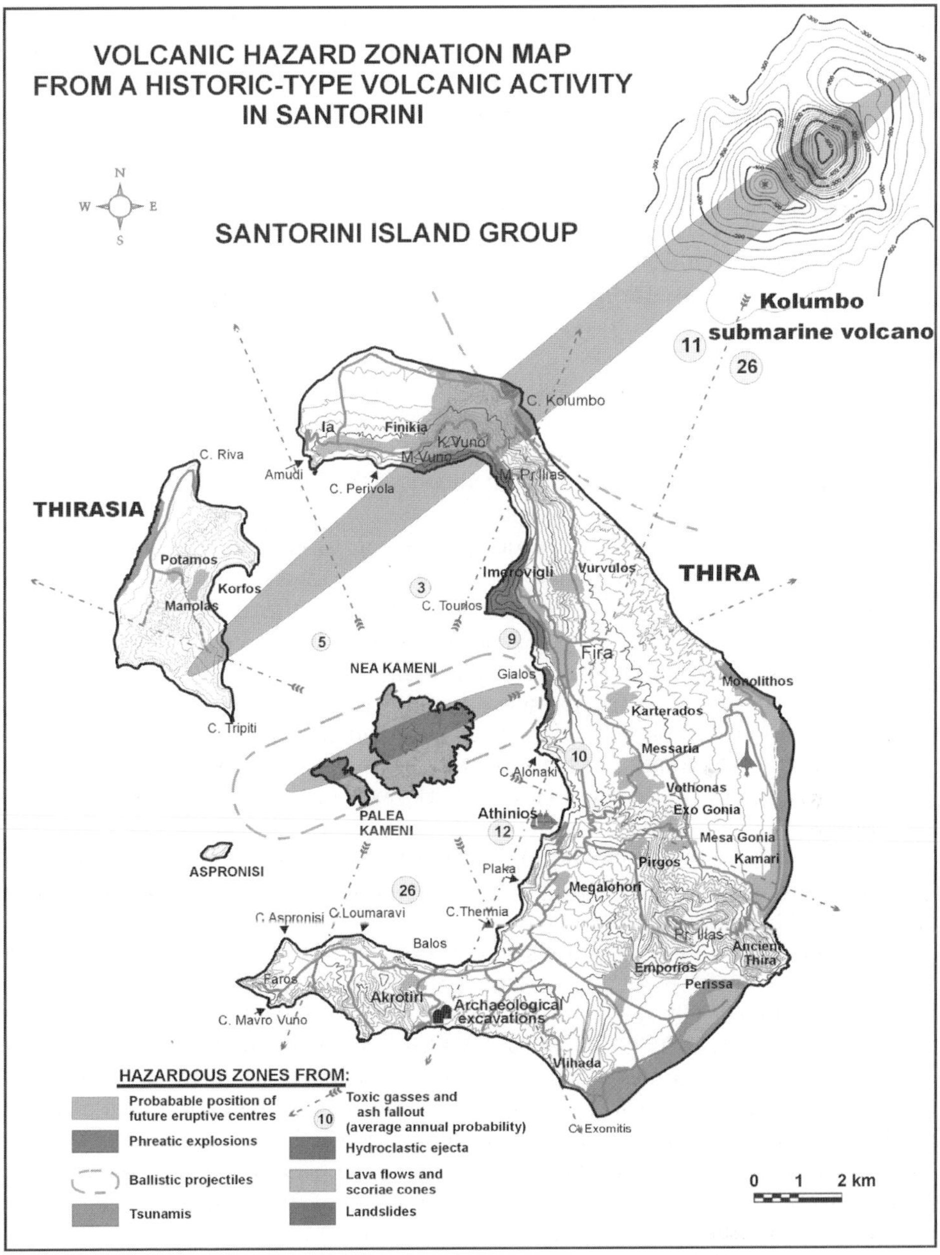

Fig. 7. Volcanic hazard zonation map of the Santorini island group.

Regarding the *possibility to forecast* the next eruption on Santorini, the data available on the post-Minoan activity do not permit an accurate *long-term forecasting* of the next volcanic unrest (Papadopoulos, 1990; Papazachos, 1990; Papadopoulos and Orfanogiannaki, this volume). Precursory phenomena, which have been registered since the post-Minoan activity lead to the conclusion that a *short-term forecasting* for the onset of such a type of activity is possible with an integral and efficiently operating monitoring network (Vougioukalakis, 1994; Fytikas et al., 1990, 1998).

1.2. Monitoring and present state of the active centers

Regarding the monitoring state of the active and potentially active volcanic centers, Santorini is the only area in which has been set up an integrated monitoring system and a Volcano Observatory.

The Santorini Volcano Observatory and the monitoring networks were established in 1994 within the auspices of a research program founded by the E.U (Santorini Volcano Laboratory) The Institute for the Study and Monitoring of the Santorini Volcano (I.S.M.O.SA.V.) was founded in the summer of 1995 as a non-profit organization, whose primary aim is to maintain the operation of the Volcano Observatory and the monitoring networks (http://ismosav.santorini.net).

Since 1995 ISMOSAV has succeeded in keeping in function the most important monitoring networks with the active support of the University of Thessaloniki, the Geodynamic Institute of the National Observatory of Athens, and the Institute of Geology and Mineral Exploration (Vougioukalakis et al., 2003).

The today operating monitoring network consists of :

- A continuous recording seismic network with 5 telemetered stations.
- A ground deformation network with a) a continuous recording wireless sea level stations of 4 tide-gauges; b) a periodically measured GPS network of 21 stations; c) a periodically measured EDM network with one central and 10 peripheral stations.
- A continuous chemical monitoring of the CO_2 flux in one station and a periodic monitoring of the hot springs and fumaroles composition in the broader area as well as the soil gas composition.
- A continuous thermal monitoring in two selected sites and a periodic monitoring of the temperatures in the fumaroles, hot springs and deep wells.

The Santorini Volcano is seismically monitored by the Geodynamic Institute of the National Observatory of Athens in co-operation with the Laboratories of Geophysics of the University of Thessaloniki and Athens. They have installed a radio-linked seismological network consisting of 5 stations in the islands of Santorini. All analogue signals are transmitted via antennas to the central station at the top of the hill of Profitis-Ilias. Despite of the unfavorable weather conditions (very strong wind etc.) the quality of the signals is fairly good. Satellite transmission of the registered data to the Laboratory of Thessaloniki has been obtained for the last period.

A number of earthquakes are recorded, of $0,5 \leq M_D \leq 4,2$ (M_D: Duration's Magnitude) and of focal depth varying between 1km and 50 km. It is very clear that two clusters of epicenters are located in the area under study. The first cluster is located in the Caldera of Santorini Isl. and is associated with the volcanic process of the caldera. The second one is located near the northern edge of the Santorini Isl. at the Kolumbo Reef, and is connected with the volcanic process at this volcano. An attempt is also made to estimate the orientations of the active fault planes by applying the "principal component method".

The seismic activity during this period remains in very low level in the caldera of Santorini and in a highest level at the Kolumbo Reef.

A network of electronic measuring stations for the monitoring of the sea level changes has been installed in the island. No significant changes related to any sign of volcanic reactivation have been registered during the monitored period.

For the global control of any topographic changes in the Santorini area, two geodetic networks have been installed. The first one is using DGPS measurements in 21 selected sites, covering the whole Santorini island group area. The second one consists of a radial EDM network with a central point at the Nea Kameni and 10 stations on Thira and Thirasia. Both networks registered a small-scale gradual inflation of the northern part of the caldera, indicating active tectonic processes and a probable magma intrusion along the active tectonic lines of this area.

The continuous CO_2 soil flux monitoring station, installed on the Nea Kameni most active area, register main flux values per day, which range between 5 και 50 ppm/sec, while the maximum registered values do not exceed the 150 ppm/sec. These values are considered as characteristic for the repose time period of the volcano.

Periodic sampling (2-3 times per year) and analysis of the major and some minor chemical species is carrying out for the Santorini hot springs.

Periodic sampling (2-3 times per year) and analysis of the main and minor constituents is also carrying out for the Santorini hot gasses (Nea Kameni fumaroles and Ag. Nikolaos bubbles).

Nea Kameni fumaroles consist of a strong portion of heated atmospheric air, and CO_2. Minor constituents (CH_4, H_2, CO) are present in variable proportions, reflecting different extents of mixing between atmospheric air and a CO_2 rich end member.

Palea Kameni gasses (Ag. Nikolaos cove bubbles) are essentially made up of CO_2 (99,9 % vol.) followed by N2, and O2. CH_4 and CO content are very low (< 10 ppmv) (Chiodini et al., 1998).

In all the monitored period, there are no significant changes in the chemical composition of both hot gasses and waters, which could indicate any deep feeding process. A further decrease of the total gas outflow and the minor gas constituents in the last two years, could indicate ongoing self-sealing processes for the uppermost Kameni area.

Periodically is monitored the *Radon (^{222}Rn) content* in the Thira-Thirasia soil gasses as well as in the fumarolic gasses and the hot springs (Radon dissolved in the water and Radon content of the bubble gas fraction).

The Radon content register long term oscillating periods. The Radon increase in soil gasses seems to be related to an "excitation" of the region after local seismic events of magnitude (M_D) larger than 4.

On Santorini, thermal monitoring includes continuous registration of the Nea

Kameni fumarolic temperature and the soil temperature on the CO_2 flux measurement area, as well as on the Palea Kameni (Ag. Nikolaos cove) hot spring. Periodic temperature registration (3-4 times per year) of the Nea Kameni (Afroessa cove) and Thira (Plaka, Athermi Christou) hot springs, as well as that of the deep well S_2 on south Thira is performed.

Fumarolic gasses have a temperature ranging from 60°C to 97°C, depending mainly from the altitude (the highest sites in altitude register the highest temperature values).

There are no stable sites or open vents where the maximum temperature values can be registered. The system is unstable, as fumarolic gasses outflow mainly through pyroclastic deposits. The maximum registered temperature on Nea Kameni fumaroles present a fluctuation of about 3°C (94°C ~ 97°C).

Since 2002, is noted a considerable diminution of the volume of the outflowing gasses from the Nea Kameni fumaroles, causing a rarest registration of the max temperature value. The most probable cause for this change is considered a self-sealing process of the volcanic rocks through their alteration from the circulating hot fluids, in a period of low tectonic activity.

On Nisyros, at the end of 1995, an intense seismic activity began in the island and lasted about 3 years (Vougioukalakis et al., 1998; Sachpazi et al., 2002). Thousands of local, shallow, low magnitude seismic events have been manifested, located mainly in the area between Nisyros and Yali islands, with characteristics that do not permit the discrimination between tectonically and volcanic triggered events. The same period ground deformation was manifested by the opening of two fractures on North Nisyros and Yali, along active faults previously mapped on land and underwater (Fig. 8).

Similar periods of intense seismic activity are common to the past: violent seismic shocks preceded and accompanied the eruptions of 1871 - 1873. Local swarms were also felt in Nisyros from time to time like those of April 1887 preceding by 5 months the last hydrothermal eruption of 1887 (Galanopoulos, 1953). Other swarms felt in Nisyros at the beginning of the century, in 1953 (Bornovas, 1953) and in 1970 (Stiros and Vougioukalakis, 1996) were not accompanied by any change in the state of the volcano (Vougioukalakis et al., 1998).

The differential interferometry study (Sachpazi et al., 2002) documents that the recent episode of Nisyros volcano unrest in years 1995-1997 resulted in crustal deformations of the order of 140 mm. Ground GPS measurements in the same area (Lagios et al., 1998) in the year 1997 concluded that the area was submitted to a general uplift trend. Nevertheless, there is some controversy between GPS and SAR results: GPS data support much higher rates of deformation.

It is interesting to point out that the northwest edge of Nisyros island is a tectonic block which suffered the maximum uplift in the geological history of the island. In this part outcrop the uplifted submarine basement of the island (Di Paola, 1974) and have been also detected large and rapid uplift movements in the last 3-4.000 years (Stiros, 2000; Stiros et al., this volume).

Considerable anomalies were also detected during the soil gas survey of the area during 1995-1998. The Radon content registered in the fumaroles and soil gasses of the

Fig. 8. 3D topographic representation of the Nisyros island group. The 1995 fractures are indicated with the black hatched lines.

caldera floor was very high during the whole crisis (4.500 pCi/L in soil gasses). After this period it progressively declined, arriving at less than 300 pCi/L in the 2000 soil gas survey. Increased fumarolic activity was also observed from June to September 1997. Intense fumarolic activity appeared specifically in the south flanks of the Megalos Polyvotis hydrothermal crater. The intensification of the fumarolic activity and the outflow of melted sulphur and hot mud on Stefanos crater floor took place one day after the occurrence of the two strongest earthquakes (Ml 5.3 and 5.2, August 28th 1997) since the beginning of the unrest phase earthquakes. Chiodini et al. (2002) noted also increasing H_2S/CO_2 ratios and decreasing CH_4/CO_2 ratios for several of the active fumaroles and interpreted these chemical changes as an increasing contribution of sulphur-rich, oxidizing magmatic fluids into the hydrothermal system below Nisyros island.

The study of spatial and temporal character of seismic activity together with the monitoring of ground deformation by INSAR Interferometry during the 1995-1998 episode (Sachpazi et al., 2002) support the existence of a magma chamber below the NW part of Nisyros. The inflation of this chamber could induce the unrest episode observed in 1995-1998 resulting in important ground deformations (total uplift of 140 mm) and also in intense seismic activity. The observed faulting in the island, which has been the source of serious house damages is then probably induced as a secondary effect of the pressure increase above the magma inflation.

Between November 2001 and December 2002 a considerably long fissure (>400 m long) trending N-S, has been manifested on the central caldera flat floor (Figs. 9, 10) (Vougioukalakis, 2003; Galanopoulos and Koletis, this volume). Fracturing developed in two distinct short time periods, with no vertical or lateral movement. This fissure was created by the collapse of the caldera floor soft sediment cover in a maximum visible depth of 15-20 m. No changes have been detected in the state of the volcano before, during and after that period. All the observed, registered or measured parameters have been unchanged in respect to the post 1995-1998 seismic crisis state of the volcano.

This passive fissuring not seems to be related to a volcano reactivation. The most probable scenario for the creation of the fissure is that of the collapse of the upper fine sediment layer (slow rate caldera feeling deposits, < 50m thick) in the discontinuous lower, highly permeable open framework, coarse formation (rapid rate caldera feeling deposit, between 50-100 m depth bsl).

The abnormally high precipitation rate during this period, combined with the denudation of the caldera walls vegetation and probable very shallow low magnitude local seismic events, could trigger the collapse of the upper sediment cover in the deepest preexisting open spaces. The numerous 1995-1998 shallow intracaldera earthquakes probably enhanced the existing discontinuities and open framework in the lower caldera feeling deposits, facilitating the large scale – quasi linear collapse.

A gravity monitoring network was installed in June 2001 on Nisyros volcano (Di Filippo and Toro, this volume). The extent of the variations recorded between 2001 and 2003 largely exceeded the measuring errors, suggesting that such variations could be ascribed to deeper mass variations.

Since April 2003, has been installed on the fumaroles of the youngest hydrothermal

Fig. 9. The fissure opened on Nisyros caldera floor in December 2001.

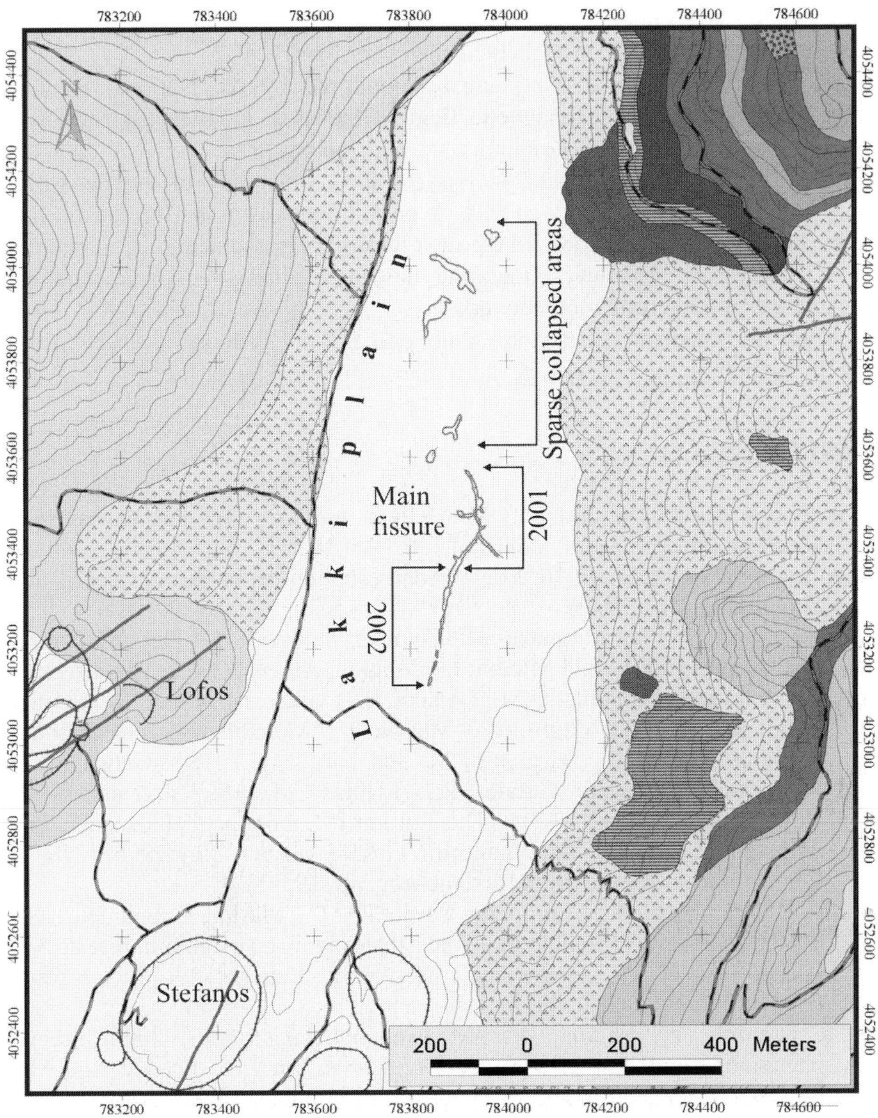

Fig. 10. The map of the main fissure and the related collapse structures in Lakki area, created during 2001-2002 winter time on Nisyros caldera floor.

crater (Mikros Polyvotis, 1887), a system for continuous monitoring of fumarolic gases. Several physicochemical and gas parameters such as the concentration of CO_2, H_2S and Rn in the fumarolic emissions, as well as the temperature of the fumarolic gas and its pressure are measured in intervals of seconds and transferred to a remote station by digital telemetry (Teschner et al., this volume). Not any significant variations for volcano reactivations have been registered since April 2003.

Since April 2002 is also continuously monitored by the National Earthquake Planning Organisation, the geomagnetic field of the area, in three (3) sites, two on Nisyros and one on Yali island (Foundulis, D., pers. comm.).

Continuously is also monitored on Nisyros the temperature of the fumaroles on Stefanos and Kaminakia craters (100°C). A periodic survey is performed on the hot fluids and soil gasses composition. In the next two years is planned to be in operation a Volcano Observatory structure, integrating the above monitored parameters with a seismic and a ground deformation network.

On Milos and Methana there is only a periodic but not systematic survey of the hot fluids composition, in the hot springs and fumaroles of the islands

REFERENCES

Barberi, F. and Carapezza, M.L., 1994. Helium and CO_2 soil gas emissions from Santorini (Greece). Bull. Volcanol., 56: 335-342.

Bond, A. and Sparks, R.S.J., 1976. The Minoan eruption of Santorini, Greece, J.Geol. Soc. London, 132: 1-16.

Bornovas, J., 1953. On the earthquakes of Nisyros in January 1953. Unpublished report, Institute of Geology and Mineral Exploration, Athens, Greece. (In Greek).

Chiodini, G., Cioni, R., Di Paola, G.M., Dotsika, E., Fytikas, M., Guidi, M., Leonis, C., Lyberopoulou, V., Magro, G., Marini, L., Meletlidis, S., Michelot, J.L., Poutoukis, D., Raco, B., Russo, M. and Virgili, G., 1998. Geochemistry of Santorini fluids. In Casale R., Fytikas M., Sigvaldasson, G. and Vougioukalakis, G.E., (Eds) "The European Laboratory Volcanoes", Proceedings of the 2d Workshop, Santorini, Greece – 2 to 4 May 1996. EUR 18161 EN, European Commission, Luxembourg, pp. 193-232.

Chiodini, G., Brombach, T., Caliro, St., Cardellini, C., Marini, L. and Dietrich, V., 2002. Geochemical indicators of possible ongoing volcanic unrest at Nisyros Island (Greece). Geophysical Research Letters 29(16): 1759-1762, doi:10.1029/2001 GL014355.

Cioni, R., Gurioli, L., Sbrana, A. and Vougioukalakis, G.E., 2000. Precursory phenomena and destructive events related to the Late Bronze Age Minoan (Thera, Greece) and AD 79 (Vesuvius, Italy) Plinian eruptions; inferences from the stratigraphy in the archaeological areas. In "The Archaeology of Geological Catastrophes", The Geological Society of London, Special Publication No 171, pp. 123-141

Dimitriadis, I.M., Panagiotopoulos, D.G., Papazachos, C.B. and Hatzidimitriou, P.M., 2004. Recent seismic activity (1994-2002) of the Santorini Volcano using data from a local seismological network. This Volume.

Di Filippo, M. and Toro, B., 2004. Gravity monitoring of Nisyros volcano activity: 2001-2003 preliminary results. This volume.

Di Paola, G.M., 1974. Volcanology and petrology of Nisyros island (Dodecanese, Greece). Bull. Volcanol., 38: 944-987.

Doumas, Ch., 1983. Thira: Pompeii of the ancient Aegean, London 1983. Thames and Hudson.

Druitt, T.H., Mellors, R.A., Pyle, D.M. and Sparks, R.S.J., 1989. Explosive volcanism on Santorini, Greece. Geol. Mag., 126: 95-126.

Druitt, T.H. and Francaviglia, V., 1992. Caldera formation on Santorini and the physiography of the islands in the late Bronze Age. Bull. Volcanol., 54: 484-493

Druitt, T.H., Edwards, L., Mellors, R.M., Pyle, D.M., Sparks, R.S.J., Lanphere, M., Davies, M. and Barreirio, B., 1999. Santorini Volcano. Geol. Soc. Mem., 19: 165.

Fouqué, F., 1879. Santorin et ses eruptions. Masson et cie, Paris.

Francalanci, L., Fytikas, M. and Vougioukalakis, G.E., 1994. Volcanological and geochemical evolution of Kimolos and Polyegos centers, Milos Island Group, Greece. IAVCEI Congress, Settembre 1994, Ankara, Turkey, Book of Abstracts.

Francalanci, L., Varekamp, J.C., Vougioukalakis, G., Defant, M.J., Innocenti, F. and Manetti, P., 1995a. Crystal retention, fractionation and crustal assimilation in a convecting magma chamber, Nisyros Volcano, Greece. Bull. Volcanol., 56: 601-620.

Francalanci, L., Vougioukalakis, G. and Eleftheriadis, G., 1995b. The intracaldera post-minoan volcanics of Santorini, Greece: preliminary petrographic and geochemical new data. In: F. Barberi, R. Casale and M. Fratta (edts.), The European Laboratory Volcanoes, Workshop Proceedings, European Commission, European Science Foundation, Luxembourg, pp. 184-188.

Francalanci, L., Vougioukalakis, G.E., Pinarelli, L., Petrone, C. and Eleftheriadis, G., 1995c. Interaction between mafic and acid magmas: the case study of the post Minoan activity of the Santorini volcanic field, Greece. Plinius, vol 14, 166-167.

Francalanci, L., Vougioukalakis, G., Eleftheriadis, G., Pinarelli, L., Petrone, C., Manetti, P. and Christofides, G., 1998. Petrographic, chemical and isotope variations in the intracaldera post-minoan rocks of the Santorini volcanic field, Greece. In Casale R., Fytikas M., Sigvaldasson G. & Vougioukalakis G.E. (Eds) "The European Laboratory Volcanoes", Proceedings of the 2d Workshop, Santorini, Greece – 2 to 4 May 1996. EUR 18161 EN, European Commission, Luxembourg, pp. 175-186.

Francalanci, L., Fytikas, M. and Vougioukalakis, G.E., 2003. Kimolos and Polyegos volcanoes, South Aegean Arc, Greece: volcanological and magmatological evolution based on stratigraphic and geochemical data. International Conference, "The South Aegean Active Volcanic Arc: Present Knowledge and Future Perspectives" (SAAVA 2003), Milos, Greece, Book of Abstracts, pp. 25-26.

Francalanci, L., Vougioukalakis, G.E., Perini, G. and Manetti, P., 2004. A West-East traverse along the magmatism of the South Aegean volcanic arc in the light of

volcanological, chemical and isotope data. This volume.
Fytikas, M., Giuliani, O., Innocenti, F., Kolios, N., Manetti, P., Mazzuoli, R., 1986a. The Plio-Quaternary volcanism of Saronikos area (western part of the active Aegean volcanic Arc). Ann. Geol. Pays Hell., 33: 23-45.
Fytikas, M., Innocenti, F., Kolios, N., Manetti, P., Mazzuoli, R., Poli, G., Rita, F. and Villari, L., 1986b. Volcanology and petrology of volcanic products from the island of Milos and neighbouring islets. J. Volcanol. Geotherm. Res., 28: 297-317.
Fytikas, M., Kolios, N. and Vougioukalakis, G.E., 1990. Post-Minoan volcanic activity of the Santorini volcano. Volcanic hazard and risk. Forecasting possibilities. In Hardy D A, Keller J, Galanopoulos V P, Flemming N C, Druitt T H. (Eds), Thera and the Aegean World III. The Thera Foundation, London,vol 2, 183-198.
Fytikas, M. and Vougioukalakis, G.E., 1993. Volcanic Structure and Evolution of Kimolos and Polyegos (Milos island group).Greek Geol. Soc. Bull. 1993, vol XXVIII/2, 221-237 (in Greek with English summary).
Fytikas, M. and Vougioukalakis, G., 1995. Volcanic hazard in the Aegean Islands. In: T. Horlick-Jones, A. Amendola and R. Casale, Natural Risk and Civil Protection.. E & FN Spon, Brussels, pp. 117-130.
Fytikas, M., Vougioukalakis, G.E., Dalampakis, P. and Bardintzeff, J.M., 1998. Volcanic hazard assessment and civil defense planning on Santorini. In: Casale, R., Fytikas, M., Sigvaldasson, G. and Vougioukalakis, G.E. (Eds) "The European Laboratory Volcanoes", Proceedings of the 2d Workshop, Santorini, Greece – 2 to 4 May 1996. EUR 18161 EN, European Commission, Luxembourg, pp. 339-351.
Galanopoulos, A., 1953. Katalog der Erdbeben in Griechenland fur die Zeit von 1879 bis 1892. Ann. Geol. Pays Hellen., 5: 144-229.
Galanopoulos, D. and Kolettis, G., 2004. Investigating the formation of a superficial fracture on Nisyros Island, Greece with the DC resistivity method. This volume.
Georgalas, G., 1953. L' éruption du volcan de Santorini en 1950. Bull. Volcanol., 13: 39-55.
Georgalas, G. and Papastamatiou, J., 1953. L' eruption du volcan du Santorini en 1939-41. L' éruption du dome Fouqué. Bull. Volcanol., 13: 3-18.
Geotermica Italiana, 1983. Nisyros 1 geothermal well final report. Pisa.
Gorceix, M.H., 1873. Sur la récente eruption de Nisyros. C R Ac Sc 77, 1039
Gorceix, M.H., 1874. Phenomenes volcaniques de Nisyros. C R Ac Sc 78, 444-446.
Heiken, G. and Mc Coy, F., 1984. Caldera development during the Minoan eruption, Thera, Cyclades, Greece. J. Geophys. Res. 89, B - 10, 8441 - 8462.
Heiken, G. and McCoy, F., 1990. Precursory activity to the Minoan eruption, Thira, Greece. In: Hardy D A, Keller J, Galanopoulos V P, Flemming N C, Druitt T H. (Eds), Thera and the Aegean World III. The Thera Foundation, London,Vol. 2. pp 79-88.
Keller, J., 1971. The major volcanic events in recent Mediterranean volcanism and their bearing on the problem of Santorini ash-layers.- Acta 1st Internat. Sci.

Congress on the Volcano of Thera, Athens, pp. 152-169.

Keller, J., 1980. Prehistoric pumice tephra on Aegean islands.- In C. Doumas (Ed.) "Thera and the Aegean World II", vol 2, pp. 49-56.

Keller, J., Gillot, P.Y., Rehren, T. and Stadlbauer, E., 1989. Chronostratigraphic data for the volcanism in the eastern Hellenic Arc: Nisyros and Kos.- TERRA abstracts, 1: 354.

Keller, J., Rehren, T. and Stadlbauer, E., 1990. Explosive volcanism in the Hellenic Arc: A summary and review. In: Hardy D A, Keller J, Galanopoulos V P, Flemming N C, Druitt T H. (Eds), Thera and the Aegean World III. The Thera Foundation, London, vol 2, pp. 13-26.

Ktenas, K., 1927. L' eruption du volcan des Kammenis (Santorini) en 1925, II, Bull. Volcanol., 4: 7-46.

Lagios, E., Chailas, S., Giannopoulos, I., and Sotiropoulos, P., 1998. Surveillance of the Nisyros volcano. Establishment and remeasurement of GPS and Radon networks. Bull. Geol. Soc. Greece, 32: 215-227 (in Greek with English abstr.).

Limburg, E. and Varekamp, J.C., 1991. Young Pumice deposits on Nisyros, Greece. Bull. Volcanol., 54: 68-77.

Marini, L., Principe, C., Chiodini, G., Cioni, R., Fytikas, M., and Marinelli, G., 1993. Hydrothermal eruptions of Nisyros (Dodecanese, Greece). Past events and present hazard J. Volcanol. Geotherm. Res., 56: 71-94.

Martelli, A., 1917. Il gruppo eruttivo di Nisyros nel mare Egeo. Mem. Soc. Geol. Ital. Sc., Serie 3a, 20: 258-309.

Papadopoulos, G.A., 1990. Deterministic and stochastic models of the seismic and volcanic events in the Santorini Volcano. In Hardy D A, Keller J, Galanopoulos V P, Flemming N C, Druitt T H. (Eds), Thera and the Aegean World III. The Thera Foundation, London,vol. 2, 151-159.

Papadopoulos, G.A. and Orfanogiannaki, K., 2004. Long-term prediction of the next eruption in Thera Volcano from conditional probability estimates. This volume.

Papazachos, B.C., 1990. Long and Short Term Prediction of Volcanic Eruptions in Santorini. In Hardy D A, Keller J, Galanopoulos V P, Flemming N C, Druitt T H. (Eds), Thera and the Aegean World III. The Thera Foundation, London,vol. 2, 224-228.

Papazachos, B.C., Dimitriadis, S.T., Panagiotopoulos, D.G., Papazachos, C.B. and Papadimitriou, E.E., 2004. Deep structure and active tectonics of the Southern Aegean Volcanic Arc. This volume.

Pe-Piper, G. and Piper, D.J.W., 2002. The igneous rocks of Greece, The anatomy of an orogen. Beitrage regionalen geologie der erde. Gebruder borntraeger. 573 pp.

Pe-Piper, G. and Piper, D.J.W., 2004. The South Aegean active volcanic arc: relationships between magmatism and tectonics. This volume.

Principe, C., Arias, A. and Zoppi, U., 2003. Hydrothermal explosions on Milos: from debris avalanches to debris flows deposits. In "The South Aegean Active Volcanic Arc: Present Knowledge and Future Perspectives" international conference, Milos 2003. Book of abstracts 95.

Pyle, D.M., 1990. New estimates for the volume of the Minoan eruption. In: Hardy D

A, Keller J, Galanopoulos V P, Flemming N C, Druitt T H. (Eds), Thera and the Aegean World III. The Thera Foundation, London, 2, pp. 113-121.

Reck, H., 1936. Santorini. - Der Werdergang eines Inselvulcans und sein Ausbruch 1925 - 1928. Dietrich Reimer, Berlin, 3 vols. MANCA

Sachpazi, M., Kontoes, Ch., Voulgaris, N., Laigle, M., Vougioukalakis, G.E., Sikioti, O., Stavrakakis, G., Baskoutas, J., Kalogeras, J. and Lepine, Cl., 2002. "Seismological and SAR signature of unrest at Nisyros caldera, Greece". Journal of Volcanology and Geothermal Research, 116(1-2): 19-23.

Seymour, K.St. and Vlassopoulos, D., 1989. The potential for future explosive volcanism associated with dome growth at Nisyros, Aegean volcanic arc, Greece. J. Volcanol. Geotherm. Res., 37: 351-364.

Smith, P.E., York, D., Chen, Y. and Evensen, N.M., 1996. Single crystal 40Ar/39Ar dating of a Late Quaternary paroxysm on Kos, Greece: Concordance of terrestrial and marine ages. Geophys. Res. Letters, 23: 3047-3050.

Sparks, R.S.J. and Wilson, C.J.N., 1990. The Minoan deposits: An Overview of their characteristics and interpretation. In Hardy D A, Keller J, Galanopoulos V P, Flemming N C, Druitt T H. (Eds), Thera and the Aegean World III. The Thera Foundation, London, vol. 2, 89-99.

Stewart, A.L. and McPhie, J., 2003. Facies architecture of the submarine-to-subaerial volcanic succession on Milos, Greece. In "The South Aegean Active Volcanic Arc: Present Knowledge and Future Perspectives" international conference, Milos 2003. Book of abstracts 24.

Stiros, S.C. and Vougioukalakis, G.E., 1996. The 1970 Yali (SE edge of the Aegean volcanic arc) earthquake swarm: surface faulting associated with a small earthquake. Annales Tectonicae, vol. X, no 1-2: 20-30.

Stiros, S.C., 2000. Fault pattern of Nisyros Island volcano (Aegean Sea, Greece): structural, coastal and archaeological evidence. *In*: McGuire, W.J., Griffiths, D.R., Hancock, P.L. & Stewart, I.S. (eds) *The Archaeology of Geological Catastrophes*. Geological Society, London, Special Publications, 171, 385-397.

Stiros, S.C., Chasapis A. and Kondogianni, V., 2004a. Geodetic evidence for slow, small scale inflation of the Santorini caldera. This volume.

Stiros, S.C., Pirazzoli, P.A., Fontugne, M., Arnold, M. and Vougioukalakis, G.E., 2004b. Late-Holocene coastal uplift in the Nisyros volcano (SE Aegean Sea): Evidence for a probable new phase of slow shallow intrusive activity. This volume.

Strabo. Geographica, Lib. I, 3, 59.

Teschner, M., Vougioukalakis, G.E., Faber E., Poggenburg, J. and Chatzigiannis, G., 2004. Real time monitoring of gas-geochemical parameters in Nisyros fumaroles. This volume.

Traineau, H. and Dalabakis, P., 1989. Mise en evidence d´une éruption phreatique historique sur l´ile de Milos (Grece). C.R. Acad. Sci. Paris, t. 308, Serie II: 247-252.

Vougioukalakis, G., 1989. Geological map of Nisyros island, 1:25.000 scale, IGME.

Vougioukalakis, G., 1993. Volcanic stratigraphy and evolution of Nisyros island. Bull. Geol. Soc. Greece, XXVIII: 239-258.

Vougioukalakis, G.E., 1994. Volcanic hazard estimation of Santorini, Aegean Sea, Greece. In F. Barberi, R. Casale, and R. Fantechi (Eds) "The Mitigation of Volcanic Hazards", Proceedings of the course, EUR 16804 EN, 1996, 471-484.

Vougioukalakis, G., Mitropoulos, D., Perissoratis, C., Andrinopoulos, A. and Fytikas, M., 1994. The submarine volcanic centre of Kolumbo, Santorini Greece. Bull. Geol. Soc. Greece, XXX/3: 351-360.

Vougioukalakis, G., Francalanci, L., Sbrana, A. and Mitropoulos, D. 1995. The 1649-1650 Kolumbo submarine volcano activity, Santorini, Greece. In: Barberi, F., Casale, R. and Fratta, M., (edts): "The European Laboratory Volcanoes, Workshop Proceeding", European Commission, European Science Fondation, Luxembourg, p. 189-192.

Vougioukalakis, G.E., 1998. Blue Volcanoes: Nisyros. Nisyros Regional Council, Nisyros, 78 pp.

Vougioukalakis, G., Sachpazi, M., Perissoratis, C. and Lyberopoulou, Th., 1998. The 1995-1997 seismic crisis and ground deformation on Nisyros volcano, Greece: a volcanic unrest?. 6th Int. Meeting on Colima volcano, Abstracts Vol.

Vougioukalakis, G.E., (2002). Hellenic Aegean Active Volcanoes: Contrasting perspectives for sustainable development. Insula (International journal of Island Affairs) 11/1, dossier Volcanic Islands, July 2002, 7-11

Vougioukalakis, G.E., Fytikas, M., Panagiotopoulos, D., Sachpazi, M., Laopoulos, Th., Stiros, S.C., Lagios, E. 2003. Institute for the study and monitoring of the Santorini Volcano: Monitoring the Santorini volcano. Meeting on "The South Aegean Volcanic Arc :Present knowledge and future perspectives", Milos island, Greece. Book of abstracts 97.

Vougioukalakis, G.E., 2003. Passive fissuring in long-dormant caldera volcanoes: The Nisyros case. Meeting on "The South Aegean Volcanic Arc: Present knowledge and future perspectives", Milos island, Greece. Book of abstracts 105.

Washington, H.S., 1926. Santorini eruption of 1925. Bull. Geol. Soc. Am., 37: 349-384.

Recent seismic activity (1994-2002) of the Santorini volcano using data from local seismological network

I.M. Dimitriadis*, D.G. Panagiotopoulos, C.B. Papazachos, P.M. Hatzidimitriou, E.E. Karagianni and I. Kane

Geophysical Laboratory, School of Geology, Aristotle University of Thessaloniki, P.O. Box 352-1, GR-54124, Thessaloniki, Greece

ABSTRACT

The South Aegean Active Volcanic Arc consists of a chain of five volcanic centers, the most active of which is the Santorini Volcano. A local radio-linked seismological network is installed on the island consisting of five permanent and four temporary stations. The temporary stations have been in operation periodically during the period 1994-1996 and two of them were installed on adjacent islands. All stations are equipped with vertical-component short period seismometers. During the period 1994-2002 a significant number of earthquakes has been recorded, with local (duration) magnitudes, M_D, up to 5.0 and focal depths varying between 0 km and 35 km. Two clusters of epicenters have been located in the broader area of the Santorini Volcano. The first cluster is located in the caldera of the volcano and is associated with the volcanic process in the Kameni Island. The second (larger) cluster is located near the northern edge of the Santorini Island at the Kolumbo Reef and is connected with the volcanic process at this reef. These clusters can be appropriately associated with the two main tectonic features (faults) in the area under study. The first one (N60°E direction) corresponds to the continuation of the Amorgos fault in the area, while the secondary tectonic line (EW direction) is probably related with the southern edge of a submarine graben, which is located between the islands Amorgos and Santorini. Using the data set of the best-located earthquakes, recorded during the period 1994-2002, an attempt has been made to derive an appropriate equivalent 1-D earth model for the area under study, in order to improve the accuracy of the determined hypocenters, as well as to obtain a preliminary knowledge of the volcano structure.

Keywords: Santorini Volcano; Seismic activity; hypocentral estimation; 1-D velocity model.

---

* Corresponding author: e-mail: danis@lemnos.geo.auth.gr

1. INTRODUCTION

1.1 Geographic and Geotectonic Setting of the Santorini Volcano

The Hellenic arc is one of the dominant tectonic features of the southern Aegean area. It separates the Aegean Sea from the Mediterranean Sea and has the main properties of a typical island arc (Papazachos and Comninakis, 1971; McKenzie, 1978; Le Pichon and Angelier, 1979, 1981). It consists of an outer (southern) "sedimentary arc", which is a link between Dinaric Alps and the Turkish Taurids, and of an inner (northern) "volcanic arc" with Quaternary volcanoes. Between the sedimentary and the volcanic arc is the Cretan Sea, with a maximum water depth of about 2 km. On the convex side of the arc (eastern Mediterranean) a system of troughs and trenches (Ionian, Pliny and Stravo) is found, which is called the Hellenic Trench with a maximum water depth of about 5 km. A well-defined Bennioff zone of intermediate depth earthquakes, dipping from the Mediterranean Sea to the Aegean Sea and associated with this arc, was first identified by Papazachos and Comninakis (1969).

Santorini is one of the five volcanic centers, which make up the South Aegean Active Volcanic Arc (SAAVA) (Fig. 1). The volcanic centers include three Quaternary Volcanoes (Santorini, Nisyros and Methana), solfataras and fumaroles fields (Sousaki, Egina, Milos and Kos) (Georgalas, 1962; Fytikas et al., 1985). The volcanoes, along

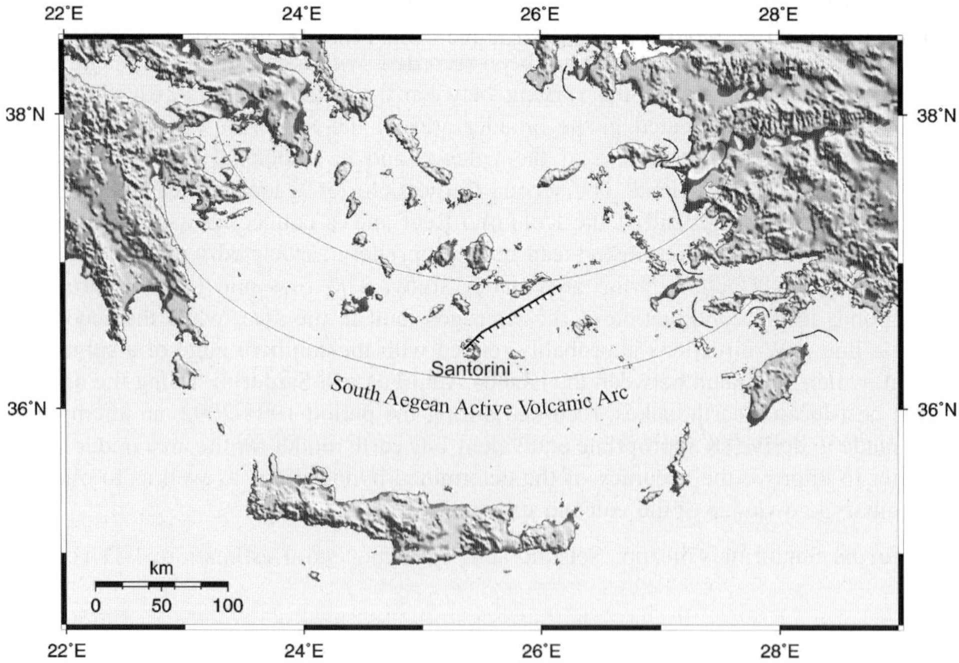

Fig. 1. The Amorgos Fault and the South Aegean Active Volcanic Arc (SAAVA).

with the clusters of epicenters of shallow and intermediate depth earthquakes, form the five seismovolcanic clusters (Sousaki, Methana, Milos, Santorini and Nisyros) (Papazachos and Panagiotopoulos, 1993). These are connected with tectonic zones of weakness, which follow a direction N60^{0}E. The most active volcanic center is the Santorini Volcano, which has erupted at least seven times in the last 500 years (1573, 1650, 1707, 1866, 1925, 1939 and 1950) (Papazachos, 1989; Fytikas et al., 1990).

The main tectonic feature in the area of Santorini is the Amorgos fault with a NE-SW direction, where the large earthquake in 1956 with M=7.5 occurred (Shirokova, 1972; Papazachos et al., 2001; Pavlides and Valkaniotis, 2003) (Fig. 1). Another important tectonic feature in the same area is the graben northern of Santorini, which has an approximately ENE-WSW direction (Perissoratis, 1996).

1.2 Geology of Santorini

Santorini is a group of five islands (Thera, Therasia, Aspronisi, Palaea Kameni and Nea Kameni). Thera, Therasia and Aspronisi enclose a sea-flooded caldera. Apart from a small non-volcanic basement found in the southeastern part of Thera, these islands are composed of volcanic rocks. Volcanism in the area of Santorini started about 2 million years ago with the built up of Hristiana volcanic centers, the extrusion of dacitic vents on the Akrotiri peninsula and continued to produce different kinds of lavas and pyroclasts (Friedrich, 1994, 2000). However, the most characteristic type of activity over the last 200.000 years has been the cyclic construction of shield volcanoes interrupted by large explosive and destructive events like the Minoan eruption (Bond and Sparks, 1976; Heiken and McCoy, 1984, 1990; Druitt et al., 1989, 1999). In particular, a thick layer of white pumice and tephra, which has been laid down by the Minoan Eruption, covers a great part of Santorini (Fig. 2).

Non-volcanic rocks are exposed at the Profitis Ilias Hill and at other parts of the main island (Kilias et al., 1996; Mountrakis et al., 1996). The oldest volcanic rocks can be found on the Akrotiri peninsula, while the youngest volcanic rocks can be found on the Palaea and Nea Kameni islands, which formed during eruptions in historic times (Druitt and Francaviglia, 1992; Seidenkrantz and Friedrich, 1992).

2. SEISMIC MONITORING AT THE SANTORINI VOLCANO

2.1 Local Seismological Network

Within the framework project of the seismic monitoring of the Santorini Volcano, the Geodynamic Institute of the National Observatory of Athens in cooperation with the Geophysical Laboratories of the University of Athens and Thessaloniki installed a local radio-linked seismological network consisting of eight stations on the islands of Santorini and Ios in 1994 (Panagiotopoulos et al., 1996). In particular, the stations of COLUMBO, AKROTIRI, CENTRAL and OIA have been installed on the Thera island, the stations of RIVA and KERA on the Therasia island, the station of KAMENI on the Nea Kameni island and the station of IOS on the Ios island (Fig. 3). The stations of IOS,

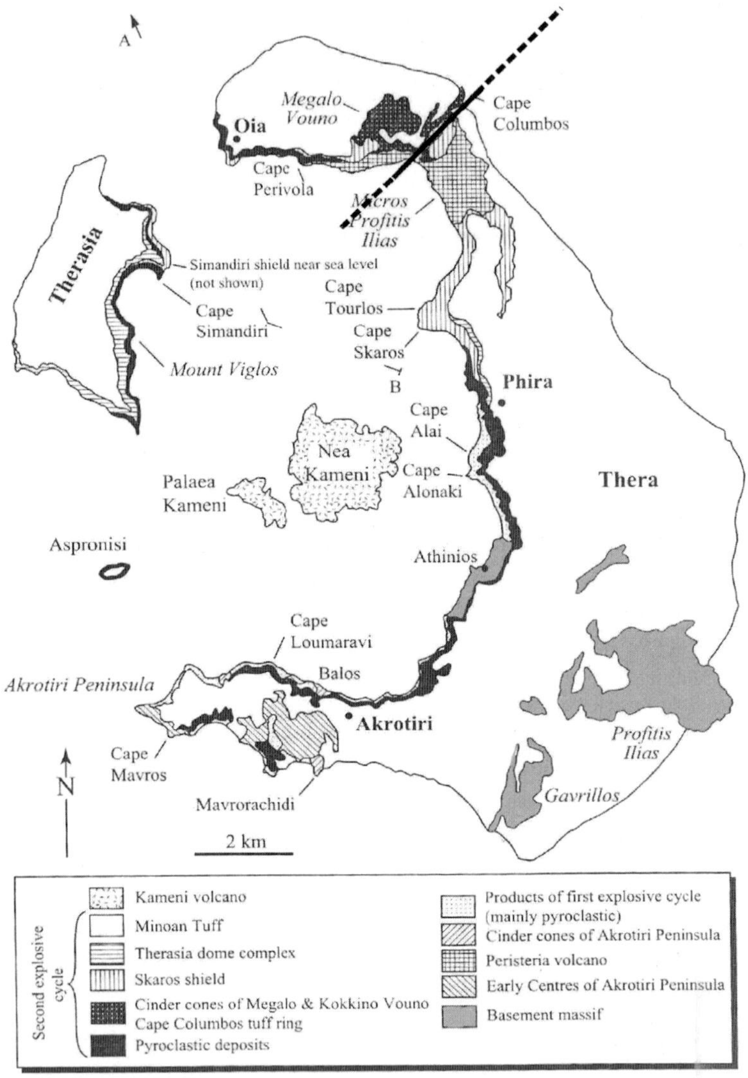

Fig. 2. Geological map of the Santorini Islands; the heavy line represents the continuation of Amorgos fault (modified from Druitt et al., 1999).

COLUMBO and RIVA have been in operation during the period 1994-1996. The other five stations (AKROTIRI, CENTRAL, KERA, KAMENI, OIA) are in operation until today. All stations are equipped with vertical-component short period seismometers and the analogue seismic signals are transmitted via antennas to the central station, which is situated on the highest point of Thera (Profitis Ilias Hill, 500 m altitude). The main station (observatory) is a relatively small construction that insures the antenna and the

Table 1. The initial local velocity model used for the estimation of the events.

Depth (km)	V_P (km/s)	V_S (km/s)
0,0	4,00	2,25
1,0	6,00	3,37
24,0	6,60	3,71
32,0	7,70	4,33
40,0	8,10	4,55

computerised recording unit and analysing of the seismological data. The recording and analysing system of the signals consists of three units: the conversion system of analogue to digital signals, the server and the terminal (workstation).

2.2 Estimation of the epicenters and hypocenters

The main objective of this paper is the precise hypocenter estimation of local earthquakes that occurred in the broader area of the Santorini Volcano during the period of 1994-2002. The estimation of the epicenters was performed using the computer program HYPOELLIPSE (Lahr, 1989, 1999) with an initial local velocity model derived from a large-scale 3-D model (Papazachos and Nolet, 1997) for the area under study (Table 1).

During the period 1994-2002, 1076 events have been recorded, with local (duration)

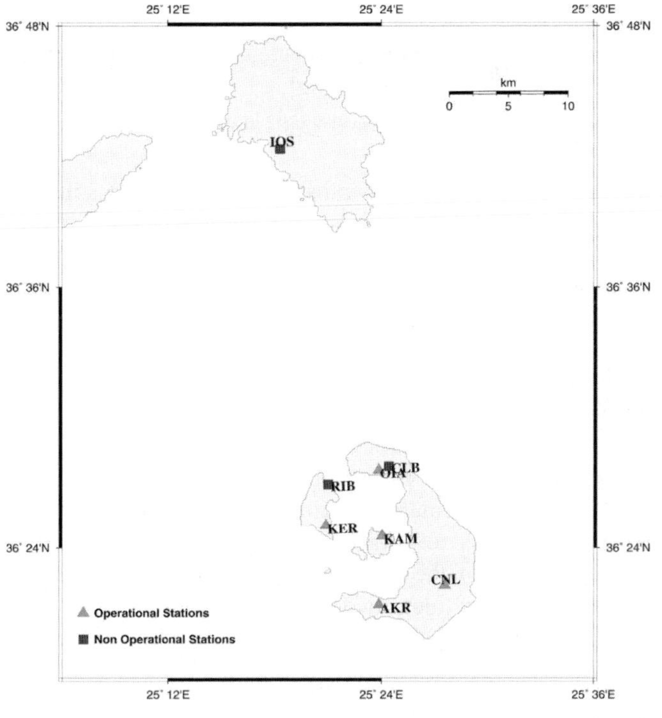

Fig. 3. Map of the local seismological network installed on the Santorini and adjacent islands.

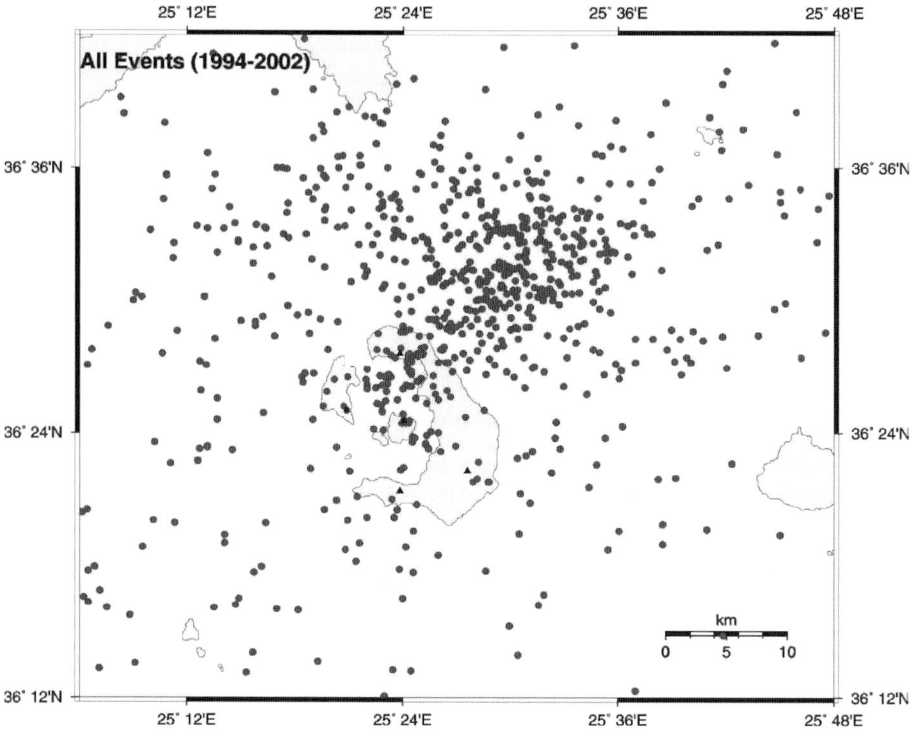

Fig. 4. Epicenters of local earthquakes of the period 1994-2002 in the area under study.

magnitudes, M_D, up to 5.0 and focal depths up to 35 km. Despite the poor spatial coverage of the recording stations and the simplified earth model used, two clusters of epicenters have been located in the broader area of the Santorini Volcano. As is shown in Fig. 4, a large cluster of epicenters is located near the northeastern edge of the island at Kolumbo Reef, while a small cluster of epicenters is located in the caldera of the volcanic center.

In order to separate the best-located events due to the uncertainty of the determined hypocenters and the lack of P and S phases, earthquakes with: ERH (minimum horizontal error) < 20km, ERZ (minimum depth error) < 25km, RMS (minimum residual error) < 0.5, phases number per event > 6, Azimuthal Gap < 340^0, the minimum focal depth used < 35km, and, finally, a minimum epicentral distance < 30km were selected (Fig. 5). The final data set satisfying these conditions consists of 157 events. Fig. 6 shows the distribution of the epicenters of the best-located earthquakes, where two very clear tectonic lines are identified. The main one has a NE-SW direction and corresponds to the Amorgos Fault, while the secondary tectonic line has an approximately E-W direction and is related with a graben northern of Santorini.

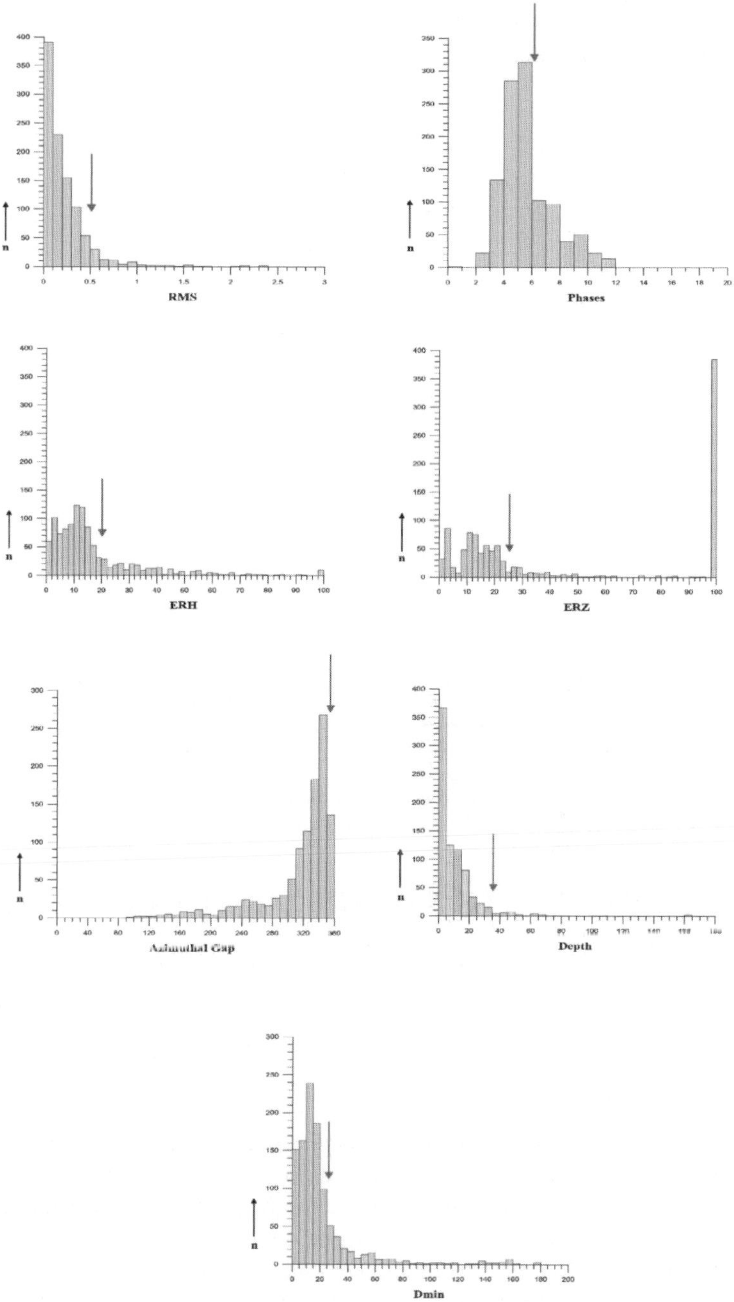

Figure 5. Histograms of the seismic parameters along with the confined parameters (black arrows) used for the selection of the best-located events.

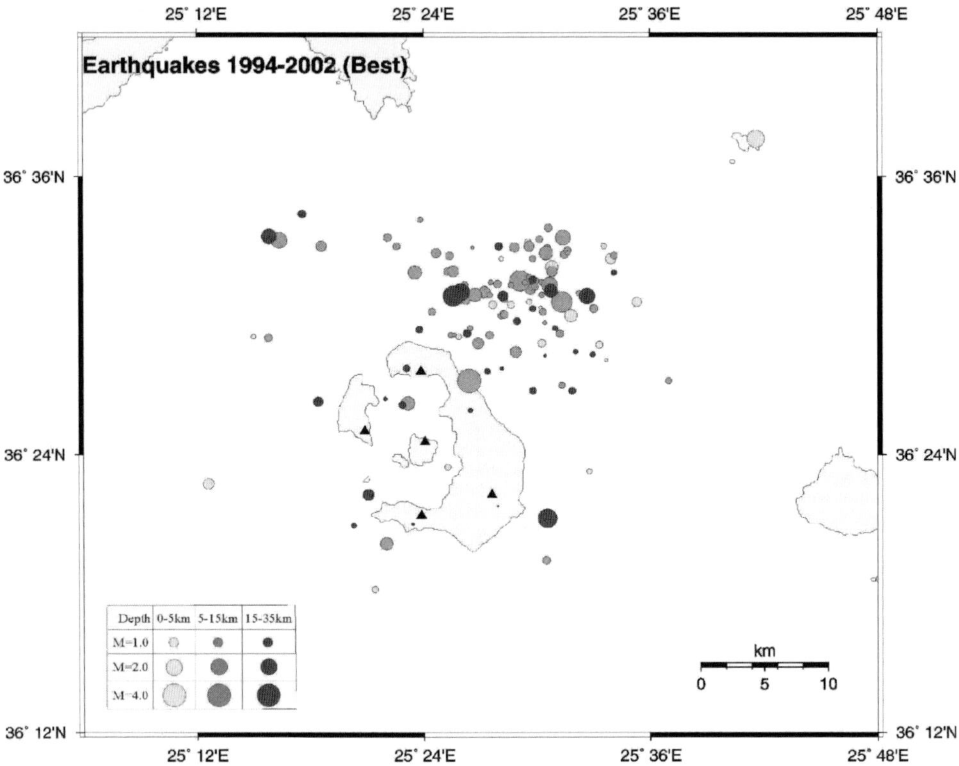

Fig. 6. Epicenters of the best-located earthquakes of the period 1994-2002 in the area under study.

3. MAGNITUDE ESTIMATION OF THE LOCAL EARTHQUAKES

Durations of the recorded waveforms were used in order to estimate the local magnitude of all earthquakes. The equation, which was used for this estimation, is the following one:

$$M_D = c_1 \log T + c_2 \log D + c_3 \quad (1)$$

where M_D is the duration magnitude, T is the duration of the waveform of arrival of P waves up to the time where the recorded pp amplitude is less than 2 mm and D is the epicentral distance. For the constants c_1 and c_2 we adopted the values, 1.97 and 0.0012 respectively, determined for the broader Greek area (Kiratzi, 1984). Constant, c_3, is calculated for every station used for the magnitude estimation. In order to estimate the value c_3, the recordings of 28 strong earthquakes of the broader area of Santorini from the Greek National Networks and the National Networks of adjacent countries were used. The presence of an active extensional tectonic regime in the area under study is

REFERENCES

Aki, K. and Lee, W.H.K., 1976. Determination of three-dimensional velocity anomalies under a seismic array using first P arrival times from local earthquakes: A homogeneous initial model. Journal of Geophysical Research, 81: 4381-4399.

Bond, A. and Sparks, R.S.J., 1976. The Minoan eruption of Santorini, Greece. Journal of the Geological Society of London, 132: 1-16.

Druitt, T.H., Mellors, R.A., Pyle, D.M. and Sparks, R.S.J., 1989. Explosive volcanism on Santorini, Greece. Geophysical Magazine, 126: 95-126.

Druitt, T.H. and Francaviglia, V., 1992. Caldera formation on Santorini and the physiography of the islands in the late Bronze Age. Bulletin of Volcanology, 54, No. 6: 484-493.

Druitt, T.H., Davies, M.S., Edwards, L., Sparks, R.S.J., Mellors, R.M., Pyle, D.M., Lanphere, M. and Barreirio, B., 1999. The Santorini Volcano. Geological Society Special Memoir, 19, Geological Society Pub House, London.

Friedrich, W.L., 1994. Feuer im Meer. Spektrum Akademischer Verlag, Heidelberg, Berlin.

Friedrich, W.L., 2000. Fire in the Sea: Volcanism and the Natural History of Santorini. Cambridge University Pr., Cambridge.

Fytikas, M., Innocenti, F., Manetti, P., Mazzuoli, R., Peccerillo, A. and Villari, L., 1985. Tertiary to Quaternary evolution of volcanism in the Aegean region. In: The Geological Evolution of the Eastern Mediterranean, Special Publ. Geol. Soc., 17: 687-699.

Fytikas, M., Kolios, N. and Vougioukalakis, G., 1990. Post-Minoan Volcanic Activity of the Santorini Volcano. Volcanic hazard and risk, forecasting possibilities. In: Hardy, D.A., (Editor), Thera and the Aegean World III, 2. The Thera Foundation, London, pp. 183-198.

Georgalas, G.C., 1962. Catalogue of the Active Volcanoes of the World including solfataras fields, Greece.

Heiken, G. and McCoy, F., 1984. Caldera Development During the Minoan Eruption, Thera, Cyclades, Greece. Journal of Geophysical Research, 89: 8441-8462.

Heiken, G. and McCoy, F., 1990. Precursory activity to the Minoan Eruption, Thera. In: Hardy, D.A., (Editor), Thera and the Aegean World III, 2. The Thera Foundation, London, pp. 79-88.

Karagianni, E.E., Panagiotopoulos, D.G., Panza, G.F., Suhadolc, P., Papazachos, C.B., Papazachos, B.C., Kiratzi, A., Hatzfeld, D., Makropoulos, K., Priestley, K. and Vuan, A., 2002. Rayleigh wave group velocity tomography in the Aegean area. Tectonophysics, 358: 187-209.

Kilias, A.A., Mountrakis, D.M., Tranos, M.D. and Pavlides, S.B., 1996. The prevolcanic metamorphic rocks of Santorini island: structural evolution and kinematics during the Tertiary (South Aegean, Greece). Proceedings of 2nd Workshop on European Laboratory Volcanoes, May 2-4 1996, Santorini, Greece, pp. 23-36.

Kiratzi, A., 1984. Magnitude scales for earthquakes in the broader Aegean area. Ph.D. Thesis (in Greek), University of Thessaloniki, Greece.

Kissling, E., Ellsworth, W.L., Eberhart-Phillips, D. and Kradolfer, U., 1994. Initial reference models in local earthquake tomography. Journal of Geophysical Research, 99: 19'635-19'646.

Kissling, E., Kradolfer, U., and Maurer, H., 1995. VELEST User's Guide (Version 3.1). Institute of Geophysics, ETH Zurich.

Lahr, J.C., 1989. HYPOELLIPSE/Version 2.0: A computer program for determining local earthquakes hypocentral parameters, magnitude, and first-motion pattern. U.S. Geological Survey Open-File Report, 89-116, p. 92.

Lahr, J.C., 1999. HYPOELLIPSE/Version 1.0: A computer program for determining local earthquake hypocentral parameters, magnitude, and first-motion pattern (Y2K Compliant Version). U.S. Geological Survey Open-File Report, 99-23.

Le Pichon, X. and Angelier, J., 1979. The Hellenic arc and trench system: a key to the Neotectonic evolution of the eastern Mediterranean area, Tectonophysics, 60: 1-42.

Le Pichon, X. and Angelier, J., 1981. The Aegean Sea, Phil. Trans. R. Soc., London, 300: 357-372.

McKenzie, D.P, 1978. Active tectonics of the Alpine-Himalayan belt: the Aegean Sea and surrounding regions, Geophysical. J. R. astr. Soc., 55: 217-254.

Mountrakis, D.M., Pavlides, S.B., Chatzipetros, A., Meletlidis, S., Tranos, M.D., Vougioukalakis, G. and Kilias, A.A., 1996. Active deformation of Santorini. Proceedings of 2nd Workshop on European Laboratory Volcanoes, May 2-4 1996, Santorini, Greece, pp. 13-22.

Panagiotopoulos, D.G., Stavrakakis, G., Makropoulos, K., Papanastasiou, D., Papazachos, C.B., Savvaidis, A.S. and Karagianni, E.E., 1996. Seismic monitoring at the Santorini volcano. Proceedings of 2nd Workshop on European Laboratory Volcanoes, May 2-4 1996, Santorini, Greece, pp. 311-324.

Papazachos, B.C. and Comninakis, P.E., 1969. Geophysical features of the Greek island arc and eastern Mediterranean ridge. Proceedings of the C.R. des Séances de la Conference Reunite a Madrid, 16, pp. 74-75.

Papazachos, B.C. and Comninakis, P.E., 1971. Geophysical and tectonic features of the Aegean arc. Journal of Geophysical Research, 76: 8517-8533.

Papazachos, B.C., 1989. Long and short term prediction of the volcanic eruptions in Santorini. In: Proceedings of 3rd International Congress "Thera and the Aegean World III", Volume 2, September 4-8 1989, Santorini, Greece, pp. 125-129.

Papazachos, B.C. and Panagiotopoulos, D.G., 1993. Normal faults associated with volcanic activity and deep rupture zones in the southern Aegean volcanic arc. Tectonophysics, 220: 301-308.

Papazachos, B.C., Mountrakis, D.M., Papazachos C.B., Tranos, M.D., Karakaisis, G.F. and Savvaidis, A.S., 2001. The fault which have caused the known major earthquakes in Greece and surrounding region between the 5th century BC and today. Proceedings of 2nd National Conference Anti-Seismic Engineering and

Technical Seismology, November 28-30 2001, Thessaloniki, Greece, pp. 17-26.

Papazachos, C.B. and Nolet, G., 1997. P and S deep structure of the Hellenic area obtained by robust nonlinear inversion of travel times. Journal of Geophysical Research, 102: 8349-8367.

Pavlides, S.B. and Valkaniotis, S., 2003. Tectonic regime of Santorini-Amorgos area, Proceedings of International Conference "The South Aegean Active Volcanic Arc: Present Knowledge and Future Perspectives", September 17-20 2003, Milos Island, Greece. Book of abstracts, p. 76.

Perissoratis, C., 1996. The Santorini volcanic complex and its relation to the stratigraphy and structure of the Aegean arc, Greece. Marine Geology, 128: pp. 37-58.

Seidenkrantz, M.S. and Friedrich, W.L., 1992. Santorini, Part of The Hellenic Arc: Age Relationship of Its Earliest Volcanism. In: Seidenkrantz, M.S. (Editor), Foraminiferal Analyses of Shelf Areas. Stratigraphy, Ecology and Taxonomy. Ph.D. Dissertation, University of Aarhus, pp. 41-65.

Shirokova, E.I., 1972. Stress pattern and probable motion in the earthquake foci of the Asia-Mediterranean seismic belt. In: L.M. Balakina et al. (Editors), Elastic Strain Field of the Earth and Mechanisms of Earthquake Sources. Nauka, Moscow, p. 8.

Wessel, P. and Smith, W., 1995. New version of the Generic Mapping Tools. Eos. Trans. Am. Geophys. Union 76, p. 329.

The South Aegean Active Volcanic Arc
M. Fytikas and G.E. Vougioukalakis

Geodetic evidence for slow inflation of the Santorini caldera

S. Stiros[1,*], A. Chasapis[2] and V. Kontogianni[1]

[1]Lab. of Geodesy, Dept. of Civil Eng., Patras University, Patras 26500, Greece.

[2]Institute of Geology and Mineral Exploration (IGME), 70 Mesoghion st., Athens 11527 Greece.

ABSTRACT

Santorini (Thera) is a volcanic island complex dominated by a partly submerged caldera defined by the islands of Thera and Therasia and the islet of Aspronisi. Santorini is the most important and most active volcano in the Aegean, famous for the Minoan (3600 years old) eruption which buried the ancient but very "modern" town of *Akrotiri*. Volcanic activity in Santorini was in many cases associated with ground deformations, subsidence and uplift, caused by magma flow at relatively shallow depths. Ground deformations therefore can be regarded as precursors of a future volcanic paroxysm.

A geodetic monitoring system aiming at an early identification of a future dilation of the caldera as a result of magma inflation was established in 1994, in the framework of an interdisciplinary project for the surveillance of this volcano. The geodetic monitoring system consisted of a radial EDM network with a central point at the Nea Kameni islet and 10 stations in the Thera and Therasia islands. Between June 1994 and May 2003 eight epochs of baseline measurements at a centimeter-level accuracy were made. A small-scale (up to 10 cm), gradual inflation of the northern part of the caldera (between Nea Kameni and Therasia), possibly associated with magma ascent along a dike has been inferred from these epochs of measurements.

Keywords: Santorini (Thera) volcano, geodetic monitoring, caldera inflation

1. INTRODUCTION

Volcanic activity in Santorini was in many cases associated with ground deformations, coastal subsidence and uplift; the best examples were the formation of the intra-calderic islets of Nea Kameni and of Palea Kameni in the historic times. Such ground movements are caused by magma flow at relatively shallow depths and can be regarded

* Corresponding author: e-mail: stiros@upatras.gr

as precursors of a future volcanic paroxysm. For this reason, geodetic monitoring of the volcano was included in an interdisciplinary surveillance project started in 1994, in the framework of the European Union DG XII Environment Project and of the Institute for the Study and the Monitoring of the Santorini Volcano, succeeding earlier, smaller-scale projects. Initially, this project was limited to EDM measurements, but in October 2000 a GPS network was established on the same benchmarks. The results and conclusions from this geodetic subproject are the subject of this article.

2. MAGMA FLOW AND SURFACE DEFORMATIONS

Experience from numerous active volcanoes and theoretical considerations indicate that any upward or downward movements of magma in vents or chambers produce elastic stresses which are reflected in changes of the micro-topography and of horizontal distances and elevations of benchmarks around the volcanic centers (Fig. 1); such topography changes may range from a few centimeters to several meters (Mogi, 1958; Bonaccorso et al., 1996; Dvorak, 1997).

In the last thousand years the volcanic activity in Santorini was rather confined to the caldera (Druitt and Francaviglia, 1992; Fytikas et al., 1990) any magma movements would cause geodetically observable changes of distances among points of the caldera.

3. GEODETIC NETWORK

The topography of the Santorini island complex, a nearly circular caldera with subvertical walls with a radius of 3 - 6 km, was ideal for the establishment of a radial EDM geodetic network dedicated to the identification of baseline length changes. A central station was established at Nea Kameni and 10 peripheral stations numbered 1 to 10 were established on the caldera walls in Thera and Therasia, with baselines numbered 1 to 10 (Fig. 2; Stiros and Chasapis, 2003). This network has a nearly uniform azimuthal distribution and can easily and unambiguously control baseline length changes which could reflect possible caldera inflation-deflation processes.

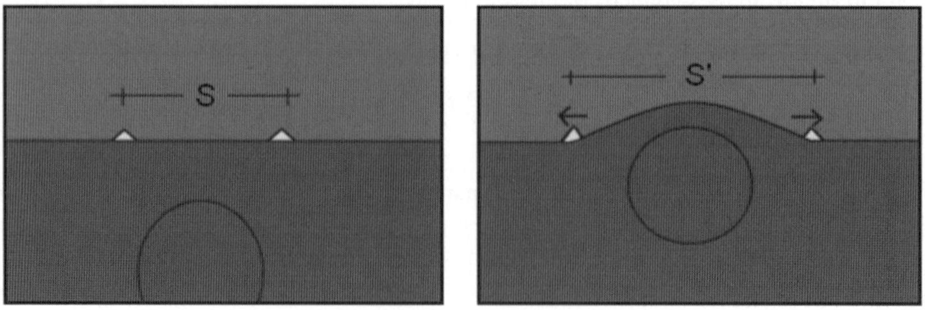

Fig. 1. Topography changes as a result of magma chamber inflation.

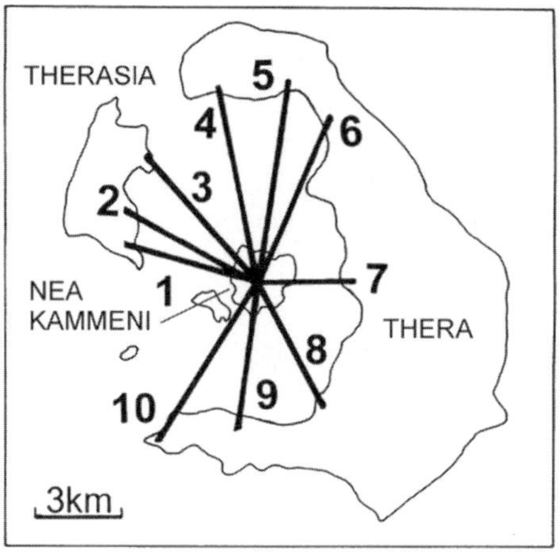

Fig. 2. Baselines for the volcano geodetic monitoring EDM network.

All stations of the network were selected among pillars of the National Triangulation Network established mostly in the 1960's. Care was taken that pillars selected are founded on stable ground (consolidated pumice deposits, stable rock masses) so they are representative of kinematics and deformation of a wider area. Baseline lengths vary between 3.2 to 6.7 km and because of the nearly uniform topography (steep cliffs of the caldera walls 100-300 m above sea-level, central station on a peak) all baselines cross a nearly uniform medium, rather free of perturbation of the atmosphere close to the ground or the water (the raypath of all baselines is practically >100 m above water/ground; Stiros and Chasapis, 2003; Fig. 3).

4. EDM SURVEYS AND RESULTS

Between June 1994 and May 2003 eight epochs of measurements were made. Measurement procedure was nearly identical in all surveys: observations were made by the same surveying party, the same instrument, an AGA 6000 laser Geodimeter and the necessary accessories (centering plates, reflectors, thermometers, barometers etc), with methods permitting a typical accuracy of 1 mm±1 mm/km for each baseline (e.g. repeated measurements when wind was blowing and the atmosphere was uniform).

Baseline length changes of the eight surveys are summarized in Fig. 4. Among the 10 baselines, those between Therasia and Nea Kameni (baselines 1, 2, 3) show maximum cumulative changes between 7 and 10 cm, while the changes of all other lines are up to ±2-3 cm, and only in two cases up to 4.5 cm and 7 cm. The observed length changes are systematic, and their amplitude higher than the corresponding uncertainly level at the 95% level, i.e. approximately 2 cm.

In addition, local ground instability effects can be ruled out for most measurements, both on the grounds of field observations and on the pattern of the network. This indicates that there is evidence for a small-scale dilation of a part of the volcano, probably because of a minor inflation of a magma chamber; probably ascent of magma along a dyke.

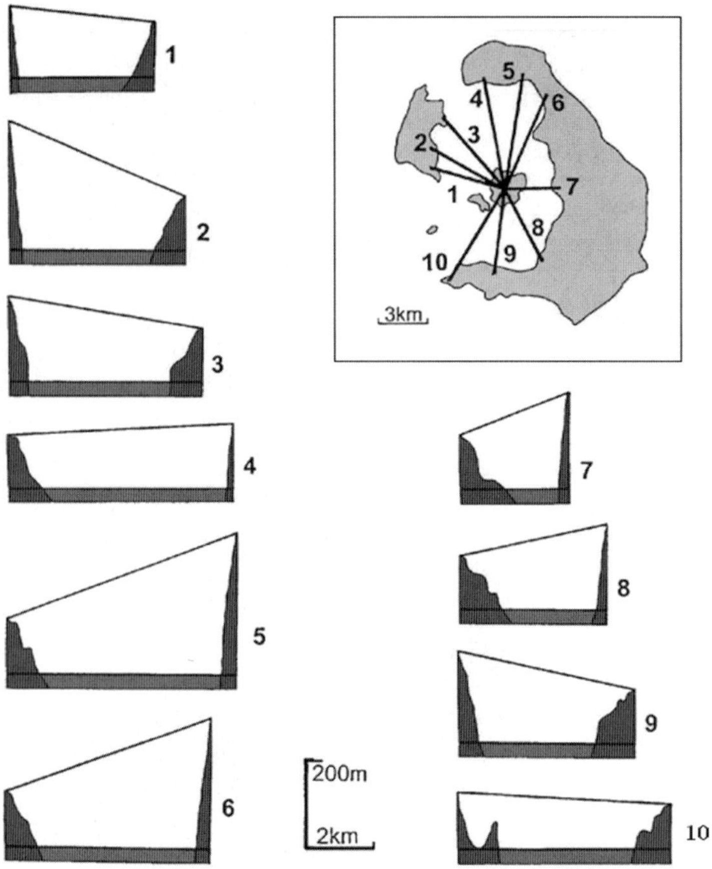

Fig. 3. Raypath of the 10 baselines.

5. GPS SURVEYS

Since October 2000, the same network was measured by GPS, in combination with the EDM survey, in order to compare EDM and GPS data in the perspective of replacing EDM measurements by GPS. No significant differences in the coordinates of the GPS stations versus time was observed, in accordance with EDM data.

6. CONCLUSIONS

The EDM surveying data indicate a statistically significant, systematic increase in the length of baselines between Nea Kameni and Therasia, shortly after the first

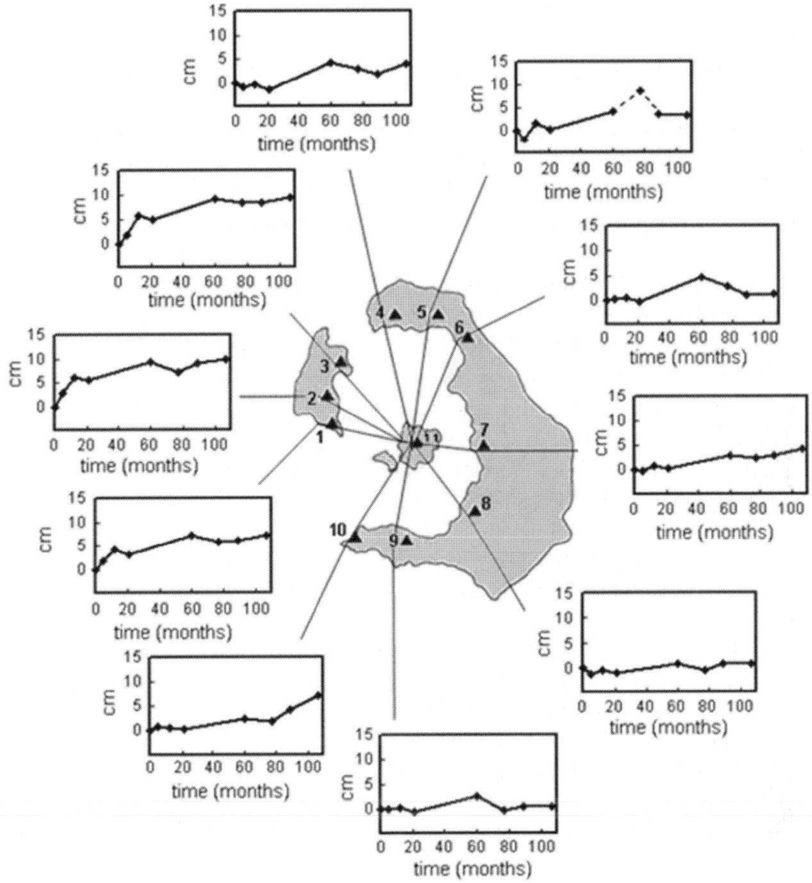

Fig. 4. Summary of changes of baseline lengths between June 1994 and May 2003. Significant changes are confined to baselines of Therasia.

measurement, but definitely no changes in the southern part of the caldera (baselines 7, 8, 9, 10 in Fig. 4). During the last years, however, no significant changes in the baseline lengths were observed. These results indicate an activity of the volcano feeding system. Whether such small-scale topography changes are premonitory phenomena of a near-future magmatic anomaly, is not possible for the moment to know.

Acknowledgements

This is a contribution to the activities of the Institute for the Study and Monitoring of the Santorini Volcano. Partial funding though EU DG XII Environment Programme is acknowledged.

REFERENCES

Bonaccorso, A., Ferruci, F., Patane, F. and Villari, L., 1996. Fast deformation processes and eruptive activity of Mount Etna (Italy). J. Geophys. Res., 101: 17467-17480.

Druitt, T.H. and Francaviglia, V., 1992. Caldera formation on Santorini and the physiography of the islands in the late Bronze Age. Bull. Volcanol., 54: 484-493.

Dvorak, J., 1997. Volcano Geodesy: The search for magma reservoirs and the formation of eruptive vents. Reviews of Geophysics, 35(3): 343-384.

Fytikas, M., Kolios, N. and Vougioukalakis, G., 1990. Post Minoan volcanic activity of the Santorini Volcano. Volcanic Hazard end Risk. Forecasting Possibilities. In: D.A. Hardy (ed.), Thera and the AEGEAN World III, vol. 2: 183-198.

Mogi, K., 1958. Relations between the eruptions of various volcanoes and the deformations of the ground surfaces around them. Bull. Earthquake Research Institute, Tokyo University, 36: 99-134.

Stiros, S. and Chasapis, A., 2003. Geodetic monitoring of the Santorini (Thera) volcano. Survey Review, 37: 287, 84-88.

Long-Term prediction of the next eruption in Thera volcano from conditional probability estimates

G.A. Papadopoulos* and K. Orfanogiannaki

Institute of Geodynamics, National Observatory of Athens, 11810 Athens, Greece

ABSTRACT

After the eruptions of small or medium Volcanic Explosivity Index that took place in Thera (Santorini) from 1925 to 1950, the volcano remains dormant. Because of this, no physical changes of some predictive value are detectable and, therefore, only probabilistic approaches are applicable for the long-term prediction of the next eruption (Papadopoulos, 1990; Papazachos, 1990). Inspired from effective methods tested in seismology, a method has been applied which is based on the lognormal distribution of the eruption inter-arrival times combined with Bayesian statistics which allows for the introduction of the "memory" element that no eruption has occurred since the last eruption time, *t*. The application was made under the assumption that the Thera eruption record is complete only from the 16th AD century onwards. Conditional probabilities for some characteristic time intervals (e.g. 1 yr, 5 yrs, 10 yrs) are calculated and compared with random (Poissonian) probabilities which do not take memory into account. The results of the method are renewable in the sense that time, *t,* is not constant but an increasing variant.

1. INTRODUCTION

The volcano of Thera (Santorini), South Aegean volcanic arc, is considered as one of the most important active volcanoes worldwide. In fact, the Late Bronge Age or "Minoan" large eruption that occurred in the 17th century BC not only attracts the great interest of scientists coming from a variety of disciplines, like volcanology, seismology, archaeology and many others, but also indicates the high volcanic potential of Thera in the long-term sense. In post-Minoan times several historical eruptions of small or moderate size were reported in the Thera island complex. The size of a particular volcanic eruption is measured by the Volcanic Explosivity Index (VEI) as it introduced by Newhall and Self (1982) which is a simple, flexible indicator of the explosive character of the eruption. Eruptions can be assigned a VEI on a scale of 0 to 8 using one

* Corresponding author: e-mail: g.papad@egelados.gein.noa.gr

or more criteria concerning the volume of ejecta, the column height, the duration of continuous blast etc. In their volume «*Volcanoes of the World*», Simkin et al. (1981) list all the historically known volcanic eruptions of Thera along with an estimation of the respective VEI. In post-Minoan times only the eruption of 197BC is assigned a VEI =4 while the VEI of the rest eleven eruption events, occurring from 19AD to 1950, varies between 2 and 3. The most recent eruption phases were observed in the time interval from 1925 to 1950. The prediction of the next Thera eruption phase is of importance from both scientific and civil protection points of view. The prediction in the short-term sense could be approached from observational data collected by the volcano monitoring systems. In about the last fifteen years several efforts have been made for instrumental monitoring of the volcano including local earthquake recording (Delibasis et al., 1990; Panagiotopoulos et al., 1996), magnetic and gravity monitoring (Lagios et al., 1990a, b) and ground temperature changes (Astaras et al., 1997). However, the volcano is dormant since 1950 and, therefore, no adequate observational data are collected up to now. As for the long-term sense, probabilistic approaches from memoryless time processes were tested in the past (Papadopoulos, 1990; Papazachos, 1990). In this paper, we apply a probabilistic approach which allows the calculation of Bayesian or conditional probabilities for the occurrence of the next eruption in a future time interval of prescribed length and compare them with the random probabilities.

2. THE DATA

All the known post–Minoan eruptions in Thera are described by Georgalas (1962) while Simkin et al. (1981) calculated the VEI for the eruptions. However, the statistics of the eruption reporting implies that it is complete only from the AD16th century onwards (Papadopoulos, 1988). The occurrence times of the known eight Thera eruptions which have occurred since the AD16th century, along with their characteristics, are listed in Table 1. From their time separation it is evident that the events of 1570 (or 1573), 1650, 1707-1711 and 1866-1870 constitute independent eruption cycles. However, the frequent events of the time interval 1925-1950 indicate that probably more than one of the listed events constitute a unique eruption cycle. On

Table 1. Data on the Thera eruptions that occurred since 16th century AD. Key: S=submarine eruption, D=dome eruption, I=island-forming eruption, E=normal explosion, F=lava flow(s), VEI=Volcanic Explosivity Index. All but the 1650 eruptions were intra-caldera events.

Date of the volcanic activity	Characters	VEI
1570 or 1573	SDIEF	3
1650 Sept. 26 – Dec. 6	SI	3
1707 May 23 – 1711 Sept. 11	SDIEF	3
1866 Jan. 26 – 1870 Oct. 15	SDIEF	2
1925 Aug. 11 – 1926 Jan.?	SDEF	2
1928 Jan. 3 – 1928 Mar. 17	DEF	2
1939 Aug. 20 – 1941 July ?	DEF	2
1950 Jan. 10 – 1950 Febr. 2	DEF	2

the basis of certain volcanological features, Papadopoulos (1986) suggested that the 1928 and 1950 eruptions were the final events of the 1925-1926 and 1939- 1941 activities, respectively. This analysis implies that six eruption cycles occurred in Thera in the examined time interval of about 490 years. Nevertheless, in order to obtain results as reliable as possible some alternative procedures have been performed; one for six eruption cycles and another for eight cycles. However, since all eruptions but the 1650 one were intra-caldera events, the calculations were repeated by taking into account only the intra-caldera eruptions. Therefore, two additional alternatives were considered, one for five eruption cycles and another for seven cycles.

3. THE METHOD

The methodology applied for the calculation of conditional probabilities has been inspired from seismology (e.g. Working Group, 1988). The assumption is made that the inter-arrival time, τ, of successive events follows the lognormal distribution, that is $\log\tau$, is distributed normally. Then, the probability density function of τ is:

$$f(\tau)=\frac{1}{\sqrt{(2\pi\sigma\tau)}}\exp\left[-(\ln\tau-\mu)^2/2\sigma^2\right] \tag{1}$$

where μ = mean of $\ln\tau$, σ = standard deviation of $\ln\tau$. The probability that the recurrence interval is shorter than τ is:

$$F(\tau)=\int_0^{\tau} f(\tau')d\tau' \tag{2}$$

The conditional probability that the next event will occur within the period from T_0 to $T_0 + T$ equals to:

$$P((T_0,T+T_0))=\frac{F(T_0+T)-F(T_0)}{1-F(T_0)} \tag{3}$$

where T_0 is the time elapsed since zero time, that is the date of the last eruption, and T is the time window within which we examine the probability of occurrence of the next eruption. In our case study, the date of our calculation is the 2nd of August 2003. Zero time is the 2nd of February 1950 and T_0 = 53.5 years. In addition, for reasons of comparison random probabilities P ($X \geq 1$) to observe at least one eruption were calculated from:

$$P(X\geq 1)=1-P(0) \tag{4}$$

where:

$$P(X)=\exp(-\lambda t)(\lambda t)^X/X! \tag{5}$$

is the probability to observe one event in a time interval of length t and λ is the mean eruption rate.

To our knowledge the particular methodology has not been tested to other volcanic centers of the world. However, other statistical approaches were extensively applied in the past. Wickman (1966a, b) introduced the application of renewal processes in testing volcanic eruption time series. Muňoz (1983), tested random and renewal processes for Chilean volcanoes, Scandone (1983) tried random processes for Etna and De la Cruz-Reyna (1993) tested random and renewal processes for Colima volcano, Mexico.

4. RESULTS

The results are summarized in Tables 2 and 3. One may observe that both the random and conditional probabilities for the occurrence of the next eruption in Thera become significant, that is exceed about 0.5 only at time intervals longer than 50 years. It is also of interest that because of the relatively low number of events the results are sensitive to the total number of events involved in the calculations. As a consequence, for a particular sample of eruption events random probabilities are lower or higher than the corresponding conditional probabilities depending on the number of events

Table 2. Random probabilities $P(X \geq 1)_6$ and $P(X \geq 1)_8$ of observing at least one eruption in Thera in the next T years 2 August 2003. $P_{c\,(6)}$ and $P_{c\,(8)}$ are the conditional probabilities to observe the next eruption within T years from 2 August 2003 given that no eruption occurred after the last event of 2 February 1950. Index 6 and 8 refer to six and eight eruption cycles, respectively.

T (yrs)	$P(X \geq 1)_6$	$P_{c\,(6)}$	$P(X \geq 1)_8$	$P_{c\,(8)}$
1	0.0139	0.0198	0.0188	0.0110
5	0.0676	0.0866	0.0906	0.0656
10	0.1306	0.1675	0.1730	0.1290
20	0.2442	0.3014	0.3161	0.2200
50	0.5034	0.5958	0.6133	0.4280
100	0.7534	0.8213	0.8504	0.6167
150	0.8775	0.9162	0.9422	0.7188

Table 3. Random probabilities $P(X \geq 1)_5$ and $P(X \geq 1)_7$ of observing at least one intra-caldera eruption in Thera in the next T years from 2 August 2003. $P_{c\,(5)}$ and $P_{c\,(7)}$ are the conditional probabilities to observe the next intra-caldera eruption within T years from 2 August 2003 given that, no eruption occurred after the last event of 2 February 1950. Index 5 and 7 refer to six and eight eruption cycles, respectively.

T (yrs)	$P(X \geq 1)_5$	$P_{c\,(5)}$	$P(X \geq 1)_7$	$P_{c\,(7)}$
1	0.0119	0.0120	0.0159	0.0107
5	0.0582	0.0605	0.0769	0.0534
10	0.1131	0.1161	0.1479	0.1059
20	0.2134	0.2166	0.2739	0.1966
50	0.4512	0.4440	0.5507	0.3802
100	0.6988	0.6699	0.7981	0.5575
150	0.8347	0.7924	0.9093	0.6587

introduced in the sample.

5. CONCLUSIONS

The applied methodology, which combines Bayesian statistics with the assumption that the eruption inter-arrival times are lognormally distributed, seems to be effective for the calculation of conditional probabilities for the occurrence of the next eruption in Thera volcano. Such probabilities are renewable in the sense that as long as no eruption takes place the conditional probability does not remain constant.

REFERENCES

Astaras, Th., Lambrinos, N. and Soulakellis, N., 1997. Multitemporal monitoring of the Santorini Volcano by use of short and thermal infrared LANDSAT-5 TM images. In: P. Marinos, G. Koukis, G. Tsiambaos and G. Stournaras (Editors), Engineering Geology and the Environment. Balkema, Rotterdam, pp. 495-503.

De la Cruz-Reyna, S., 1993. Random patterns of occurrence of explosive eruptions at Colima Volcano, Mexico. J. Volcanol. Geotherm. Res., 55: 51-68.

Delibasis, N., Chailas, S., Lagios, E. and Drakopoulos, J., 1990. Surveillance of Thera Volcano, Greece: Microseismicity Monitoring. In: D.A. Hardy, J. Keller, V.P. Galanopoulos, N.C. Flemming, T.H. Druitt (Editors), Thera and the Aegean World III. The Thera Foundation, London, pp. 199-206.

Georgalas, G.C., 1962. Catalogue of the active volcanoes of the world including solfatara fields, Part X,Greece. International Association of Volcanology, pp. 1-40.

Lagios, E., Tzanis, A., Chailas, S., and Wyss, M., 1990a. Surveillance of Thera Volcano, Greece: Monitoring of the Geomegnetic Field. In: D.A. Hardy, J. Keller, V.P. Galanopoulos, N.C. Flemming, T.H. Druitt (Editors), Thera and the Aegean World III. The Thera Foundation, London, pp. 207-215.

Lagios, E., Tzanis, A., Hipkin, R., Delibasis, N. and Drakopoulos, J., 1990b. Surveillance of Thera Volcano, Greece: Monitoring of the Local Gravity Field. In: D.A. Hardy, J. Keller, V.P. Galanopoulos, N.C. Flemming, T.H. Druitt (Editors), Thera and the Aegean World III. The Thera Foundation, London, pp. 216-223.

Muñoz, M., 1983. Eruption Patterns of the Chilean Volcanoes Villarrica, Llaima, and Tupungatito. Pure Appl. Geophys., 121: 835-852.

Panagiotopoulos, D.G., Stavrakakis, G., Makropoulos, K., Papanastasiou, D., Papazachos, C., Savvaidis, A. and Karagianni, E., 1996. Seismic monitoring at the Santorini volcano. In: 2nd Workshop on European Laboratory Volcanoes, Abstract volume, Santorini, 2-4 May 1996.

Papadopoulos, G.A., 1986. Large intermediate depth shocks and volcanic eruptions in the Hellenic arc during 1800-1985. Phys. Earth Planet. Inter., 43: 47-55.

Papadopoulos, G.A., 1988: Statistics of historical earthquakes and associated phenomena in the Aegean and surrounding regions. In: P. Marinos and G. Koukis (Editors), Proc. Intern. Symp. Engineering Geology as Related to the Study, Presentation and Protection of Ancient Works, Monuments and Historical Sites, Athens, Sept. 1988, Balkema, Rotterdam, 3, pp.1279-1283.

Papadopoulos, G.A., 1990. Deterministic and stochastic models of the seismic and volcanic events in the Santorini volcano. In: D.A. Hardy, J. Keller, V.P. Galanopoulos, N.C. Flemming and T.H. Druitt (Editors). Thera and the Aegean World III, Proc. of the 3rd Internat. Congress, Santorini, Sept. 1989, The Thera Foundation, London, 2, pp. 151- 159.

Papazachos, B.C., 1990. Long and short term prediction of the volcanic eruptions in Santorini. In: D.A. Hardy, J. Keller, V.P. Galanopoulos, N.C. Flemming and T.H. Druitt (Editors). Thera and the Aegean World III, Proc. of the 3rd Internat. Congress, Santorini, Sept. 1989, The Thera Foundation, London, 3, pp. 224-228.

Scandone, R., 1983. Problems related with the evaluation of volcanic risk. In: Tazieff, H. and Sabroux J.-C. (Editors), Forecasting volcanic events. Elsevier, Amsterdam, pp. 57-67.

Simkin, T., Siebert, L., McClelland, L., Bridge, D.,New-Hall, C. and Latter, J.H., 1981. Volcanoes of the World. Stroudsburg, Pa, USA, pp. 232.

Working Group on California Earthquake Probabilities, 1988. Probabilities of large earthquakes occurring in California on the San Andreas fault. U.S.Geological Survey, Open-File Report 88 –398, pp.62.

Late-Holocene coastal uplift in the Nisyros volcano (SE Aegean Sea): Evidence for a new phase of slow intrusive activity

S.C. Stiros[1,*], P.A. Pirazzoli[2], M. Fontugne[3], M. Arnold[3] and G. Vougioukalakis[4]

[1] Department of Civil Engineering, Patras University, Patras 26500, Greece

[2] CNRS-Laboratoire de Géographie Physique, 1 Place A. Briand, 92195 Meudon cedex, France

[3] LSCE (CNRS-CEA), Avenue de la Terrasse, 91198 Gif-sur-Yvette, France

[4] IGME, 70 Messogion st., Athens 11527, Greece

In approximately the last 20,000 years, after a period of violent explosive activity which led to the formation of a spectacular, 4 km-wide caldera and a subsequent phase of extrusion of domes up to 600 m high, the approximately 150,000 years old Nisyros volcano, at the SE edge of the Aegean volcanic arc was relatively quiescent, and its activity is assumed to have been confined to hydrothermal eruptions.

A detailed geomorphological and biological study of the coasts of Nisyros and of the nearly Kos Island, combined with radiocarbon analysis of collected samples, provided evidence of land uplift along the northern and western coast of Nisyros Island, at a minimum, though increasing rate of 1.7 mm/yr during the last 2,000-3,000 years. This uplift correlates with late Quaternary uplift deduced from coastal and volcanological data, and reflects a new tendency for topography build-up due to caldera inflation, and consequently to potential for a new volcanic unrest.

Keywords: Nisyros volcano, Aegean, Holocene uplift, caldera inflation, marine biology, radiocarbon dating.

* Corresponding author: e-mail: stiros@upatras.gr

1. INTRODUCTION

The small, nearly circular (diameter ~ 8 km) island of Nisyros is a Late Quaternary volcano at the SE edge of the South Aegean volcanic arc (Fig. 1) marked by a spectacular, ~4 km wide caldera. The latter was formed by explosive eruptions, followed by extrusions of rhyolitic-dacitic domes (Fig. 2) and lava flows. In approximately the last 20,000 years the activity of Nisyros volcano was limited to a few hydrothermal explosions, the most important of the historically registered ones occurred between 1871-1873 (Martelli, 1917; Di Paola, 1974; Keller et al., 1990; Marini et al., 1993; Vougioukalakis, 1993, 1998). According to the available reports, the recent activity was not associated with any signs of crustal deformation except for some minor fissures which opened during the last eruptions or during recent seismic paroxyms, but were thereafter closed, and by some minor seismic faults in the nearby Yali islet (Stiros and Vougioukalakis, 1996).

Recent investigations, however, revealed important fault activity and Late Holocene raised beaches (Stiros, 2000). Such findings motivated a detailed study of coastal change in Nisyros, the results of which are presented here.

The aims of this study, the first results of which are summarized in this article, are:

1. the detailed study of crustal movements deduced from recent relative sea-level changes and their correlation with longer-term movements deduced from stratigraphic data.
2. the correlation of the observed coastal movements with possible caldera and dome building processes.

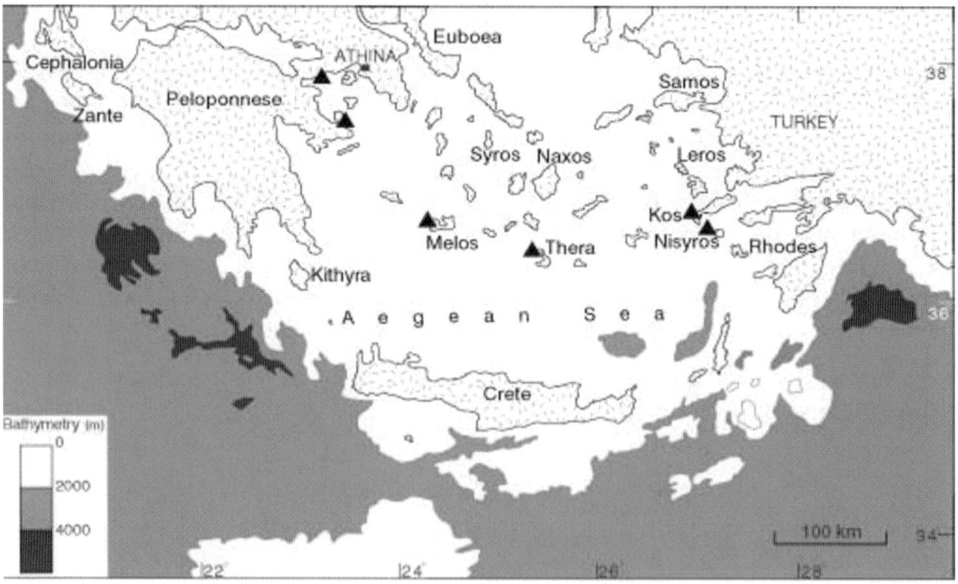

Fig. 1. South Aegan, location map. Triangles indicate the Pliocene-Quaternary volcanic centres.

3. a contribution to the assessment of the volcanic risk of this island; in particular the identification of areas where expected crustal deformation is maximum or anomalous; such areas may represent a threat for constructions and the infrastructure of the island (utilities networks, uplifted and consequently aborted harbours, etc.).

2. GEOLOGY AND VOLCANIC HISTORY

According to Di Paola (1974), Fytikas et al. (1985), Keller et al. (1990), Marini et al. (1993) and Vougioukalakis (1993) Nisyros island represents the emergent portion of an andesitic composite volcano that was build up during the last 150,000 years. Geologic formations of Nisyros consist of a succession of lava flows and domes inter-stratified with volcaniclastic layers. The oldest of these volcanics, pillow lavas and hyaloclastites, younger than 200 ka, are exposed at the NW part of the island (mostly along a strip between sites A and E in Fig. 2). The pre-volcanic basement, Mesozoic rocks and Neogene sediments, is not exposed on Nisyros. Cone-building activity associated with relatively deep (13-27 km deep) magma chambers characterises the first cycle of the geologic history of the island. The second cycle started possibly about 40 ka ago and is characterized by intense explosive activity and shallow (less than 6 km) magma chambers. As a result of this activity, the volcano cone was truncated by a summit caldera, about 4 km wide.

The north and western part of caldera depression was subsequently filled by a series of dacitic-ryodacitic domes, up to 600 m high (Fig. 2), while the southeastern part of the caldera floor, at an altitude of 110-120 m, host intense fumarolic activity and the craters of the most recent hydrothermal eruptions; the last ones of these eruptions occurred during the paroxysmal periods of 1871-1873 and 1887.

Some of these eruptions were associated with ground fissures (Marini et al., 1993; Vougioukalakis, 1998), but there is no information on other types of ground deformation accompanying volcanic events, such as coastal changes.

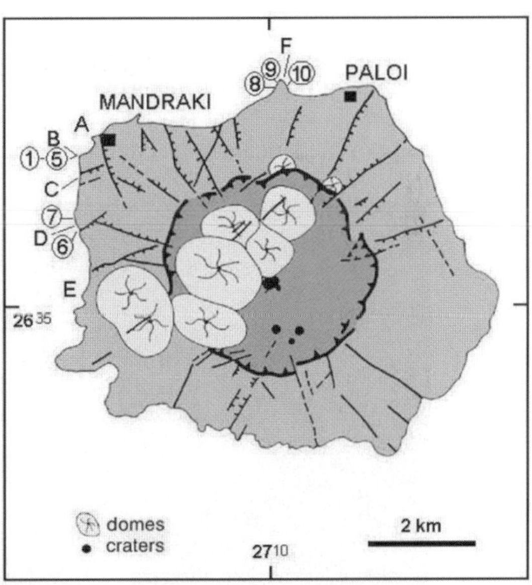

Fig. 2. Topography of Nisyros. A thick line with triangles indicates the caldera rim, inside which lava domes and craters of hydrothermal eruptions are found. Main, mostly volcanotectonic faults (lines with ticks) are also shown. Simplified after Vougioukalakis (1993). Letters indicate sites mentioned in the text, while numbers in circles the last digit of the code of samples of Table 1.

Numerous, mostly subvertical faults crosscut Nisyros island and the caldera. Their expression is clearer at the caldera rims, where vertical offsets of the order of 100-150 m are present. Vertical offsets of most of these faults decrease away from the caldera rim towards the coast, where they practically disappear ("scissors-type" faults); an evidence of their association with volcanotectonic effects (Stiros, 2000).

3. HOLOCENE EMERGENCE OF NISYROS

Evidence for Holocene coastal changes in Nisyros have been presented by Desio (1931) who mentioned east of Pali partly submerged ancient thermal baths, already mentioned by Hippocrates (4th c. BC) and Strabo (1st c. BC-1st c. AD). Di Paola (1974), Vougioukalakis (1993) and Stiros (2000) reported raised beaches at Kochlaki beach, SW of Mandraki (site B in Fig. 2).

In the framework of this project, the whole coast of Nisyros and of the southern coast of Kos Island were inspected by boat and were systematically surveyed in 1995 and 1997, while samples were collected for laboratory analysis and dating.

Our study revealed that the remains of late-Holocene emergence are confined to the northern and north-western coast of Nisyros, approximately between Pali and Cape Kanoni (site D in Fig. 2). Evidence for such an emergence provides mainly vermetid-shell and algal-crust remains still attached to the rock in growth position (Fig. 3), at elevations higher than their present-day live counterpart. For instance, a marine platform containing pottery fragments cemented by marine crusts and fossil vermetids about 20-30 cm above present sea level was observed at Pali and farther east, while at the south edge of the Kochlaki beach, SW of Mandraki (site B in Fig. 2) vermetids and

Fig. 3 Two Vermetus triqueter shells, still preserved in growth position at about 1.3 m above sea level, are visible near a centimetre scale. One of them (sample 97NI-1) has been 14C dated 1060±50 years BP. Location A in Fig 2. (Photo G80 by PAP).

algal crusts were found fixed on fragile volcanic rocks up to the height of several meters above the water. As far as the subsidence reported by Desio (1931) is concerned, we have not been able to document and evaluate it.

Sampling took place between 7 and 9 June 1997. Geographical coordinates were determined in the field using pocket Global Positioning System (GPS) devices. Elevations were measured with reference to the sea surface by means of a spirit level and a folding ruler and then corrected according to the tide predictions at the nearby stations of Leros, Rhodes and Syros, and to the meteorological conditions observed at the Kos Airport (see Stiros and Pirazzoli (2003) for details). Samples were collected from various elevations, between 0.1 m and 4.2 m above the present sea level, mainly inside rock crevices sheltered from wave action.

The results of radiocarbon dating are summarized in Table 1. Conversions of these ages into calibrated dates were obtained using the calibration correction proposed for marine samples by Stuiver and Braziunas (1993) assuming a reservoir effect of 320 ± 25 years (Stiros et al., 1992). Slightly different ages would be obtained using the reservoir effects of 390 ± 85 years for the Mediterranean, as proposed by Siani et al. (2000), or of 480 ± 72 years for the Aegean Sea (Reimer and McCormac, 2002). Yet, adaptation of any of these corrections has minor effects in the estimates of the uplift.

Most of the samples dated belong biologically to the infralittoral zone, i.e. they may have developed at any depth between sea level to 25-35 m deep (Laborel and Laborel-Deguen, 1994), and hence they can provide only minimum estimates of the uplift. Furthermore, the two *Dendropoma* samples (97NI-6 and 97NI-8) which were potentially the most precise sea-level indicators gave uncertain results: sample 97NI-6 corresponding to a vermetid formerly living at sea-level and testifying to an about 0.5 m uplift was probably contaminated by modern barnacle shells, which may live up to a few decimeters above sea level in exposed sites. Sample 97NI-8 on the other hand may not indicate a precise sea-level, for the boulder on the outer face of which it was fixed was possibly displaced by waves.

Despite these problems, the available data can be used to derive a curve reflecting the trend of land uplift in northwestern Nisyros: samples covering a wide time and elevation range have been collected form a single site (Fig. 4), and computed ages define an almost linear trend. Hence, it can be assumed that these samples developed below, but quite close to the sea level.

Based on this reasonable hypothesis, the available data were plotted in a graph to estimate a lower-bound curve for the observed relative sea-level change (Fig. 5, dotted curve). Such a minimum sea-level curve suggests an apparent, quasi-linear uplift trend during the last 2000 years, at a minimum rate of 1.7 mm/yr. Some evidence for acceleration of the uplift in the last 1500-2000 years also exists.

The uplift trend defined mainly in the Mandraki area can probably be extrapolated to whole of the NW part of the island, for farther east of Pali, a platform containing sherds has been cemented by marine crusts about 20-30 cm above present sea level; at its outer border, undated slightly elevated vermetid formations are still preserved in growth position.

Table 1. List of dated biological indicators of sea-level change from Nisyros and Kos Islands. For sample location see Fig. 2.

Nisyros										
Lat. N	Long. E	Sample	Elevation (m)	Material	Estimated palaeo -MSL elevation (m)	14C age	Laboratory number	$\delta^{13}C$ ‰	1-sigma (p = 0.66) calibrated interval	2-sigma (p = 0.95) calibrated interval
36°36.47'	27°07.64'	97NI-1	+1.3±0.2	V. triqueter	≥ +1.1	1060±50	GifA-97353		1230-1310 AD	1180-1350 AD
		97NI-2	+1.7±0.2	Vermetids or serpulids	≥ +1.5	1130±50	GifA-97354		1160-1270 AD	1070-1300 AD
		97NI-3	+3.6±0.2	Marine crust	≥ +3.4	2050±60	Gif-10939	2.76	160-340 AD	100-420 AD
		97NI-4	+4.2±0.2	Algal crust	≥ +3.8	2970±70	GifA-97355		920-790 BC	1020-750 BC
		97NI-5	+0.4±0.2	Vermetids	≥ +0.2	530±50	GifA-97356		1660-1730 AD	1620-1845 AD
36°35.78'	27°07.58'	97NI-6	+0.5±0.2	*Dendropoma* (a)	≥ 0	220±70	Gif-10940	1.03	modern	modern
36°35.93'	27°07.55'	97NI-7	+1.4±0.2	V. triqueter	≥ +1.2	1490±50	GifA-97357		780-900 AD	710-975 AD
36°37.12'	27°10.04'	97NI-8	+0.3±0.2	*Dendropoma* (b)	≥ +0.1	820±60	Gif-10941	2.00	1410-1480 AD	1340-1530 AD
36°37.23'	27°10.03'	97NI-9	+0.1±0.2	vermetids	≥ -0.1	290±65	Gif-10942	2.19	modern	modern
36°37.30'	27°10.10'	97NI-10	+0.1±0.2	vermetids	≥ -0.1	464±55	Gif-10943	1.52	1690-1870 AD	1660-1950 AD

Kos, Kardamaina										
Lat. N	Long. E	Sample	Elevation (m)	Material	Estimated palaeo -MSL elevation (m)	14C age	Laboratory number	$\delta^{13}C$ ‰	1-sigma (p = 0.66) calibrated interval	2-sigma (p = 0.95) calibrated interval
36°46.48'	27°08.11'	97 KO-1	+0.5±0.2	Shell debris	+0.5±0.2	25,350±860	Gif-10944	2.22	>23,700	>23,700

Fig. 4. View of locality A. Marine crusts can be found in situ in crevices on this rock up to 5 m in elevation. Scale is 2 metres. (Photo G78 by PAP).

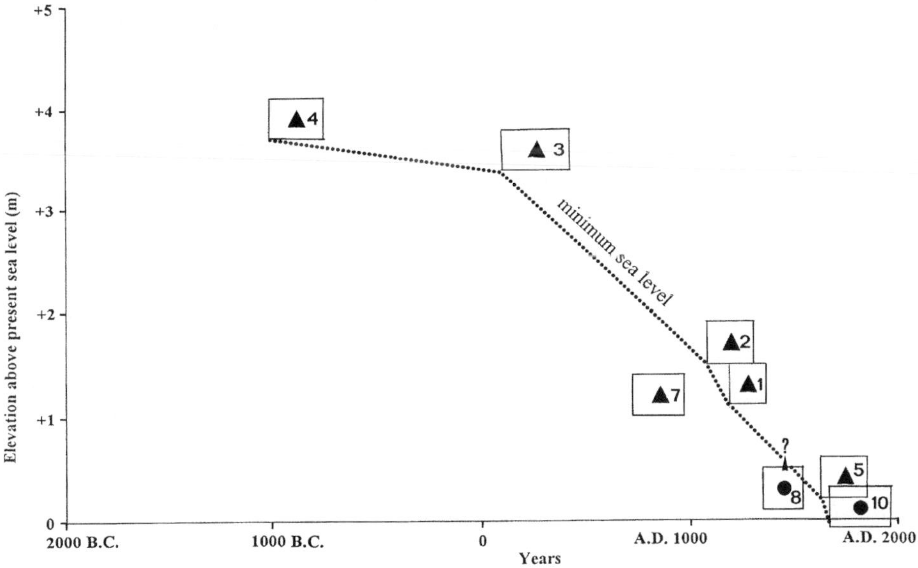

Fig. 5. Plot of observed relative sea-level drop versus age for the dated samples of Table 1. A dashed line indicates the minimum relative sea-level curve at Nisyros during the last three millennia. The numbers inside the boxes correspond to the last digits of the sample codes of Table 1. The size of the boxes indicates the uncertainty margins in sea-level reconstruction (2-sigma calibrated intervals for radiocarbon dates). The two samples indicating modern material are omitted.

4. EVIDENCE OF LATE-QUATERNARY UPLIFT

At site A there is some geomorphological evidence of a fossil marine bench, 3-4 m high, while at sites C and E of Fig. 2 layers of rounded pebbles and tuffites, testifying fossil Pleistocene beaches, up to 15 m high can be observed on the subvertical sea-cliffs. Given the geologic history of the island (see above), this uplift is up to a few thousand years old.

In the island of Kos, just south of Kardamena village, at the coast nearest to Nisyros, beachrock formation showing an about 0.5 m emergence has been observed. This is the only place where evidence of "recent" emergence could be found on the island of Kos. A sample of marine shell debris collected from the inner part of a beachrock slab has been dated by radiocarbon 23,350±860 years BP. This age should of course be interpreted as a minimum one, the real age probably belonging to an undefined time during the last interglacial. This dating proves, anyway, that the recent uplift observed in Nisyros does not extend as far as Kos Island.

5. DISCUSSION AND CONCLUSIONS

Our data indicate that Holocene land uplift is confined to the NW part of Nisyros, and is most likely taking its maximum value in the wider Mandraki area. Interestingly, the area of Holocene uplift correlates with that of Late Quaternary uplift deduced from coastal data presented above, as well as from volcanological data: The volcanic stratigraphy of the island reveals that the oldest volcanic rocks (submarine andesitic lavas and pyroclasts; Di Paola, 1974; Vougioukalakis, 1993), outcrop at the area of maximum Holocene uplift (NW coast), while lava domes are confined to the NW part of the caldera. Furthermore, the maximum amplitude in the throw of volcanotectonic scissors faults has been observed in the Palaiokastro area, close to Mandraki (Stiros, 2000).

It is therefore reasonable to conclude that the Holocene coastal data indicate that the topography build-up activity is continuing and has resumed in the last 3,000 years, and has probably accelerated in the last 1000 years (Fig. 5). The only plausible explanation is an inflation of this active volcano due to magma intrusions. Given the volcanic history of the island, an implication if this conclusion is that, the topography build-up of NW Nisyros, especially in the area where the major inhabitation center of the island is located, testifies to an increasing potential for a major, future tectonic anomaly.

Acknowledgements

We are indebted to M. Pachos, former Mayor of Nisyros and J. Koulakis former member of the municipality council for their enthousiastic support of this project and to the Municipality of Nisyros for covering the cost of the radiocarbon dating.

REFERENCES

Desio, A., 1931. Le isole italiane dell'Egeo. Memorie descrittive della Carta geologica d'Italia, vol. 24, Roma, 546 p.

Di Paola, G., 1974. Volcanology and petrology of Nisyros island (Dodecanese, Greece). Bull. Volcanol., 38: 944-987.

Fyticas, M., Innocenti, F., Manetti, P., Peccerillo, A. & Villari, L. 1985. Tertiary to Quaternary evolution of volcanism in the Aegean region. In: D. Dixon and A. Robertson (Editors), The geological evolution of the Eastern Mediterranean, Geol. Soc. Lond, Special Publication 17: 687-699.

Laborel, J. and Laborel-Deguen, F., 1994. Biological indicators of relative sea-level variations and of co-seimic displacements in the Mediterranean region. Journal of Coastal Research, 10: 395-415.

Marini, L., Principe, C., Chiodini, G., Cioni, R., Fytikas, M. and Marinelli, G., 1993. Hydrothermal eruptions of Nisyros (Dodecanese, Greece). Past events and present hazard. J. Volcan. and Geotherm. Res., 56: 71-94.

Martelli, A., 1917. Il gruppo eruttivo di Nisiro nel mare Egeo. Mem. Soc. Ital. delle Scienze (detta dei XL), Ser.3, T.XX, Roma.

Reimer, P.J. and McCormac, F.C., 2002. Marine radiocarbon reservoir corrections for the Mediterranean and Aegean Seas. Radiocarbon, 44: 159-166.

Siani, G., Paterne, M., Arnold, M., Bard, E., Métivier, B., Tisnerat, N. and Bassinot, F., 2000. Radiocarbon reservoir ages in the Mediterranean Sea and Black Sea. Radiocarbon, 42(2): 271-280.

Stiros, S., 2000. Fault pattern of Nisyros island volcano (Aegean Sea, Greece): structural, coastal and archaeological evidence. In: McGuire, B. et al. (Editors), The Archaeology of geological catastrophes, Geological Society of London, Special Publication 171: 385-397.

Stiros, S.C. and Pirazzoli, P.A., 2003. Impact of short-wavelength sea-level oscillations on coastal biological zoning: evidence from Nisyros Island (Aegean Sea), and implications for the use of the Biological Mean Sea Level as a Geodetic Datum. Journal of Coastal Research, in press.

Stiros, S. and Vougioukalakis, G., 1996. The 1970, Yali (SE edge of the Aegean volcanic arc) earthquake swarm: surface faulting associated with a small earthquake. Annales Tectonicae, 10: 20-30.

Stiros, S.C., Arnold, M., Pirazzoli, P.A., Laborel, J., Laborel, F. and Papageorgiou, S., 1992. Historical coseismic uplift on Euboea Island, Greece. Earth and Planetary Science Letters, 108: 109-117.

Vougioukalakis, G., 1993. Volcanic stratigraphy and evolution of Nisyros Island. Bull. Geol. Soc. Greece, 28: 239-258.

Vougioukalakis, G., 1998. Blue volcanoes: Nisyros. Nisyros Regional Council, 78pp. and map.

Investigating the formation of a superficial fracture on Nisyros Island, Greece with the DC resistivity method

D. Galanopoulos[1] and G. Kolettis[2]

[1]General Secretariat for Civil Protection, 2 Evangelistrias Str., GR-10563, Athens, Greece

[2]P.O.Box 9, GR-19013, Anavissos, Greece

ABSTRACT

During January 2003 a geophysical study was undertaken on Nisyros volcanic island in order to investigate the formation of a 600 m long superficial fracture. The fracture is distinguished by a main branch of an approximately N-S direction and smaller side branches of E-W, NE-SW and NW-SE directions. The phenomenon was initiated during November 2001 within the caldera of the island and is still active with the fracture gradually extending southwest, towards Stefanos hydrothermal crater. The study comprised electrical resistivity measurements with the direct current (DC) method along two parallel profiles of E-W direction, close to the north and south ends of the fracture, respectively. Each profile included 4 Half-Schlumberger soundings with a maximum AB/2 of 130 m. The field measurements were carried out with ATLAS 350, a DC resistivity meter designed and developed by the authors of this paper. The collected data are proved to be of very good quality and similar to those of previous DC resistivity surveys in the same area. The data processing resulted in electrical models, which are characterized by rather low resistivities (5-400 ohm-m) and delineate lateral electrical discontinuities at depths greater than 10 m. The latter could be attributed to fractured zones, which do not have any superficial manifestations. The derived electrical models are compatible with the local geology and support the idea that the formation of the fracture is directly related with subsidence phenomena within the top 100 m depths.

Keywords: Electrical resistivity, DC method, Schlumberger, Nisyros, Hellenic volcanic arc

1. INTRODUCTION

Nisyros is a small volcanic island located on the southeastern part of the Aegean Sea (Fig. 1a). Nisyros has a nearly round shape (Fig. 1b) with a diameter of about 8 km and an area of approximately 50.6 km^2.

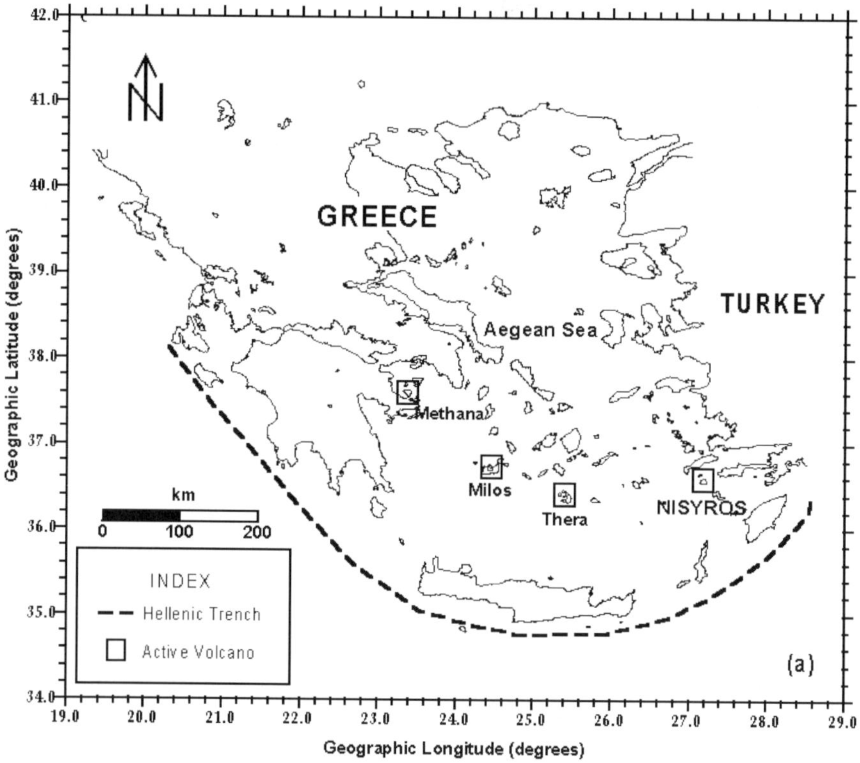

(a)

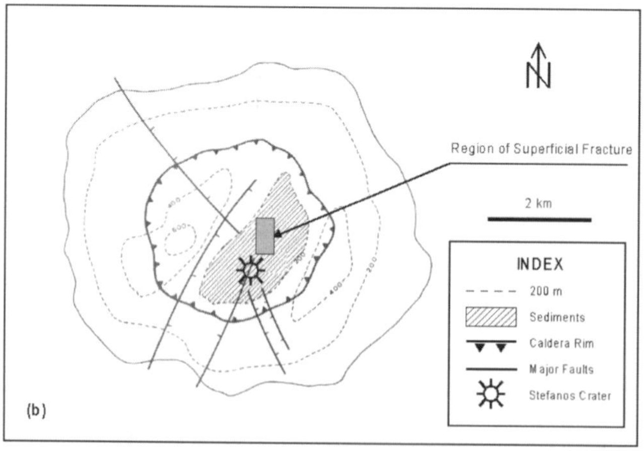

(b)

Fig. 1. (a) Hellecic Volcanic Arc: geographic location of Nisyros Island, (b) Sketch map of Nisyros Island showing the main geotectonic features and the region of superficial fracture.

The dominant feature is a large caldera of 4 km diameter, located at the central part of the island. The heights of the caldera rim above sea level vary from 250 to 600 m (Vougioukalakis, 1998).

Nisyros, like Santorini and Milos islands and Methana peninsula belongs to the group of active volcanoes of Greece located along an island arc (Fig. 1a), known either as South Aegean Active Volcanic Arc (SAAVA) (Fytikas et al., 1987), or Hellenic Volcanic Arc (HVA) (Galanopoulos, 1993). The formation of volcanoes along this arc is associated with the relative motion of large lithospheric plates and subduction phenomena (McKenzie, 1978; Angelier et al., 1982; Fytikas et al., 1984).

Di Paola (1974), Papanikolaou and Lekkas (1990) and Vougioukalakis (1993; 1998) are among the various researchers who have thoroughly studied the geology and volcanism of Nisyros. The high temperatures (450°C) observed at depths of 1.8 km, the various superficial manifestations like hot springs, fumaroles and large hydrothermal craters and the results from recent tectonic (Papanikolaou and Nomikou, 2001), seismic (Papadopoulos et al., 1998), GPS (Lagios et al., 1998: Lagios, 2000), interferometric (Parcharidis and Lagios, 2001; Ganas et al., 2002) and Landsat (Ganas and Lagios, 2003) studies provide significant evidence that Nisyros is a dormant active volcano.

During November 2001, a fracture with a total length of about 400 m was developed within the upper geological formations of the caldera fill. According to Vougioukalakis (2002, 2003) the development of the fracture is probably due to the subsidence of loose superficial sediments over existing fractures and voids of the underlying geological formations. A prolonged period of rainfall and two local seismic events probably initiated the occurrence of the phenomenon. During December 2002, the subsidence phenomena were repeated and the fracture was extended about 200 m to the southwest.

The only geophysical investigations with the DC resistivity method prior to the present study were those undertaken by the Institute of Geology and Mineral Exploration (Voutetakis, 1974; Nathanael, 1983). We present the results of a preliminary electrical survey consisting of 8 Half-Schlumberger soundings undertaken during January 2003. The survey was carried out with a DC resistivity meter designed and developed by the authors of this paper. The objective of this study was the investigation of the formation of the 600 m long superficial fracture.

2. THE SUPERFICIAL FRACTURE

The fracture is sited at the center of the eastern part of the caldera (Fig. 1b) and has a rather complex shape (Fig. 2) which comprises a main branch of an approximately N-S direction and smaller side branches of E-W, NE-SW and NW-SE directions.

The most complex shape is encountered in the northern section of the fracture, which was developed during November 2001 and for reasons of simplicity will be called in this paper 'fracture 2001'. This section is distinguished by six side branches AB, BC, BF, DE, FG and FH, correspondingly (Fig. 2). The more recent section includes only one branch and coincides with the southern part of the fracture. This section was developed during December 2002 and for reasons of simplicity will be called 'fracture 2002' (Fig. 2). The average width and maximum depth of the trench that

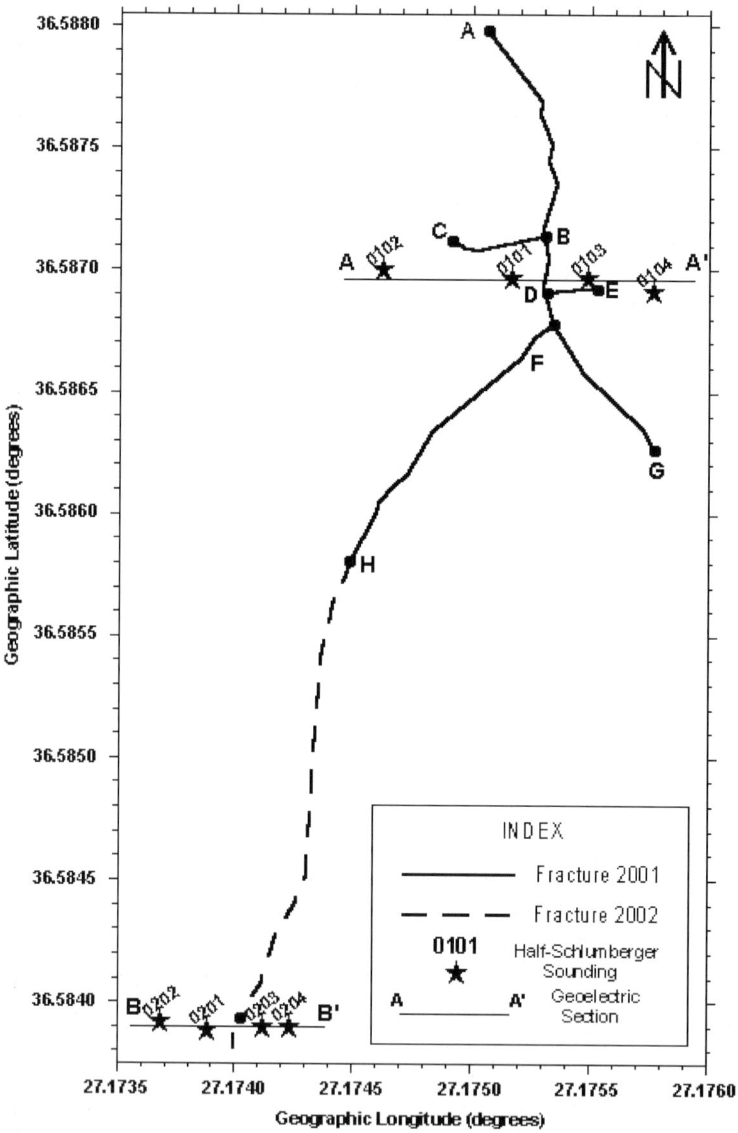

Fig. 2. Map of the region of the superficial fracture showing the locations of the Half-Schlumberger soundings.

has been formed along ‘fracture 2001’ are 0.7-1 m and 15 m and along ‘fracture 2002’, 1-2 m and 11 m, respectively (Vougioukalakis, 2002, 2003).

The fracture extends over an area of 200 m x 400 m. The top 100 m of the caldera fill in this area comprise two geological formations (Vougioukalakis, 1993, 2002).

Fig. 3a demonstrates a simplified section of the geological stratigraphy in this area. The upper geological formation is 25-35 m thick, has low porosity and consists of fine-grained sediments, such as clays and sand, and pebbles. These are products of hydrothermal explosions and erosion, or lacustrine deposits. The lower formation is 40-60 m thick and consists of boulders and large blocks, which collapsed to the inner part of the caldera during or immediately after the last phase of the explosion that formed the caldera itself. The formation has very large porosity (100%) due to the existing voids between the boulders and large blocks (Figs. 3a, b).

According to Vougioukalakis (2002, 2003) the development of the fracture is most likely due to the subsidence of the fine-grained sediments of the upper geological formation into existing fractures and voids of the lower geological formation of the caldera fill. The phenomenon was probably initiated and speed up by prolonged periods of rainfall and local seismic events with surface magnitude less than 3.5 R, thus not detectable by the Seismic Network of the Geodynamic Institute of the National Observatory of Athens, however reported by local people.

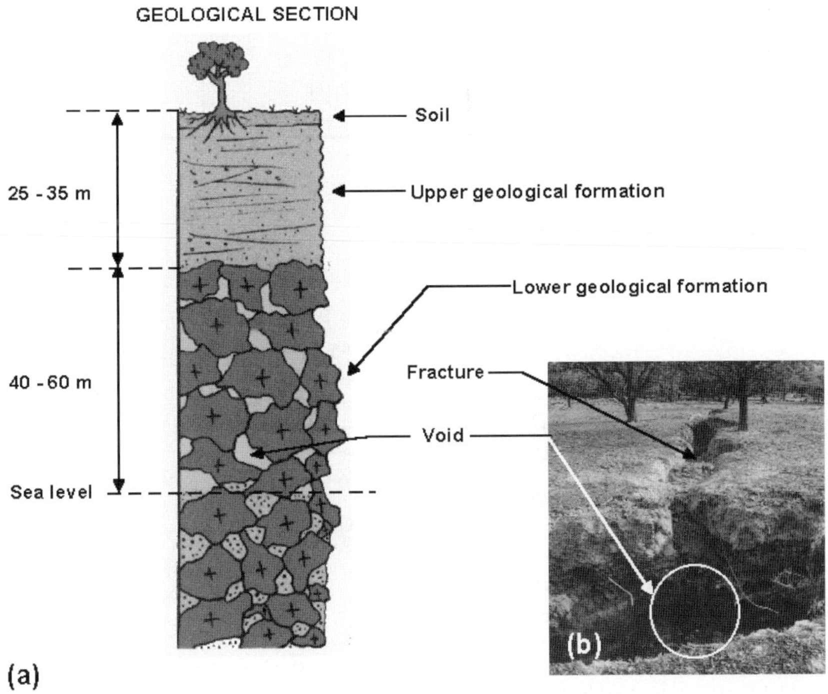

Fig. 3. (a) Geological section of the Nisyros Caldera fill in the region of the superficial fracture (after Vougioukalakis, 2002), (b) A subsurface void below branch DE of 'fracture 2001'.

3. METHODOLOGY

A suitable geophysical method for delineating shallow fracture zones and voids is the DC resistivity method. The method (Telford et al., 1981) employs an artificial source of direct electric current I, which, is injected into the ground by using a pair of metal electrodes, called current electrodes A and B. The resulting potential difference ΔV in the region of the electric current flow is measured by using a second pair of metal electrodes, called potential electrodes M and N (Fig. 4a). The ratio ΔV/I multiplied by an appropriate coefficient K which depends on the geometry of the electrode array in use, gives the electrical resistivity ρ of rocks.

One of the most common electrode configurations is that introduced by Schlumberger. In this case $K=\pi(AB/2)^2/MN$, where AB and MN are the distances between the two current and the two potential electrodes, respectively. The electrical

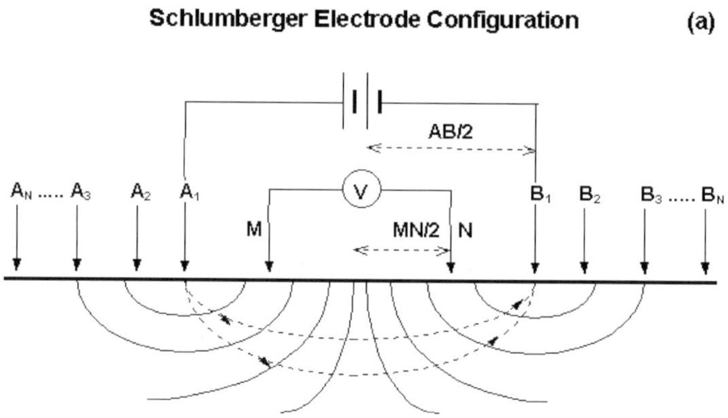

(a)

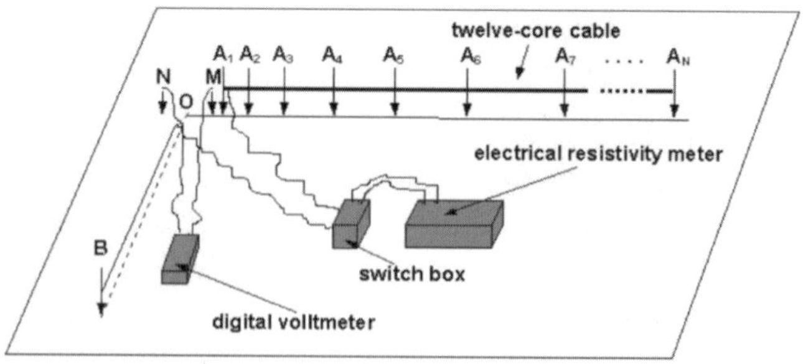

(b)

Fig. 4. (a) The Schlumberger electrode configuration (b) Set-up of the field apparatus and the Half-Schlumberger electrode configuration.

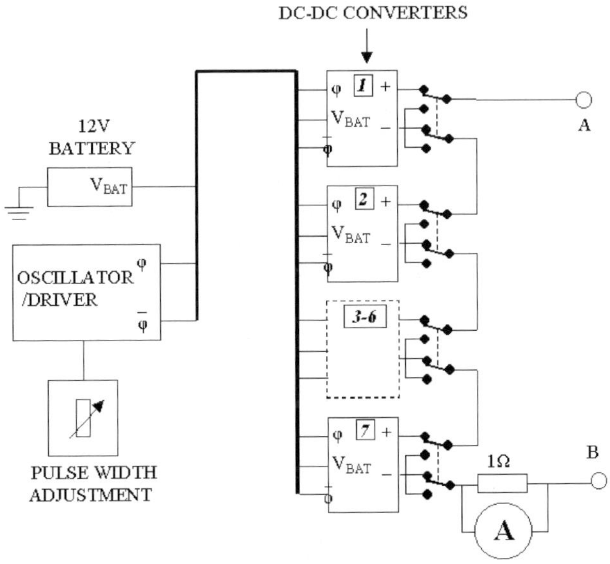

Fig. 5. The block diagram of ATLAS 350 dc resistivity meter.

resistivity ρ of rocks can be determined as a function of depth by increasing successively the distance between the two current electrodes (Fig. 4a). For a total number of N distances between the current electrodes, the electrical resistivity ρ_n corresponding to each distance is $\rho_n=K_n(\Delta V_n/I_n)$, where n=1,2,3, ... N, ΔV_n, I_n and K_n the potential difference, electric current and geometric coefficient relative to each distance.

In this study the electric current and potential difference were measured by using a version of the Schlumberger electrode configuration known as the half-Schlumberger array. According to this version, one of the two current electrodes e.g. the current electrode B is placed at a fixed point, a large distance (about 2-3 times the maximum AB/2) away from the center O of the array and in a way that OB is perpendicular to OA (Fig. 4b). Then electrode A is moved successively at the prescribed positions A_1, A_2, A_3,...,A_N, where N is the total number of measurements. The electrical resistivity ρ_n which corresponds to each of the distances $(A_nB)/2$ is calculated by using the above mentioned equation, however, the potential difference ΔV is now doubled in order to balance for the loss of half of the voltage gradient measured at the potential electrodes.

4. APPARATUS AND ELECTRICAL RESISTIVITY MEASUREMENTS

The electrical resistivity measurements were carried out with ATLAS 350 (Fig. 5), a dc resistivity meter designed and developed by the authors of this paper. The system is powered by a 12V, 7Ah rechargeable Pb battery and consists of an oscillator with adjustable output pulse width, which drives 7 switch-controlled DC-DC converters, 12

V to 50 V each. ATLAS 350 provides a maximum voltage output of 350 V, at intermediate steps of 50, 100, 150, 200, 250, and 300 V and a maximum electric current of 800 mA. The system includes a built-in LED milliammeter for the measurement of current and an external, high resolution, digital voltmeter for the measurement of potential difference.

The self potential (SP) of the Earth was taken into account during each measurement of I and ΔV, by subtracting its value from the corresponding value of ΔV. The ratio (ΔV-SP)/I was then multiplied by the appropriate geometric coefficient K giving $\rho=2K(\Delta V-SP)/I$. SP is the potential difference measured between the potential electrodes M and N in the absence of any artificial current flow. This potential difference is due either to bioelectric activity in vegetation, varying electrolytic concentrations in ground water or to the flow of natural electric currents (telluric currents) within the Earth, which are induced by the varying magnetic field of the Earth as a result of various magnetospheric and ionospheric phenomena (Telford et al., 1981).

The field measurements were carried out during January 2003 and comprised 8 Half-Schlumberger soundings, which were conducted along two parallel profiles of E-W direction (Fig. 2). The first profile crosses the northern part of the fracture, while the second one is located about 5 m south of its southern end. Each profile included 4 Half-Schlumberger soundings with a maximum AB/2 of 130 m.

5. CONTOUR SECTIONS AND ONE-DIMENSIONAL MODELING

The apparent electrical resistivity data are proved to be of very good quality (Figs. 6a, b) and similar to those of previous DC resistivity surveys in the same area (Voutetakis, 1974; Nathanael, 1983). The data were modeled one-dimensionally by using the computer program IPI developed by Bobachev et al. (1999) in the Department of Geophysics of Moscow State University. Tables 1 and 2 demonstrate the model parameters and rms% values for each Half-Schlumberger sounding. It was proved that a five layer electrical model was necessary in order to obtain a low rms% value per site. The contour sections of Figs. 7a, b and the one-dimensional model sections of Figs. 8a, b, provide the derived geo-electric structure, below the region of the superficial fracture.

The contoured sections (Figs. 7a, b) illustrate the variation of apparent electrical resistivity with AB/2 below profiles AA' and BB'. Considerably low resistivities (<125 ohm-m) are observed below both the profiles, while below the location of the superficial fracture a lateral electrical discontinuity seems to be present in every section. An additional lateral electrical discontinuity exists below the locations of soundings 0203 and 0204.

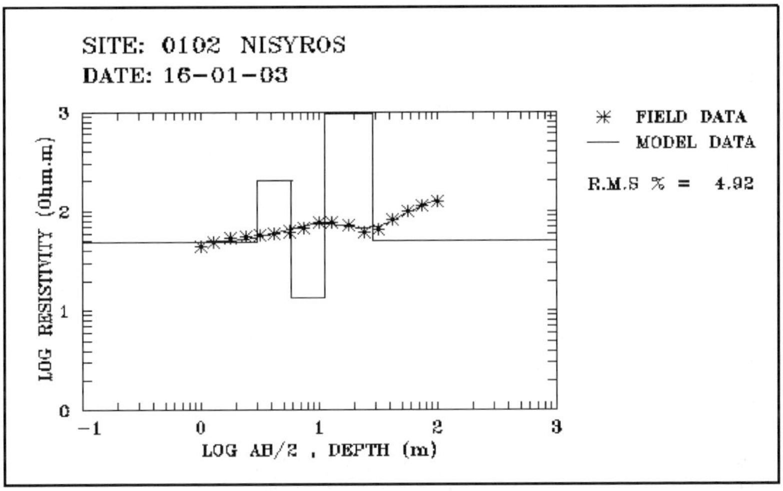

(a)

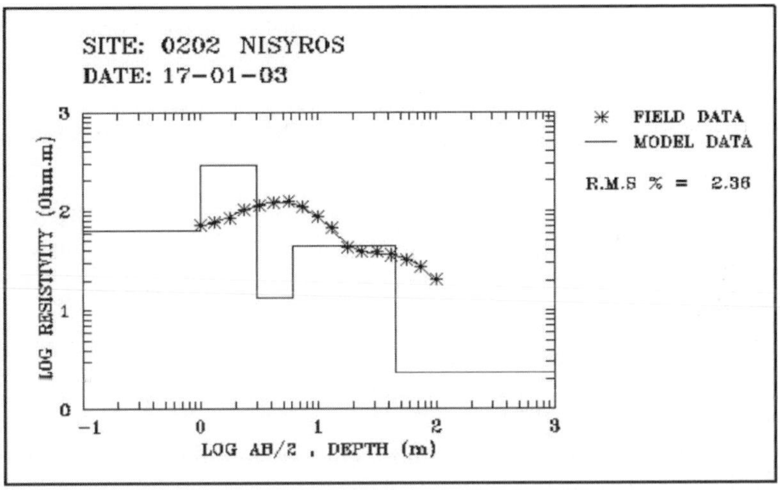

(b)

Fig. 6. (a) Apparent electrical resistivity data and one dimensional model for sounding 0102. Solid curve indicates model response (b) Apparent electrical resistivity data and one dimensional model for sounding 0202. Solid curve indicates model response.

The one-dimensional model (1D) sections (Figs. 8a, b) are products of the combination of the one-dimensional models of Tables 1 and 2 and provide the distribution of electrical resistivity with depth below profiles AA' and BB'. Both sections are distinguished by relatively low electrical resistivities (2-400 ohm-m). The prominent feature of section AA' is a lateral electrical discontinuity under the location of the superficial fracture. The discontinuity is encountered at a depth of 10 m and

Table 1. Values of the one dimensional models of section AA' (ρ = electrical resistivity of layer d = depth to the top of layer).

Layer No	SECTION AA' – ONE DIMENSIONAL MODELS							
	0102		0101		0103		0104	
	ρ (ohm-m)	d (m)	ρ (ohm-m)	d (m)	ρ (ohm-m)	d (m)	ρ (ohm-m)	d (m)
1	49.5	3.0	27.0	0.1	42.8	0.8	58.9	0.7
2	206.0	5.9	130.0	6.9	65.0	6.5	176.0	1.4
3	13.6	11.1	7.64	11.4	7.93	9.1	39.7	8.3
4	960	28.5	342.0	36.0	23.2	46.1	18.3	39.1
5	50.6	-	48.5	-	328.0	-	204.0	-
rms%	4.92		3.71		4.53		2.80	

Table 2. Values of the one dimensional models of section BB' (ρ = electrical resistivity of layer d = depth to the top of layer).

Layer No	SECTION BB' – ONE DIMENSIONAL MODELS							
	0202		0201		0203		0204	
	ρ (ohm-m)	d (m)	ρ (ohm-m)	d (m)	ρ (ohm-m)	d (m)	ρ (ohm-m)	d (m)
1	63.0	1.0	56.8	0.8	224.0	1.4	389.0	0.7
2	292.0	3.0	330.0	3.4	159.0	2.5	246.0	2.7
3	13.5	6.2	9.02	5.5	147.0	4.1	24.4	4.1
4	44.9	44.9	30.6	79.5	40.4	100.0	38.6	83.1
5	2.7	-	2.9	-	3.8	-	3.7	-
rms%	2.36		2.59		3.34		4.18	

seems to extent at greater depths. A similar lateral electrical discontinuity is encountered below section BB' at an average depth of 70 m. A second lateral electrical discontinuity is observed below the locations of soundings 0203 and 0204 at an average depth of 90 m.

6. DISCUSSION

The derived electrical models are compatible with the local geology (Vougioukalakis, 1993, 2002). In particular, the observed electrical resistivities (2-400 ohm-m) are similar to those predicted (McNeil, 1980) for the volcanic origin sediments (clays, sands and pebbles) and rocks, which are expected to the maximum depth (130 m) of investigation. Although the 1D sections AA' and BB' include geo-electric layers of similar resistivity, the layers are not in the same order with respect to depth. This observation leads to the conclusion that the subsurface structure below the region of the superficial fracture consists of the same geological layers but with different layering order. Since the predicted resistivity values for clays (1-100 ohm-m) and sands (10-800 ohm-m) overlap, it is rather difficult to suggest a direct relation of volcanic sediments and rocks with specific geo-electric layers.

The observed lateral geo-electric discontinuities probably correspond to faults,

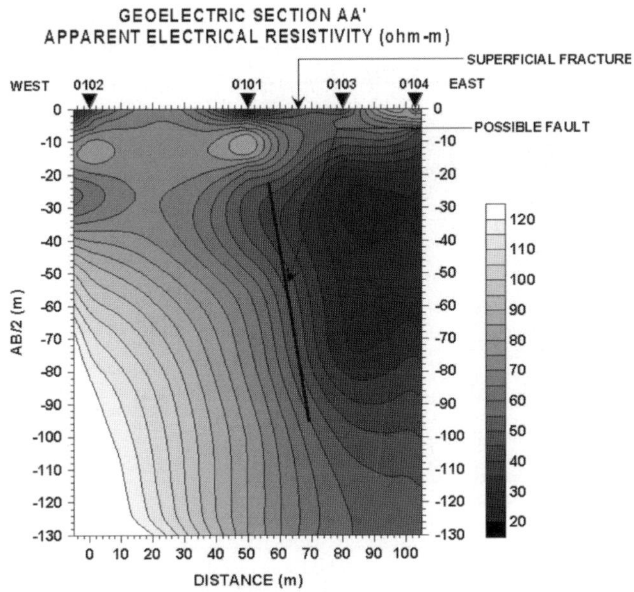

(a)

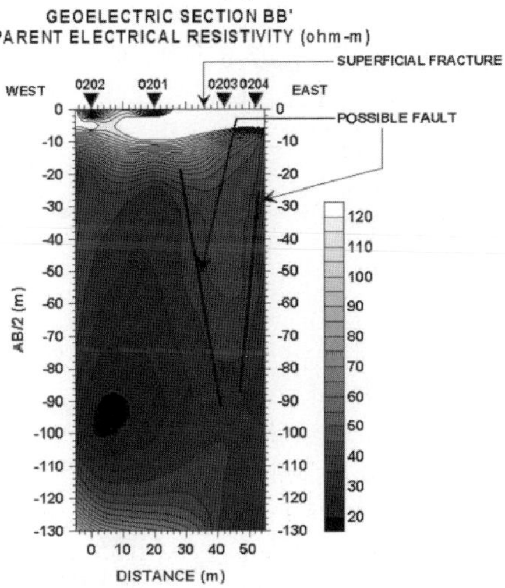

(b)

Fig. 7. (a) Apparent electrical resistivity variation with AB/2 below profile AA' (b) Apparent electrical resistivity variation with AB/2 below profile BB'.

which do not exhibit any superficial manifestations. These faults seem to correlate with the locations of the fracture and maybe extension to the north of the NE-SW fault zone, observed south of the investigated area (Fig. 1b). In the case of geo-electric section BB',

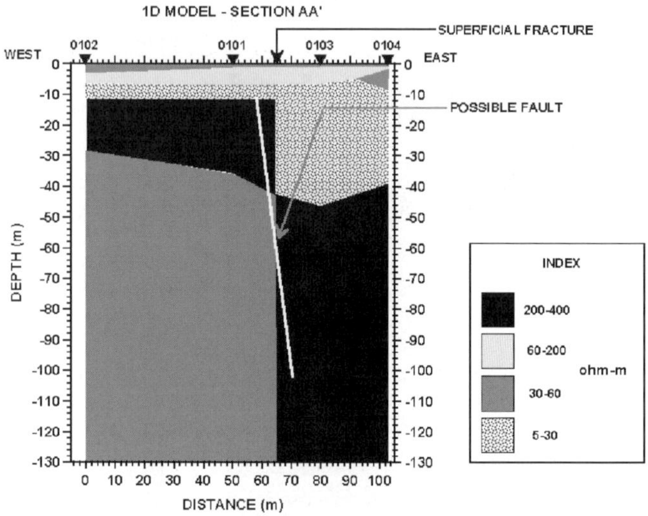

(a)

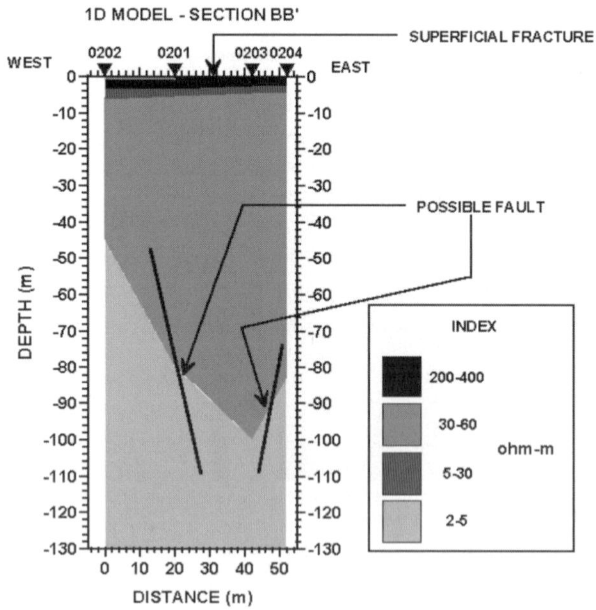

(b)

Fig. 8. (a) One dimensional model section below profile AA' (b) One dimensional model section below profile BB'.

the lateral discontinuities below the eastern part of the section give the impression of a small tectonic depression. Following the above analysis, it seems possible that the subsidence of the fine-grained sediments of the upper geological formation into existing fractures and voids of the lower geological formation of the caldera fill is due to recent seismic activity (Vougioukalakis, 2002) connected to the forth mentioned fault zones.

Since the undertaken research work was a preliminary study of the formation of the 600 m long superficial fracture, a combination of further and more detailed geological, geophysical and geo-chemical investigations is needed to provide a more thorough study of the phenomenon.

Acknowledgements

The authors would like to express their sincere thanks to Mr I. Georgakopoulos, General Secretary for Civil Protection and the Ministry of Interior, Public Administration and Decentralization of Greece for undertaking this project. The authors are thankful to Major P. Katsimatidis and the Municipality of Nisyros for providing facilities during fieldwork. Dr D. Galanopoulos would like to thank Dr G. Vougioukalakis for his friendship, valuable help and many fruitful discussions during and after fieldwork.

REFERENCES

Angelier, J., Lyberis, N., le Pichon, X., Barrier, E. and Huchon, P., 1982. The tectonic development of the Hellenic Arc and the Sea of Crete: a synthesis. Tectonophysics, 86: 159-196.

Bobachev, A.A., Modin, I.N., Pervago, E.V. and Shevnin, V.A., 1999. IPI-1D: Minimal (Free) programs for 1D VES data interpretation. www.geol.msu.ru/deps/geophys/rec_labe.htm

Di Paola, G.M., 1974. Volcanology and petrology of Nisyros Island (Dodecanese, Greece) Bull.Volcanol., 38: 944-987.

Galanopoulos, D., 1993. Preliminary magnetotelluric studies along the Hellenic Volcanic Arc: implications for the collision between the African plate and the 'Aegea" microplate. Phys. Earth Planet Inter., 81: 139-153.

Ganas, A., Lagios, E., Dietrich, V.J., Vassilopoulou, S., Hurni, L. and Stavrakakis, G., 2002. Interferometric mapping of ground deformation in Nisyros volcano, Aegean Sea, during 1995-1996. Proc.6th Pan-Hellenic Geographical Congress. Thessaloniki, Greece, Vol.II, pp.135-141.

Ganas, A. and Lagios, E., 2003. Landsat 7 Night Imaging of the Nisyros volcano, Greece. International Journal of Remote Sensing, 24 (7): 1579-1586.

Fytikas, M., Innocenti, F., Manetti, P., Mazzuoli, R., Peccerillo, A. and Vilari, L., 1984. Tertiary to Quaternary evolution of volcanism in the Aegean region. Geol. Soc. London Spec. Publ., 17: 687-699.

Fytikas, M., Innocenti, F., Kolios, N., Manetti, P. and Mazzuoli, R., 1987. The Plio-Quaternary volcanism of Saronikos area (western part of the Active Aegean

Volcanic Arc), Ann.Geol.de pays Hellen., 23-44.

Lagios, E., Chailas, S., Giannopoulos, J. and Sotiropoulos, P., 1998. Surveillance of Nisyros Volcano: establishment and remeasurement of GPS and radon networks. Bull.Geol.Soc.Greece. 32(4), 215-227 (In Greek with English abstract).

Lagios, E., 2000. Intense crustal deformation rates on Nisyros Island, Greece, deduced from GPS studies, may foreshadow a forthcoming volcanic event. In: Balassanian, S., Cisternas, A., Melkumyan, M. (Editors), Proc. 2nd Int. Conf. On Earthquake Hazard and Seismic Risk Reduction. Kluwer, Dordrecht, pp.249-259.

McKenzie, D.P., 1978. Active tectonics of the Alpine-Himalayan belt: the Aegean Sea and the surrounding regions. Geophys.J.R.Astron.Soc., 55: 217-254.

McNeil, J.D., 1980. Electrical Conductivity of Soils and Rocks. Technical Note TN-5. Geonics Ltd, Canada.

Nathanael, E., 1983. Geoelectric survey on Nisyros Island. Technical Report. Institute of Geology and Mineral Exploration, Greece (In Greek).

Papadopoulos, G.A., Sachpazi, M., Panopoulou, G. and Stavrakakis, G., 1998. The volcanoseismic crisis of 1996-97 in Nisyros, SE Aegean Sea, Greece. Terra Nova, 10: 151-154.

Papanikolaou, D., and Lekkas, E., 1990. Geological structure and evolution of the Nisyros Volcano. Bull.Geol.Soc.Greece. 25(1): 405-419 (In Greek with English abstract).

Papanikolaou, D., and Nomikou, P., 2001. Tectonic structure and volcanic centers at the eastern edge of the Aegean volcanic arc around Nisyros Island. Bull.Geol.Soc.Greece. 34(1): 289-296 (In Greek with English abstract).

Parcharidis, I. and Lagios, E., 2001. Deformation in Nisyros Volcano (Greece) using Differential Radar Interferometry. Bull.Geol.Soc.Greece. 34(4), 1587-1594 (In Greek with English abstract).

Telford, W.M., Geldart, L.P., Sheriff, R.E. and Keys, D.A, 1981. Applied Geophysics. Cambridge University Press.

Vougioukalakis, G., 1993. Volcanic stratigraphy and evolution of Nisyros Island. Bull.Geol.Soc.Greece. 28(2): 239-258 (In Greek with English abstract).

Vougioukalakis, G., 1998. Blue Volcanoes: Nisyros. Published by Nisyros Regional Council, Nisyros, Greece.

Vougioukalakis, G., 2002. The fracture within the Nisyros caldera and the state of the volcano during December 2001. Technical Report. Institute of Geology and Mineral Exploration, Greece (In Greek).

Vougioukalakis, G., 2003. The extension of fracture within the Nisyros caldera and the state of the volcano during January 2003. Technical Report. Institute of Geology and Mineral Exploration, Greece (In Greek).

Voutetakis, S., 1974. Geoelectric research on Nisyros: contribution to the solution of geothermal problems. Technical Report. Institute of Geology and Mineral Exploration, Greece (In Greek).

Gravity monitoring of Nisyros volcano activity: 2001-2003 preliminary results

M. Di Filippo and B. Toro

Dipartimento di Scienze della Terra, Università di Roma "La Sapienza". Italia

ABSTRACT

Topographic monitoring of gravity benchmarks, based on a differential-mode, dual-frequency GPS, started in 2002 on Nisyros volcano.

A monitoring network was installed in June 2001. The sites were selected in parts of the island experiencing different seismic-tectonic processes, in order to detect their relative movements.

In June 2002, monitoring with the microgravity network was repeated for the first time. The comparison of the resulting values with those recorded in 2001 showed a maximum variation of +0.034 mGal in the active volcanic area, with an increasing trend towards South. In 2003, the monitoring was repeated. The resulting G variations were comparable to those observed in 2001-2002. However, with the installation of a new station in 2002, the G variations proved to be confined to the most active volcanic areas.

The extent of the variations recorded between 2001 and 2003 largely exceeded the measuring errors, suggesting that such variations are to be ascribed to mass variations of the volcanic complex.

Keywords: Net gravity monitoring, Nisyros, Aegean volcanic arc, eruption forecasting

1. LOCATION

The island of Nisyros, together with the islets of Strongili, Pahia, Pyrgousa and Yali, is located in the eastern edge of the S.A.A.V.A.

The island is a volcanic massif resting on a calcareous basement (Martelli, 1917; Di Paola, 1975; Vougioukalakis, 1998). The emergence of the island is more recent, as it started about 160,000 years ago, as a result of a number of volcanic events ending approximately 15000 years ago. The island has a persistent, albeit mild and latent, volcanic activity with historic hydrothermal eruptions (last eruptive stages: 1871-1873 and 1887).

Between 1996 and 1999, Nisyros was affected by a strong seismic activity (Papadopoulos et al., 1998).

Based on the findings from the deformation monitoring network, which was put in place by the Geophysical Laboratory of the National and Kapodistrian University of Athens as a part of a research programme of the European Union (GEOWARN), Nisyros has been experiencing soil deformations in the past few years.

2. GRAVITY FIELD WORK AND DATA REDUCTION

In June 2001, a microgravity network was installed on the Nisyros island, with a view to monitoring gravity variations and report changes, if any, in elevation and/or density induced by geodynamic activity.

Topographic monitoring of gravity benchmarks based on a differential-mode, dual-frequency GPS has begun in 2002.

The sites were selected in parts of the island experiencing different seismic-tectonic processes, in order to detect their relative movements. 12 gravity benchmarks (featured by stability over time and ease of access) were put in place. For each measuring station, a monograph with its exact location of the station and photographs was prepared.

Gravity measures were taken with a LaCoste & Romberg mod. D no. 60 gravity meter.

Each station was connected at least twice with the adjacent stations. Additional back and forth links were established (see diagram in Fig. 1).

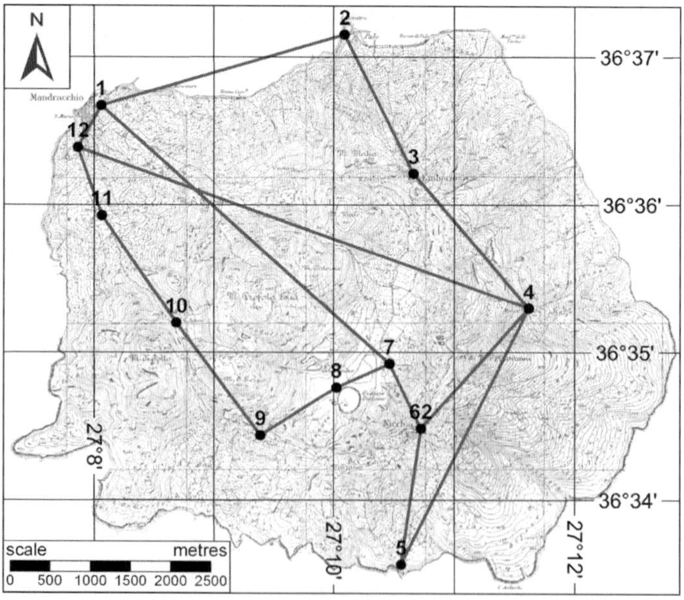

Fig. 1. Sites of the stations and microgravity network.

The daily variations were computed with ETC software, using a 1.17 amplitude amplification factor. The time needed to complete each loop never exceed 90 minutes. Instrumental drift was distributed over the measurements involved as a linear function of time; on average, it was roughly equal to 0.020 mGal.

The network so obtained was offset with standard compensation methods. The following statistical parameter was selected as measurement accuracy indicator:

$$\overset{N}{\underset{1}{\eta}} = \left(\Sigma_i \left|d_i\right|\right) / N \qquad (1)$$

where N is the number of stations and $|d_i|$ denotes the maximum deviation between the values observed at the i-th station. In our study the computation yielded 0.012 mGal.

In 2001, the microgravity network was connected to a station located near the castle of the Kos harbour, in order to determine the gravity variations of the entire island of Nisyros with respect to Kos. The G delta between Kos and station no. 1 of Nisyros was equal to $+1.136 \pm 0.017$ mGal.

In June 2002, microgravity measurements were replicated for the first time. The resulting values were compared to previous values, assuming station no. 1 (built-up area of Nisyros) as a reference point.

To improve the understanding of gravity variations and limit the areas showing evident gravity variations, a new gravity monitoring station was installed near the Ag. Panteleimon church of Avlaki (gravity station no. 5).

In 2003, in the same seasonal period as in previous years, microgravity network monitoring and GPS-based topographic surveying were repeated.

3. RESULTS

The comparison of the 2001-2002 data indicated a maximum variation of +0.034 mGal (Fig. 2, showing the gravity variation lines with a clear upward trend towards S).

The comparison of the 2002-2003 data always showed a positive variation of gravity (+0.032 mGal), of the same extent as in the period 2001-2002; such variation proved to be limited to the area with more evident hydrothermal activity (Stefanos crater, Laki area, Fig. 3).

The overall gravity variation recorded between 2001 and 2003 in the Stefanos crater area was equal to +0.060 mGal, Fig. 4; the extent of the variations between 2001 and 2003 largely exceeded the measuring errors. From 2002 to 2003, no changes were recorded in the elevation of the network stations.

These findings suggest that the observed gravity variations are to be attributed to mass variations of the volcanic complex.

The application of the Gauss theorem (flow theorem) to the gravity variations enabled to determine the extent of the mass variations responsible for the gravity increase; such variations, which amount to about 6500 tons, are likely to be related to the inflow of magma into the subsoil of the area with the gravity increase.

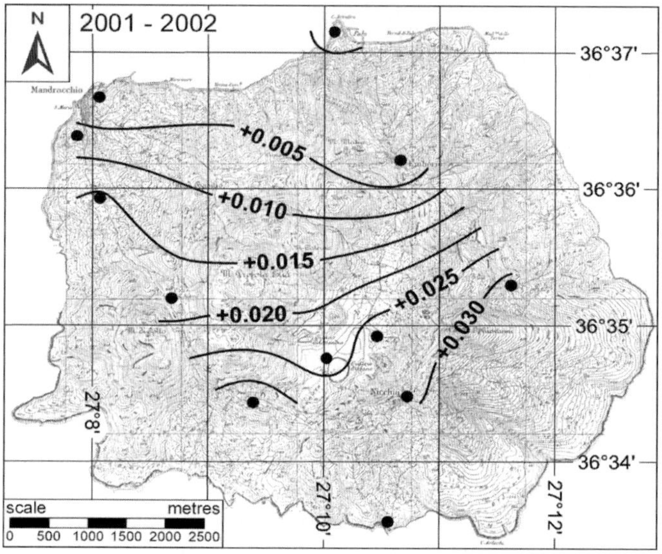

Fig. 2. Contour gravity map of variations between 2001 and 2002 (the contour interval is 0.005 mGal).

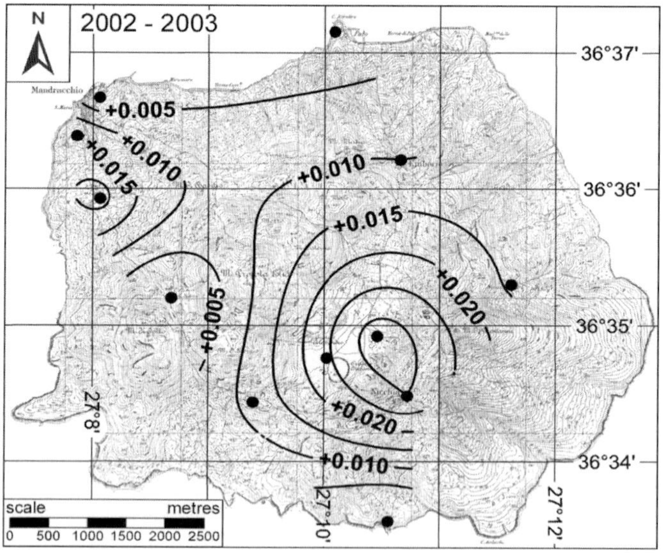

Fig. 3. Contour gravity map of variations between 2002 and 2003 (the contour interval is 0.005 mGal).

This finding infers magma movements in the part of the island with evidence of more recent volcanic activity. These phenomena are to be regarded as possible precursors of renewed volcanic activity.

In 2004, microgravity network monitoring is planned to be repeated at Nisyros and to be extended to areas not yet covered by it.

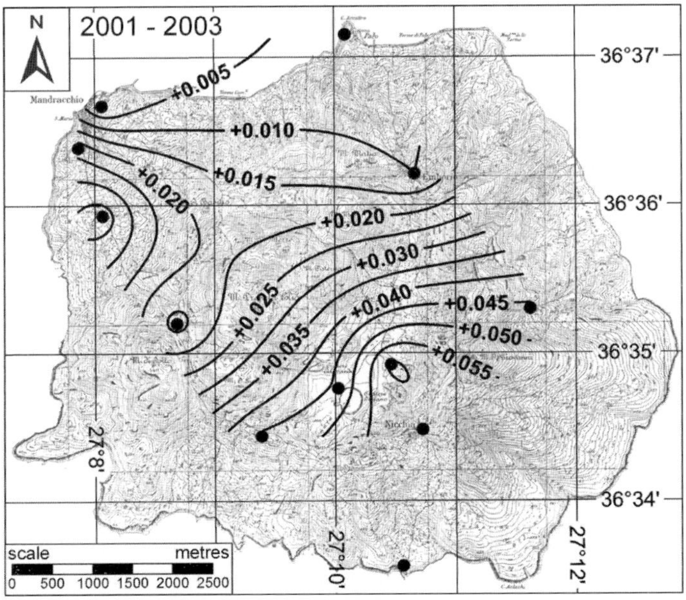

Fig. 4. Total gravity changes for 2001 – 2003 (contour interval 0.005 mGal).

REFERENCES

Papadopoulos, G., Sachpazi, M., Panopoulou, G. and Stavrakakis, G., 1998. The volcanoseismic crisis of 1996-1997 in Nisyros, SE Aegean Sea, Greece. In: 23rd general assembly of the European Geophysical Society; Part 1, Society symposia, Solid Earth geophysics and geodesy. Annales Geophysicae, 16, Suppl. 1; pp. 195. 1998.

Di Paola, G.M., 1975. Volcanology and petrology of Nisyros Island (Dodecanese, Greece). Bulletin Volcanologique, 38 (4): pp. 944-987. Heidelberg.

Martelli, A., 1917. Il Gruppo eruttivo di Nisiro nel mare Egeo. Mem. Soc.ital. Sc. Detto di XL Serie 3a, T.XX.

Vougioukalakis, G.E., 1998. Blue vulcanoes: Nisyros. Ed. Nisyros Regional Council, Greece.

Real time monitoring of gas-geochemical parameters in Nisyros fumaroles

M. Teschner[1,*], G.E. Vougioukalakis[2], E. Faber[1], J. Poggenburg[1] and G. Hatziyannis[2]

[1] Federal Institute for Geosciences and Natural Resources (BGR), Hannover, Germany.

[2] Institute for Geology and Mineral Exploration (IGME), Athens, Greece

ABSTRACT

In this paper the installation and operation of a system for continuous monitoring of fumarolic gases are described. Several physicochemical and gas parameters such as the concentration of CO_2, H_2S and Rn in the fumarolic emissions, as well as the temperature of the fumarolic gas and its pressure are measured in intervals of seconds and transferred to a remote station by digital telemetry. Variations in the monitored parameters which were observed during a short flurry of seismic activity in close vicinity to Nisyros are also reported.

Keywords: Nisyros, fumaroles, volcanic gas composition, real-time monitoring, geochemistry

1. INTRODUCTION

Many of the phenomena observed by effusive, explosive and eruptive behaviour of volcanoes are initiated, influenced and controlled by different complex processes. They are related to the transport of fluids in the magmatic and in the near-surface hydrothermal systems. These processes generate a broad variety of chemical and physical signals on different time scales which may be used as input for monitoring and quantifying changes in the volcano's activity or for modelling the dynamic processes which produce them (Martinelli, 1997).

Whereas geophysical methods are widely introduced as surveillance tools, geochemical monitoring is used only infrequently and lacks some general acceptance by the community of volcanologists. In the past sampling and analysing of volcanic gases from fumaroles have been performed mostly discontinuously with time intervals of weeks or even months between measurements. Any short term variation in geochemical

[*] Corresponding author: email: manfred.teschner@bgr.de

parameters will be missed. Undoubtedly these sampling frequencies are too low to allow efficiently comparing gas data and e.g. seismic information. Therefore, continuous monitoring systems, when successfully developed and correctly applied, will improve the understanding of processes in volcanic systems.

Optical techniques for gas analysis at remote locations have been introduced since long. Using a correlation spectrometer (COSPEC) technique SO_2 flux has routinely been measured at various volcanoes (e.g. at Kilauea (Hawaii) since 1979 (Sutton et al., 2001), at Mt. Etna (Italy) since 1987 (Caltabiano et al., 1994) or at Soufrière Hills (Montserrat) since 1995 (Young et al., 1998)). The application of various optical techniques to monitor fumarolic gases has recently been reviewed by De Natale et al. (2001). However, with these techniques SO_2 concentration can only be measured within plumes. Thus, fumaroles are difficult to monitor individually by optical techniques.

Few instrumental systems for continuous gas monitoring have been discussed in the literature. Investigation of gases from a well located at the foot of the active cone on Vulcano Island, Italy, have been presented by Toutain et al. (1992) with data for CO_2, He and ^{222}Rn. Japanese scientists report on a monitoring system for volcanic gases extracted from an observation well in the vicinity of Izu-Oshima volcano (Shimoike and Notsu, 2000). They also review other papers on gas-monitoring systems.

Because of the presence of hot water vapour and the variable contents of corrosive components like CO_2, SO_2, H_2S or HCl in the volcanic fluids, analytical equipment may be damaged in short time. Only limited technical information is available in the literature and from manufacturers on suitable monitoring equipment which can be directly installed to active fumaroles. Zimmer and Erzinger (1998, 2003) and Zimmer et al. (2000) applied a gas chromatographic system which has been operated continuously at the summit of Merapi volcano. A system which analysed fumarolic gases pumped through a pipe to a station composed of a gas chromatograph, a mass spectrometer and several other physical instruments was described by Faber et al. (1998).

For direct, on-site monitoring of gases like CO_2, H_2S, Rn and of physical parameters like fumarolic pressure lightweight and corrosion-resistant instruments with low power consumption have not been available. Here we present information on a system which was briefly described by Faber et al. (2000) and – after installation and operation for a long period at Galeras volcano, Colombia – in more detail by Faber et al. (2003). The basic components of this monitoring system have been developed in BGR laboratories, commercialization is not excluded.

2. GEOLOGICAL SETTING AND LOCATION OF FUMAROLES

Nisyros volcanic island, built up during the last 150 ka, lies at the eastern end of the south Aegean active volcanic arc (Francalanci et al., 1995). The last magmatic activity of Nisyros is of unknown age (>15ka). However, hydrothermal eruptions were frequent in historical times. They affected the southern part of the Lakki plain, presently the site of widespread fumarolic activity. The last hydrothermal eruptions were recorded for the

Fig. 1. View of recent hydrothermal craters and place of installation of monitoring station.

19th century and they formed the craters of Polyvotis and Phlegethon (1871–1873) and Polyvotis Mikros (1887) (Fig. 1).

It seems likely that seismic shocks played a fundamental role in triggering at least the last hydrothermal eruptions in the 19th century. Again a strong seismic crisis was observed on Nisyros island in 1996 –1997 and, fortunately, this crisis was not followed by any hydrothermal eruption. Chiodini et al. (2002) noted increasing H_2S/CO_2 ratios and decreasing CH_4/CO_2 ratios for several of the active fumaroles and interpreted these chemical changes as an increasing contribution of sulphur-rich, oxidizing magmatic fluids into the hydrothermal system below Nisyros island. Considering the historical information about hydrothermal eruptions, the recent changes in the fumarolic gas composition and the physical phenomena affecting Nisyros may be interpreted as long-term precursors of a new period of volcanic unrest possibly culminating in a magmatic eruptive phase.

3. MONITORING SYSTEM

We decided to connect the gas extraction device directly to a fumarole, as the hot vapour and gases escaping the fumaroles seem to be linked in a short way to regions influenced by the magmatic body of the volcano and/or to its overlaying hydrothermal systems. In the fumarole gases we believe to sense changes in temperature and gas composition at depth more rapidly than by analysing diffusive emanating soil gases.

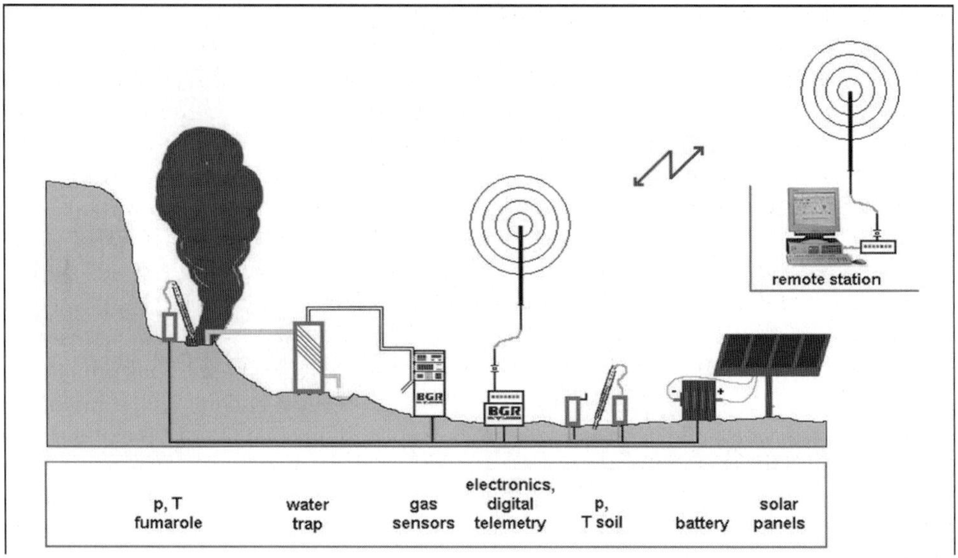

Fig. 2. Sketch of the instrumental set up of the gas monitoring system.

No real time monitoring system for any volcanologically important parameter has been installed up to now on Nisyros island. Our multi-parameter station has been installed close to one of the active fumaroles of the youngest hydrothermal intra-caldera crater (Polyvotis Mikros, 1887; Fig. 1). The installation of the equipment was started during April 2003 and was supplemented during August and November 2003. The operation of the monitoring system is performed in the frame of a BGR-IGME research project, in straight collaboration with the municipality of Nisyros.

The Nisyros monitoring station includes the following main components (Fig. 2):

- physical sensors to measure temperature of fumarolic gases and of surrounding soil
- physical sensors to measure fumarolic and atmospheric pressure
- a system to remove water vapour
- gas-geochemical sensors for the measurement of CO_2, H_2S and $^{220}Rn/^{222}Rn$ (Rn sensor temporarily disconnected). Electrical signals from all gas-geochemical sensors have to be considered as proxies for the variation of concentration over time, exact calibration will be performed later.
- an electronic system including A/D-converters, interconnected by a digital bus
- power supply using solar panels and back-up batteries
- digital telemetry (868 MHz) to a remote station in Emborios (a small village on the caldera rim). From here connection to BGR or IGME offices by standard telephone line or GSM is used.
- software to control all components of the monitoring system and to store measured data (software developed by BGR)

4. DISCUSSION AND CONCLUSIONS

Fluctuations of instrumental records may have various reasons. Instabilities due to technical reasons have to be detected first and afterwards to be corrected which may be a difficult task. But this is a prerequisite to detect and interpret fluctuations which are caused by geological, volcanological, meteorological or other "natural" events. Fig. 3 shows a data set recorded between 4 – 25 August 2003. The temperature of the fumarolic gas shows some pronounced scattering during short time periods, usually with durations of several hours (which are not due to instrumental problems). The bottom graph of Fig. 3 indicates some earthquakes recorded on 5, 10, 11 and 15 August 2003, together with their magnitude and a weighted distance to Nisyros monitoring station, but a correlation to fumarole temperature scattering is not obvious. It may be speculated whether the temperature fluctuation in the morning of the 4. August 2003 is linked to an earthquake which occurred about 63.9 km away from the monitoring site at midnight. A fluctuation with a similar magnitude was found around noon of 8. August 2003, but the next seismic events in close vicinity have been recorded about 2 and 3 days later. There are other, smaller fluctuations in the temperature record where the time off-set to a seismic event is only one day. A strong earthquake with a magnitude over 6 on the Richter scale was recorded on 14. August 2003. The distance to the hypocenter

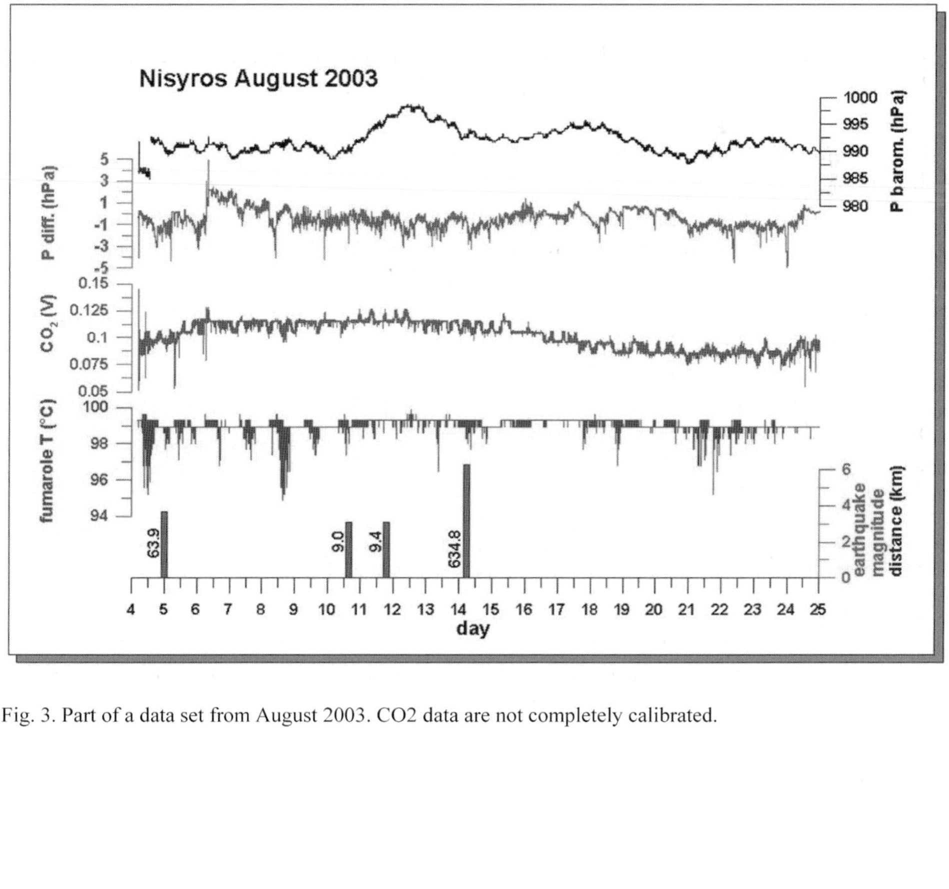

Fig. 3. Part of a data set from August 2003. CO2 data are not completely calibrated.

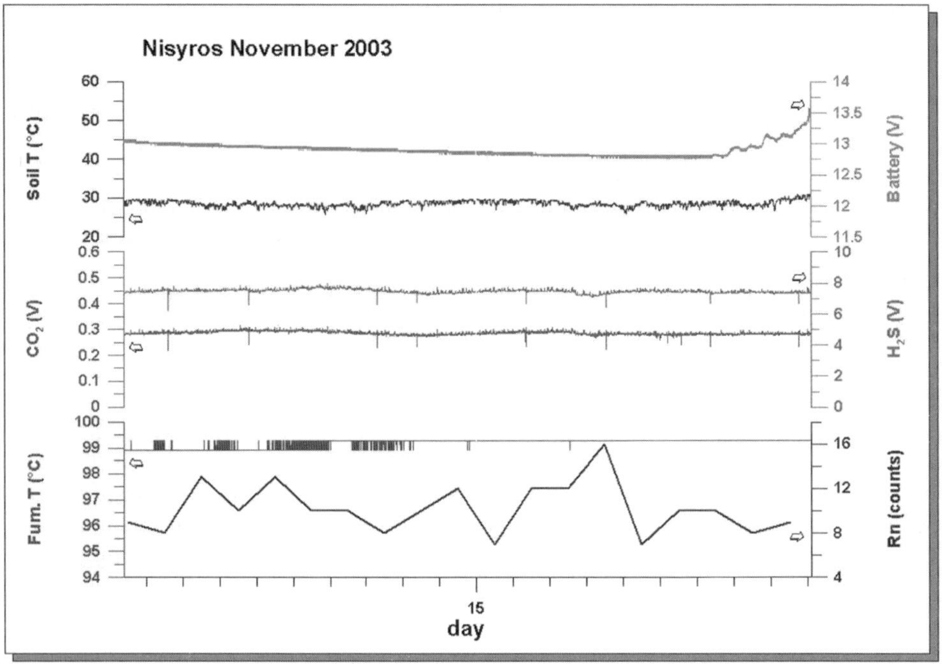

Fig. 4. Part of a data set from November 2003. Data for 222Rn/220Rn measurements show an integration time of 60 minutes. CO2 and H2S data are not yet exactly calibrated.

was over 600 km, so it is not surprising that we do not find a good proxy in the fumarole temperature record.

The pressure of the fumarolic gas is very low and does not exceed a few hecto-Pascal. It seems to fluctuate without a clear correlation to the seismic events in the vicinity of the monitoring station or to changes in the fumarole temperature record. During August 2003 these fluctuations of the fumarole pressure were independent of the atmospheric pressure.

The concentration of CO_2 in the total gas mixture shows a slight modulation over some weeks. Until exact calibration on site the data of this sensor have to be taken as proxy only. The CO_2 concentration in the fumarolic gas mixture seems to be somewhat higher during periods of seismic unrest. However, a direct correlation to an earthquake in the vicinity of the monitoring station was not yet established.

Fig. 4 shows a one day record of CO_2, H_2S, Rn and temperature signals obtained after the installation of a more sensitive CO_2 sensor and additional equipment in mid November. Although the on-site calibration of the total flow system has not yet been completed the concentrations estimated are in good agreement with gas geochemical data for Nisyros fumaroles given by Brombach et al. (2003) (e.g. 7867 ppm CO_2 in the total fumarolic gas mixture for a sample taken during February 2001). The same applies to the H_2S concentration. For the sample mentioned above Brombach et al. (2003) report a value of 1819 ppm H_2S.

Radon and Thoron concentrations are very low in the fumarolic gas of Polyvotis Mikros. These data were recorded with an integration time of one hour (Fig. 4). Fluctuations observed are small and do not correlated with changes in other gas components, temperature or atmospheric pressure. These first results point to the necessity that we have to collect time series for longer periods to judge on a possible correlation of seismic events occurring in the South Aegean region and gas geochemical or physical parameters.

Acknowledgements

We kindly acknowledge the continuous support and encouragement of IGME and BGR staff for this monitoring project. Severe thanks go to the municipality of Nisyros in Mandraki which supported the project by providing manpower during the phase of installation of the monitoring station and by balancing some running expenses. We are thankful that part of the equipment installed as well as travel expenses could be financed by a BGR project with the German Ministry of Economics and Labour, grant number BMWi VI A 2-27/01. Financial support of the IGME team was supplied by the 3rd Framework EU IGME project "Continuous monitoring of the Hellenic geothermal fields".

REFERENCES

Brombach, T., Caliro, St., Chiodini, G., Fiebig, J., Hunziker, J.C. and Raco, B., 2003. Geochemical evidence for mixing of magmatic fluids with seawater, Nisyros hydrothermal system, Greece. Bulletin of Volcanology, 65: 505-516.

Caltabiano, T., Romano, R. and Budetta, G., 1994. SO_2 flux measurements at Mount Etna (Sicily). Journal of Geophysical Research, 99: 12809-12819.

Chiodini, G., Brombach, T., Caliro, St., Cardellini, C., Marini, L. and Dietrich, V., 2002. Geochemical indicators of possible ongoing volcanic unrest at Nisyros Island (Greece). Geophysical Research Letters, 29(16): 1759-1762, doi:10.1029/2001GL014355.

De Natale, P., Gianfrani, L. and De Natale, G., 2001. Optical methods for monitoring of volcanoes: techniques and new perspectives. Journal of Volcanology and Geothermal Research, 109: 235-245.

Faber, E., Inguaggiato, S., Garzón-Valencia, G. and Seidl, D., 1998. Continuous gas measurements at volcanic fumaroles. Mitteilungen Deutsche Geophysikalische Gesellschaft e.V. DGG Special Volume III/1998, ISSN-Nr. 0947-1944: 83-87.

Faber, E., Poggenburg, J., Garzón, G., Morán, C. and Inguaggiato, S., 2000. Gas monitoring at volcanoes. Mitteilungen Deutsche Geophysikalische Gesellschaft e.V., DGG Special Volume IV/2000, ISSN-Nr. 0947-1944: 77-80.

Faber, E., Morán, C., Poggenburg, J., Garzón, G. and Teschner, M., 2003. Continuous gas monitoring at Galeras volcano, Colombia; first evidence. Journal of Volcanology and Geothermal Research, 125(1-2): 13-23.

Francalanci, L., Varecamp, J.C., Vougioukalakis, G., Defant, M.J., Innocenti, F. and Manetti, P., 1995. Crystal retention, fractionation and crustal assimilation in a convecting magma chamber, Nisyros Volcano, Greece. Bulletin of Volcanology, 56: 601-620

Martinelli, B., 1997. Volcanic tremor and short-term prediction of eruptions. Journal of Volcanology and Geothermal Research, 77: 159-171.

Shimoike, Y. and Notsu, K., 2000. Continuous chemical monitoring of volcanic gases in Izu-Oshima volcano, Japan. Journal of Volcanology and Geothermal Research, 101: 211-221.

Sutton, A.J., Elias, T., Gerlach, T.M., Stokes, J.B., 2001. Implications for eruptive processes as indicated by sulfur dioxide emissions from Kilauea Volcano, Hawaii, 1979-1997. Journal of Volcanology and Geothermal Research, 108: 283-302.

Toutain, J.P., Baubron, J.-C., Le Bronec, J., Allard, P., Briole, P., Marty, B., Miele, G., Tedesco, D. and Luongo, G., 1992. Continuous monitoring of distal gas emanations at Vulcano, southern Italy. Bulletin of Volcanology, 54: 147-155.

Young, S., Francis, P.W., Barclay, J., Casadevall, T.J., Gardner, C.A., Darroux, B., Davies, M.A., Delmelle, P., Norton, G.E., Maciejewski, A.J.H., Oppenheimer, C., Stix, J., Watson, I.M., 1998. Monitoring SO_2 emission at the Soufrière Hills volcano: implications from changes in eruptive conditions. Geophysical Research Letters, 25: 3681.

Zimmer, M. and Erzinger, J., 1998. Geochemical monitoring on Merapi volcano, Indonesia. Mitteilungen Deutsche Geophysikalische Gesellschaft e.V., DGG Special Issue III/1998, ISSN-Nr. 0947-1944: 89-92.

Zimmer, M. and Erzinger, J., 2003. Continuous H_2O, CO_2, ^{222}Rn and temperature measurements on Merapi Volcano, Indonesia. Journal of Volcanology and Geothermal Research, 125(1-2): 25-38.

Zimmer, M., Erzinger, J. and Sulistiyo, Y., 2000. Continuous chromatographic gas measurements on Merapi volcano, Indonesia. Mitteilungen Deutsche Geophysikalische Gesellschaft e.V., DGG Special Volume IV/2000, ISSN-Nr. 0947-1944: 87-91.

The Vani manganese deposit, Milos island, Greece: A fossil stratabound Mn-Ba-Pb-Zn-As-Sb-W-rich hydrothermal deposit

G.P. Glasby[1], C.T. Papavassiliou[1,*], J. Mitsis[1], E. Valsami-Jones[2], A. Liakopoulos[3] and R.M. Renner[4]

[1] Department of Economic Geology and Geochemistry, University of Athens, Panepistimioupoli, Zografou, Athens 157 84, Greece.

[2] Natural History Museum, Cromwell Rd, London SW7 5BD, U.K.

[3] Institute of Geology and Mineral Exploration (I.G.M.E.), 70, Messoghion St., 115 27 Athens, Greece.

[4] School of Mathematical and Computing Sciences, Victoria University of Wellington, P.O. Box 600, Wellington, New Zealand.

The Vani manganese deposit is a fossil stratabound hydrothermal deposit formed by the penetration of hydrothermal fluids through a lithified pyroclastic tuff. Two types of deposit have been recognized: "high-temperature" hydrothermal Mn deposits formed initially when the hydrothermal fluids penetrated faults and fissures within the volcaniclastic sandstone and bedded hydrothermal Mn deposits formed subsequently as the cooling hydrothermal fluids migrated along the bedding planes of the volcaniclastic sandstone. Both are late-stage, low-temperature deposits. Mineralogical analysis showed that the principal manganese minerals present are (in decreasing order of abundance) cryptomelane, pyrolusite, hollandite, ramsdellite, coronadite and romancheite with jacobsite, franklinite and hydrohetaerolite present in minor amounts. On average, the "high-temperature" hydrothermal Mn deposits appear to be marginally enriched in pyrolusite, ramsdellite and perhaps coronadite and jacobsite and depleted in haematite compared to the bedded hydrothermal Mn deposits but these variations are not statistically significant. Variations in the abundances of minerals between individual samples are much greater with pyrolusite, cryptomelane and hollandite varying between low and very abundant and ramsdellite, coronadite, romanechite and barite between

* Corresponding author: e-mail: papavas@geol.uoa.gr

absent and very abundant. However, no systematic patterns in the relative abundances of the various minerals could be observed. The compositional data also showed wide variations in element concentrations between samples. On average, the "high-temperature" deposits are significantly enriched in Mn and the bedded deposits in Na, K, Mg, Ca, Al, Ti, Fe, Zn, Zr, Nb, Ce, Hf and Th. This reflects the fact that the "high-temperature" deposits formed first when the Mn concentration in the hydrothermal fluids was higher. The bedded deposits formed subsequently and are characterized by higher concentrations of lithogenous elements derived from the associated volcaniclastic sandstone. However, no well-defined patterns of association between the ore-forming elements could be observed in the samples. Nonetheless, these data demonstrate that the Vani manganese deposit is a Mn-Ba-Pb-Zn-As-Sb-W-rich hydrothermal deposit which is similar in mineralogy and composition to the epithermal vein deposits of the southwestern United States. Based on a comparison with the JADE submarine hydrothermal field in the Okinawa Trough, it is suggested that Pb, Zn As and Sb may have been leached as chloro complexes from felsic rocks of the Aegean intracontinental Arc by deeply penetrating chloride-rich hydrothermal fluids during the formation of the Vani manganese deposit, although a magmatic contribution is possible. The high positive Eu anomalies in the deposit confirm that leaching of the divalent Eu^{2+} from the host rocks took place at temperatures greater than 250°C during this time.

Keywords: manganese deposit, hydrothermal, stratabound, Cape Vani, Milos island.

1. INTRODUCTION

The Vani manganese deposit is situated at Cape Vani in the rugged NW sector of island of Milos in the Aegean volcanic arc (Fig. 1). The deposit lies proximal to the sea and occurs between 35 m above sea level and an unknown depth below sea level. Individual beds of the deposit may be in excess of 5 m thick (Liakopoulos et al., 2001). The deposit is considered to be a fossil stratabound hydrothermal deposit formed within volcaniclastic sandstone which was mined between 1886 and 1909 and again between 1916 and 1928. Reserves of the mine were estimated to be about 2.1 million tonnes of manganese ore (Glasby et al., 2001). Two types of ore can be recognized:

- hard, dense steel grey, lustrous manganese ores formed by the migration of hot hydrothermal fluids along faults, fractures and fissures within volcaniclastic sandstone; these deposits can take the form of hydrothermal vent tubes, fissure vein fillings, massive botryoidal structures or manganese dendrites (Fig. 3a, b, c, d, e).
- black, friable, bedded manganese ores with high moisture content formed by the penetration of cooling hydrothermal fluids along the bedding planes of layers of porous volcaniclastic sandstone (Fig. 3f, g, h).

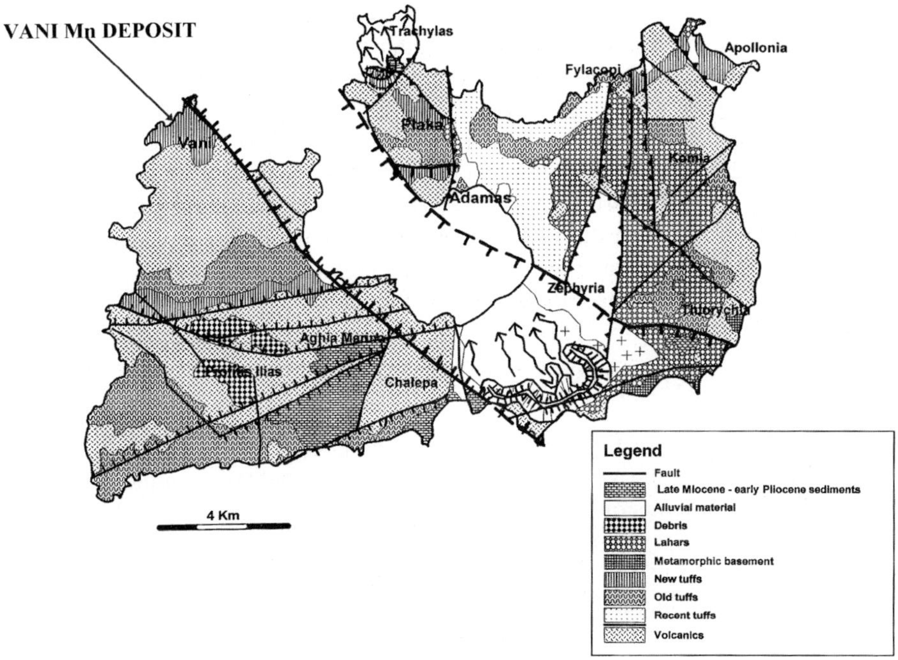

Fig. 1. Simplified geological map of Milos showing the location of the Vani area (Papanikolaou et al., 1993).

According to Glasby et al. (2001), the original tuffaceous material was deposited in a shallow submarine basin at Cape Vani in the Upper Pliocene (3.5-2.0 Ma). The tuff was subsequently compacted and lithified to produce a volcaniclastic sandstone in which vertical joints (fissures) developed in its structure. Hydrothermal fluids subsequently penetrated these fissures and migrated along the bedding plane of the sandstone. Hydrothermal manganese mineralization was therefore controlled by the distribution of these fissures and the porosity of the sandstone. Initial deposition of the hydrothermal manganese oxide minerals within the open fissure systems is thought to have occurred at higher temperatures than the subsequent deposition of the minerals within the bedded sandstone. These two types of deposit display characteristic features and have been named "high-temperature" hydrothermal Mn deposits and bedded hydrothermal Mn deposits, respectively. Differences in the mode of formation of these two types of deposit would account for the variability in their texture and morphology. Strong tectonic activity resulted in rapid uplift in the area which elevated the deposit above sea level.

Relatively few scientific publications have dealt with the mineralogy and composition of the Vani manganese deposit (e.g. Liakopoulos, 1987; Hein et al., 2000). We have now carried out a detailed sampling and analytical programme in order to elucidate the origin of this deposit. This investigation therefore builds on our previous studies on the history of mining (Glasby et al., 2001) and the nature and origin of the Vani deposit (Liakopoulos et al., 2001). The latter publication gives detailed overview of the Vani manganese deposit based largely on data presented in the thesis of Liakopoulos (1987).

The principal aim of this study has been to determine the mineralogical and compositional characteristics of the Vani manganese deposit and to discriminate the "high-temperature" hydrothermal Mn deposits from the bedded hydrothermal Mn deposits in order to establish the main controls on the mineralogy and composition of this deposit. In addition, it was hoped to establish the closest analogues of this deposit in order to gain further insights into its mode of formation.

2. SAMPLE DESCRIPTION

We undertook 5 days of field work and sampling at Cape Vani in March and April, 2000 in order to sample the Vani manganese deposit in detail. We sampled six vertical sections of the "high-temperature" hydrothermal Mn deposits (ϕ4, ϕ9, ϕ10, ϕ11, ϕ12 and ϕ13 and dump sample; 49 samples) and three vertical sections of the bedded hydrothermal Mn deposits (ϕ5, ϕ7 and ϕ8; 26 samples). Faulting in the area has post-dated the manganese mineralization and the lithological succession has been separated into tectonic blocks. Since there are no marker horizons, it was impossible to correlate these sections stratigraphically. In virtually all cases, the maganese deposits occurred within a matrix of volcanoclastic sandstone in which barite-rich veins and horizons were relatively common. Contamination of the manganese-rich samples by variable amounts of volcaniclastic sandstone and barite was therefore inevitable. In this section, we summarize the nature of the samples in each section for both the "high-temperature" hydrothermal Mn deposits and the bedded hydrothermal Mn deposits. The locations of these sections are given in Fig. 2. Because the nature of the Vani manganese deposit is very variable, we present a detailed description of the samples from each section in the APPENDIX. Photographs illustrating some of the typical characteristics of the "high-temperature" and bedded deposits are shown in Fig. 3.

3. SAMPLING

In all, 75 samples were collected in the course of this study, 49 from the "high-temperature" deposits and 26 from the bedded deposits. Because of the variable nature of the deposits, particularly of the "high-temperature" deposits, relatively small samples of material were collected for curation and analysis (100-200g). Of this, 5-10 g of material was ground to a fine powder in an agate mortar and pestle for mineralogical and chemical analysis in accordance with the sampling strategy of Usui and Mita (1995) for a hot spring manganese deposit from Hokkaido, Japan. However, it should be pointed out that Roser (1983), in his study of manganese mineralization in New Zealand, ground 200-500 g of sample in a tungsten carbide 'Tema' swing mill for 45

Fig. 2. Schematic map of the Vani manganese deposit showing the locations of the six vertical sections of the "high-temperature" hydrothermal Mn deposits (f4, f9, f10, f11, f12 and f13 and dump sample) and three vertical sections of the bedded hydrothermal Mn deposits (f5, f7 and f8).

secs and then homogenized the mixtures by means of a preliminary mixing in an agate mortar followed by 2-3 secs of swing milling.

4. ANALYTICAL METHODS

X-ray diffraction measurements were carried out on a Siemens Model 5005 X-ray diffractometer in combination with the DIFFRAC plus software package. The diffractometer was operated using Cu Kα radiation at 40 kV and 40 mA and employing

Fig. 3. Photographs showing some typical characteristics of (a) "high-temperature" and (b) bedded hydrothermal manganese deposits from Vani.

the following scanning parameters: 0.020° step size and 1.0 sec. step time. The raw files were evaluated by use of the EVA 2.2 program of the Siemens DIFFRACplus-D5005 software package. Abundances of the minerals present in each sample were estimated from the intensity of the X-ray diffraction peaks and are ascribed values in the range absent to very abundant for each mineral. However, these estimates are only qualitative and cannot be related directly to the percentage of each mineral in the sample.

Samples were digested for chemical analysis by placing 200 mg of finely-powdered sample in a high-pressure TFM Teflon vessel in an ETHOS 1600 MILESTONE

Microwave System and digesting it with a mixture of 3 ml HNO3 (65%), 2 ml HF (40%) and 0.5 ml H_2O_2 (35%). After decomposition, the cooled solution was transferred to a PTFE Teflon beaker, evaporated to dryness, redissolved with 6 ml of HNO_3 (65%) and evaporated again. The residues were then taken up with 6 ml of distilled water, 1 ml $HClO_4$ (70%) and 1 ml H_2O_2 (35%), placed in the same high-pressure Teflon vessel in the ETHOS Microwave System and digested again. After digestion and cooling, the solution was diluted to 100 ml with distilled water. Concentrated HNO_3 was added to the solution to bring its final strength to 1% HNO_3. Al, Ca, Mg, K, Na, Fe, Mn, Pb, Zn were analyzed at the University of Athens using a Perkin Elmer 1100b atomic absorption spectrophotometer. The precision and accuracy of these analyses was checked by reanalyzing Mn-ore samples of known compositions using local standards from other Mn-bearing deposits in Greece and was ± 5% for Fe, Mn, Ca, Na, K, Al and ± 10% for Pb and Zn.

The remaining elements were analyzed at the Natural History Museum, using ICP-AES (ARL 3410) and quadrupole ICP-MS (Thermo Elemental PQ3). The detection levels (in solution) for samples analysed by ICP-AES were: K: 0.3 ppm, Li: 0.01 ppm, Na: 0.5 ppm, Ba: 0.003 ppm, Sr: 0.0005 ppm, Al: 0.1 ppm, Ca: 0.0003 ppm, Y: 0.01 ppm, V: 0.05 ppm, Mg: 0.001 ppm, Cr: 0.05 ppm, Fe: 0.02 ppm, Mn: 0.005 ppm, Ni: 0.05 ppm, Cd: 0.02 ppm, Co: 0.04 ppm, Pb: 0.2 ppm, Zn: 0.01 ppm, As: 0.2 ppm, Sb: 0.2 ppm, W: 0.1 ppm and Ag: 0.02 ppm. The detection levels (in solution) for samples analysed by ICP-MS were: Y: 0.015 ppb, Zr: 0.05 ppb, Nb: 0.065ppb, La: 0.014 ppb, Ce: 0.017 ppb, Pr: 0.006 ppb, Nd: 0.04 ppb, Sm: 0.035 ppb, Eu: 0.015 ppb, Gd: 0.035 ppb, Tb: 0.005 ppb, Dy: 0.017 ppb, Ho: 0.005 ppb, Er: 0.011 ppb, Tm: 0.005 ppb, Yb: 0.015 ppb, Lu: 0.005 ppb, Hf: 0.02 ppb, Ta: 0.035 ppb, Th: 0.018 ppb, U: 0.013 ppb. The analytical precision was ± 1% for concentrations above 1 p.p.m. and ± 5% for concentrations below 1 p.p.m.

Because many of the features that we sampled during the field programme were relatively small scale such as hydrothermal vent tubes, botryoidal features and layers of lustrous Mn oxides, we tended to take relatively small samples. This may have contributed to some of the variability in the mineralogy and composition of the deposits recorded here, in particular as a consequence of the heterogeneity of the manganese minerals. Although this downgrades the quality of the data, we believe that the results presented here still present a valuable insight into the nature and mode of formation of this deposit.

5. MINERALOGY

Mineralogical data have been obtained for 75 samples of the Vani manganese deposit representing the two main types of ore by means of bulk X-ray diffraction analysis. The abundances of the various minerals in the samples and their means in each stratigraphic section are presented in Table 1 and are summarized in Table 2.

The data show that the principal manganese minerals present in the Vani manganese deposit are pyrolusite, ramsdellite, cryptomelane, hollandite and coronadite in accordance with the previous findings of Liakopoulos (1987) based on X-ray diffraction analysis, optical mineralogy and electron microproprobe studies. In contrast to the findings of Liakopoulos (1987), however, romanechite was identified as one of the principal manganese minerals whereas hydrohetaerolite was only rarely identified in this study. The presence of romanechite may have been missed in the previous study because it intergrows with hollandite on a very fine scale and is therefore not easy to identify by electron microprobe. Hydrohetaerolite, haematite, jacobsite and franklinite were also identified in a few samples of which jacobsite was most common. Previously, Liakopoulos (1987) had showed that hydrohetaerolite is a rare mineral in the Vani manganese deposit and occurs as an exsolution in some hollandite specimens. Barite was more common being very abundant in some samples. However, we failed to identify δMnO_2 in these samples as previously reported by Hein et al. (2000) even though it was specifically looked for.

Post (1999) has pointed out that pyrolusite and ramsdellite are polymorphs of MnO_2 of which pyrolusite is the more stable and abundant. Pyrolusite commonly occurs as low-temperature hydrothermal deposits. Ramsdellite is a relatively rare mineral, usually occurring as low-temperature hydrothermal deposits and commonly associated with, and probably altering to, pyrolusite. Cryptomelane ($K_{1-2}Mn_8O_{16}.xH_2O$), hollandite ($Ba_{1-2}Mn_8O_{16}.xH_2O$) and coronadite ($Pb_{1-2}Mn_8O_{16}.xH_2O$), on the other hand, are members of the hollandite group enriched in K, Ba and Pb respectively. The hollandite group commonly occurs as fibrous crystals in the oxidized zones of Mn deposits and less commonly as prismatic crystals in hydrothermal vein deposits. Romanechite is a Ba-bearing manganese mineral ($Ba_{0.66}(Mn^{4+},Mn^{3+})_6O_{12}.3.5H_2O$) which commonly intergrows with hollandite on a very fine scale. Hydrohetaerolite is a Zn-bearing manganese mineral ($ZnMn_2O_4.xH_2O$) isomorphous with hausmannite.

Although the data presented here show significant variations in mineral abundances between individual samples with the abundance of pyrolusite, cryptomelane and hollandite varying between low and very abundant and that of ramsdellite, coronadite, romanechite and barite between absent and very abundant, differences in the average abundances of minerals between individual sections appear to be much less. On average, the "high-temperature" hydrothermal Mn deposits appear to be marginally enriched in pyrolusite, ramsdellite and perhaps coronadite and jacobsite and depleted in haematite compared to the bedded hydrothermal Mn deposits as shown in Tables 1 and 2 but these differences are relatively minor considering the wide variability of the mineral abundances and are not statistically significant.

Table 1. Mineralogy of Vani manganese samples determined by X-ray diffraction.

High-temperature hydrothermal Mn deposits	Pyrolusite	Ramsdellite	Cryptomelane	Hollandite	Coronadite	Romanechite	Hydrohetaerolite	Barite	Haematite	Jacobsite	Franklinite
Phi 4											
CV1a	***	-	**	**	**	tr.	-	**	-	-	-
CVIb	****	-	**	**	**	-	-	-	-	-	-
CV2	**	-	****	**	tr.	tr.	-	**	-	-	-
CV3	**	****	***	**	**	-	-	**	-	-	-
CV49	**	tr.	*	***	****	**	-	**	-	**	-
CV50	***	-	***	**	*	*	-	****	-	*	-
CV51	**	-	**	***	*	*	-	-	-	-	-
CV52	***	*	***	***	*	**	-	-	-	*	-
CV53	***	*	**	*	-	*	-	****	-	**	-
CV54	**	*	**	**	-	**	-	-	-	**	-
CV55	**	**	**	***	*	*	-	**	-	**	-
CV56	**	*	***	***	-	*	-	**	-	**	-
CV57	****	*	**	***	-	*	-	**	-	-	-
CV58	***	*	**	**	**	**	-	***	-	-	-
CV59	**	*	***	***	*	**	-	**	-	*	-
CV60	**	tr.	***	**	tr.	*	-	*	-	**	-
CV61	*	***	**	***	*	**	-	**	-	-	-
CV62	****	*	***	***	tr.	tr.	-	*	-	-	-
CV63	***	*	***	***	*	*	-	**	-	-	-
CV64	***	**	****	***	-	**	-	**	-	*	-
CV65	***	tr.	***	***	-	**	-	***	-	*	-
CV66	****	tr.	**	**	-	**	-	-	-	*	-
CV67	***	**	***	**	-	*	-	**	***	-	-
CV68	***	*	***	**	**	-	-	**	*	-	-
CV69	***	*	***	**	***	-	-	***	-	***	-
CV70	***	*	***	***	*	**	-	**	-	**	-
CV71	**	*	***	**	*	-	*	*	*	-	-
Av.	2.7	1.0	2.6	2.4	1.0	1.1	-	1.8	0.2	0.9	-
Phi 9											
CV72	***	tr.	**	***	*	*	-	***	-	**	-
CV73	**	*	**	**	***	tr.	-	**	-	**	-
CV74	**	****	***	*	*	*	-	***	tr.	-	-
Av.	2.3	1.3	2.3	2.0	1.7	0.7	-	2.7	-	1.3	-
Phi 10											
CV90	*	-	***	***	**	**	-	-	-	***	-
CV91	***	*	***	**	**	-	-	tr.	-	-	
CV92	***	**	****	***	*	-	-	-	-	-	-
CV93	**	-' C	**	**	***	tr.	-	-	-	**	-
CV94	**	**	**	**	**	-	-	-	-	**	-
Av.	2.2	1.0	2.8	2.4	2.0	0.4	-	-	-	1.4	-

Table 1. Mineralogy of Vani manganese samples determined by X-ray diffraction (continued).

High-temperature hydrothermal Mn deposits	Pyrolusite	Ramsdellite	Cryptomelane	Hollandite	Coronadite	Romanechite	Hydroheta-erolite	Barite	Haematite	Jacobsite	Franklinite
Phi 11(1)											
CV96	***	-	***	***	tr.	**	-	*	-	*	-
CV97	**	-	***	**	-	*	-	*	-	*	-
CV98	***	tr.	***	**	-	*	-	-	-	*	-
CV99	**	**	***	**	-	**	-	-	-	-	-
Phi 11(2)											
CV100	***	**	****	**	-	*	-	-	-	-	-
CV101	**	**	***	**	-	*	-	*	-	*	-
CV102	***	**	**	***	-	*	-	**	-	-	-
Phi 11 (3)											
CV103	**	*	***	**	-	**	-	-	-	**	-
Phi 11 (4)											
CV104	***	*	***	**	tr.	*	-	-	-	**	-
Phi 11(5)											
CV105	****	****	***	****	tr.	-	-	*	-	-	-
CV106	****	****	**	***	tr.	-		*	-	-	-
Av.	2.8	1.5	2.9	2.5	-	1.1	-	0.6	-	0.7	-
Phi 12											
CV109	tr.	-	***	***	***	-	-	-	-	*	-
Phi 13											
CV110	***	*	****	***	*	-	-	-	*	-	-
Dump sample											
CV111	***	*	****	***	-	-	-	-	-	-	-
Grand Av.	2.5	1.1	2.8	2.4	0.9	0.9	-	1.3	0.1	0.9	-

Bedded Hydrothermal Mn deposits	Pyrolusite	Ramsdellite	Crypto melane	Hollandite	Coronadite	Romanechite	Hydroheta-erolite	Barite	Haematite	Jacobsite	Franklinite
Phi 5											
CV30	**	**	***	***	*	-	-	-	-	**	***
CV31	**	tr.	***	***	*	*	-	****	-	-	-
CV32	*	*	****	***	tr.	-	-	-			
CV34	*	**	***	***	tr.	-	-	***	-	-	-
CV35	**	*	***	***	**	-	-	tr.	-	-	-
Av.	1.6	1.2	3.2	3.0	0.8	0.2	-	1.4	-	0.4	0.6
Phi 8											
CV36	*	-	***	***	**	*	-	*	-	-	-
CV37	**	-	**	***	*	-	-	*	-	-	-
CV38	*	-	***	***	-	**	-	*	-	-	-
CV39	*	-	**	**	-	**	-	*	*	-	-

Table 1. Mineralogy of Vani manganese samples determined by X-ray diffraction (continued).

Bedded Hydrothermal Mn deposits	Pyrolusite	Ramsdellite	Crypto melane	Hollandite	Coronadite	Romanechite	Hydroheta-erolite	Barite	Haematite	Jacobsite	Franklinite
CV40	*	-	***	***	-	tr.	*	**	*	-	-
CV41	**	-	**	***	*	tr.	-	*	***	-	-
CV42	**	*	**	****	-	*	-	**	*	-	-
CV45	*	-	*	**	-	*	-	****	-	-	-
CV46	*	-	****	***	-	*	-	*	-	-	-
CV47	**	-	****	***	tr.	-	-	*	*	-	-
CV48	**	-	***	***	-	*	-	-	-	-	-
Av.	1.5	0.1	2.6	2.9	0.4	0.8	-	1.5	0.6	-	-
Phi 7											
CV75	*	-	*	*	tr.	-	-	*	-	*	-
CV76	*	-	*	**	*	-	*	*	-	**	-
CV77	**	*	***	**	-	-	-	-	-	-	-
CV78	*	*	***	*	-	*	-	-	-	-	-
CV79	**	-	***	**	-	-	-	-	-	-	-
CV80	*	-	**	**	-	-	-	-	-	-	-
CV82	*	-	***	***	*	*	-	*	-	-	-
CV84	*	-	**	***	**	**	-	****	-	-	-
CV87	**	*	****	***	***	-	-	-	-	-	-
CV89	**	*	***	**	*	*	*	*	-	-	-
Av.	1.4	0.4	2.5	2.1	0.8	0.5	-	0.9	-	0.3	-
Grand Av.	1.5	0.4	2.7	2.6	0.6	0.6	-	1.2	0.3	0.2	0.1

Mineral names and formulae used in this table: pyrolusite (MnO_2); ramsdellite (MnO_2); cryptomelane ($K_{1-2}Mn_8O_{16}.xH_2O$); hollandite ($Ba_{1-2}Mn_8O_{16}.xH_2O$); corandite ($Pb_{1-22}Mn_8O_{16}.xH_2O$); romanechite ($Ba_{0.66}(Mn^{4+},Mn^{3+})_6O_{12}.3.5H_2O$); hydrohetaerolite ($ZnMn_2O_4.xH_2O$); barite ($BaSO_4$); haematite ($Fe_2O_3$); jacobsite ($MnO_2.MgO.Fe_2O_4$); franklinite $(Zn,Mn,Fe)(Fe,Mn)_2O_4$
Mineral abundances: xxxx = very abundant; xxx = abundant; xx = medium; x = low; tr. = trace; - = not detected.

Table 2. Average abundance of the various minerals in the "high-temperature" and bedded hydrothermal Mn deposits calculated from the data in Table 1.

	"high-temp/ture" hydrothermal Mn deposits	bedded hydrothermal Mn deposits.
Pyrolusite	Medium	Medium
Ramsdellite	Low	Low
Cryptomelane	Abundant	Abundant
Hollandite	Medium	Medium
Coronadite	Low	Low
Romanechite	Low	Low
Barite	Low	Low
Haematite	Trace	Trace
Jacobsite	Low	Trace
Franklinite	Absent	Trace

Table 3. Major elements including Zn and Pb but excluding Ti analyzed by AAS. Ti, Cu, Sr, Ag, As, Sb and W analyzed by ICP-AES. Y, Zr, Nb, REE, Hf, Ta, Th and U analyzed by ICP-MS. Analyses of major elements (including Zn and Pb) in per cent; analyses of minor elements in p.p.m.

"High-temperature" hydrothermal Mn deposits	Na_2O	K_2O	MgO	CaO	Al_2O_3	TiO_2	MnO	Fe_2O_3	Zn	Pb	Cu	Sr	Ag	As	Sb	W
Phi 4																
CV1a	2.4	6.0	0.5	0.1	7.0	0.1	32.9	1.3	0.16	2.21	525	631	n.d.	797	n.d.	114
CV1b	2.1	6.0	0.5	0.2	6.7	0.1	34.7	1.3	0.16	2.29	490	596	n.d.	683	197	249
CV2	2.6	9.6	0.3	0.1	10.8	0.2	20.3	0.8	0.21	0.65	358	625	n.d.	427	97	382
CV3	0.3	4.5	0.1	0.1	4.6	0.1	47.0	0.3	0.58	1.91	1124	1241	n.d.	2357	178	518
CV49	5.3	6.9	0.4	2.1	7.0	0.1	37.2	1.4	0.20	3.76	811	678	n.d.	969	557	465
CV50	0.4	3.1	0.1	0.2	2.3	0.1	39.0	0.7	0.26	0.84	1177	1697	n.d.	1164	795	530
CV51	1.3	5.8	0.5	0.2	6.6	0.1	36.8	1.1	0.26	0.65	666	1015	n.d.	787	413	313
CV52	1.8	4.8	0.4	0.4	5.0	0.1	41.8	0.7	0.23	0.64	809	1013	n.d.	1122	564	760
CV53	1.9	5.1	0.3	0.1	4.9	0.1	20.3	0.4	0.17	0.14	337	1561	n.d.	368	102	249
CV54	1.6	6.4	0.2	0.2	7.5	0.1	46.0	1.1	0.25	0.09	1184	596	n.d.	1297	127	331
CV55	1.2	5.4	0.4	0.1	5.3	0.1	38.9	1.0	0.20	0.94	980	711	n.d.	1116	327	401
CV56	2.0	5.4	0.3	0.2	5.9	0.1	55.2	0.4	0.28	0.12	1910	790	n.d.	1213	335	245
CV57	2.2	5.1	0.3	0.3	5.9	0.1	43.8	0.4	0.22	0.11	1280	781	n.d.	990	288	199
CV58	1.2	6.6	0.5	0.3	7.7	0.2	31.2	0.9	0.18	1.34	793	827	n.d.	663	383	258
CV59	3.2	4.3	0.1	0.2	4.4	0.1	44.8	0.4	0.63	0.69	666	1719	n.d.	2065	140	1541
CV60	2.6	5.0	0.5	0.5	5.6	0.1	43.1	0.7	0.25	0.55	1285	941	27	1099	266	594
CV61	0.3	4.5	0.1	0.2	4.5	0.1	44.9	0.3	0.57	0.69	668	1602	n.d.	1912	141	1968
CV62	1.0	4.7	0.4	0.2	4.9	0.1	44.1	1.6	0.31	0.75	1064	1422	n.d.	1248	433	200
CV63	1.1	4.1	0.4	0.5	3.6	0.1	45.1	0.7	0.38	0.74	839	1440	n.d.	1159	151	1148
CV64	0.9	2.7	0.7	0.3	1.2	0.0	58.9	1.7	0.74	0.08	1389	3499	n.d.	1321	233	1190
CV65	3.0	6.4	0.4	0.4	7.7	0.1	26.9	2.0	0.26	0.02	206	940	n.d.	570	n.d.	337
CV66	1.4	5.7	0.4	0.1	6.0	0.1	21.4	1.8	0.24	0.03	218	1126	26	370	55	349
CV67	3.2	4.5	0.9	0.1	5.4	0.1	33.3	11.7	0.31	0.14	388	394	n.d.	645	266	23
CV68	1.1	5.7	0.4	0.2	6.6	0.2	35.0	1.2	0.70	3.72	931	939	n.d.	768	83	733

Table 3. Major elements including Zn and Pb but excluding Ti analyzed by AAS. Ti, Cu, Sr, Ag, As, Sb and W analyzed by ICP-AES. Y, Zr, Nb, REE, Hf, Ta, Th and U analyzed by ICP-MS. Analyses of major elements (including Zn and Pb) in per cent; analyses of minor elements in p.p.m (continued).

"High-temperature" hydrothermal Mn deposits	Na_2O	K_2O	MgO	CaO	Al_2O_3	TiO_2	MnO	Fe_2O_3	Zn	Pb	Cu	Sr	Ag	As	Sb	W
CV69	1.3	4.9	0.5	0.3	6.1	0.4	28.1	6.7	0.54	5.14	638	666	n.d.	2209	164	822
CV70	0.6	5.8	0.1	0.1	6.3	0.2	36.4	1.9	0.69	1.46	893	1872	n.d.	1791	164	1455
CV71	0.9	6.9	0.3	0.2	7.3	0.3	36.7	6.2	1.20	1.99	923	1565	n.d.	1660	252	974
Av.	1.7	5.4	0.1	0.1	5.8	0.1	38.0	1.8	0.15	1.20	835	1144	1	1140	249	606
Phi 9																
CV72	1.2	4.8	0.4	0.3	4.1	0.1	45.1	3.2	0.89	0.74	1608	3137	n.d.	1228	n.d.	1235
CV73	1.2	5.4	0.2	0.1	6.8	0.1	34.9	1.8	0.57	3.24	1196	964	n.d.	942	63	481
CV74	1.0	4.8	0.4	0.1	4.8	0.4	40.5	7.6	0.63	1.12	1022	705	29	2030	236	406
Av.	1.1	5.0	0.3	0.2	5.2	0.2	40.2	4.2	0.70	1.70	1275	1602	10	1400	100	707
Phi 10																
CV90	1.4	4.5	0.3	9.2	8.2	0.2	36.7	1.9	0.24	3.82	675	927	n.d.	1378	233	160
CV91	2.2	3.9	0.5	0.3	4.7	0.1	48.9	0.9	0.27	2.23	897	1094	n.d.	1726	523	433
CV92	2.7	4.3	0.5	0.4	5.2	0.1	49.7	1.3	0.23	1.64	516	864	n.d.	1874	461	267
CV93	1.4	2.4	0.5	0.5	2.2	0.1	48.7	1.9	0.31	6.58	1289	545	n.d.	2217	3679	1495
CV94	2.1	6.6	0.5	0.2	8.2	0.1	31.7	0.8	0.17	3.13	3085	1166	26	1278	205	221
Av.	1.9	4.3	0.4	2.1	5.7	0.1	43.2	1.3	0.24	3.48	1293	919	5	1695	1020	515
Phi 11(1)																
CV96	2.5	4.2	1.3	1.8	8.0	0.1	50.6	2.3	0.16	0.02	10	2418	n.d.	1072	75	352
CV97	2.4	4.0	1.1	1.9	8.7	0.1	38.2	2.1	0.12	0.01	9	2053	n.d.	736	62	294
CV98	2.0	4.0	0.8	1.7	8.8	0.1	40.7	1.9	0.11	0.02	9	1962	n.d.	838	53	353
CV99	3.9	4.7	1.6	2.2	8.6	0.1	33.3	1.1	0.07	0.02	16	2138	n.d.	1308	n.d.	253
Phi 11(2)																
CV100	1.9	5.5	0.6	1.2	7.6	0.1	46.5	1.6	0.43	0.02	38	2633	n.d.	1420	108	2420
CV101	2.7	4.2	1.2	1.9	7.4	0.1	42.7	2.2	0.48	0.03	33	3147	n.d.	1572	202	1190
CV102	1.8	4.0	0.5	1.2	5.8	0.1	48.7	1.4	0.27	0.03	40	2036	23	1242	101	2873

Table 3. Major elements including Zn and Pb but excluding Ti analyzed by AAS. Ti, Cu, Sr, Ag, As, Sb and W analyzed by ICP-AES. Y, Zr, Nb, REE, Hf, Ta, Th and U analyzed by ICP-MS. Analyses of major elements (including Zn and Pb) in per cent; analyses of minor elements in p.p.m (continued).

"High-temperature" hydrothermal Mn deposits	Na_2O	K_2O	MgO	CaO	Al_2O_3	TiO_2	MnO	Fe_2O_3	Zn	Pb	Cu	Sr	Ag	As	Sb	W
Phi 11(3)																
CV103	1.7	4.6	0.5	1.3	6.8	0.1	46.0	1.7	0.31	0.01	40	2117	n.d.	1124	77	2321
Phi 11(4)																
CV104	3.1	3.8	0.9	1.9	8.5	0.2	40.0	3.4	0.46	0.01	32	1538	n.d.	1844	196	1668
Phi 11(5)																
CV105	2.0	2.9	0.3	1.9	8.5	0.1	43.5	3.3	0.16	0.14	76	1484	n.d.	3515	719	717
CV106	1.9	2.4	0.3	1.7	8.0	0.1	48.9	3.8	0.00	0.14	353	765	n.d.	2502	1023	724
Av.	2.3	4.0	0.8	1.7	7.8	0.1	43.5	2.3	0.23	0.04	60	2026	2	1561	237	1196
Phi 12																
CV109	0.3	4.5	0.1	0.2	6.1	0.2	45.1	1.4	0.64	1.44	722	1638	n.d.	2917	529	2147
Av.	0.3	4.5	0.1	0.2	6.1	0.2	45.1	1.4	0.64	1.44	722	1638	n.d.	2917	529	2147
Phi 13																
CV110	0.8	4.9	0.5	0.3	4.9	0.1	49.0	2.8	0.07	0.02	6	1504	n.d.	367	53	917
Av.	0.8	4.9	0.5	0.3	4.9	0.1	49.0	2.8	0.07	0.02	6	1504	n.d.	367	53	917
Dump sample																
CV111	0.5	5.2	0.3	0.3	2.9	0.1	59.1	3.8	0.10	0.02	163	1131	37	1107	165	659
Av.	0.5	5.2	0.3	0.3	2.9	0.1	59.1	3.8	0.10	0.02	163	1131	37	1107	165	659
Grand Av.	1.8	4.9	0.5	0.8	6.1	0.1	40.7	2.1	0.34	1.15	701	1364	-	1326	318	774

Table 3. Major elements including Zn and Pb but excluding Ti analyzed by AAS. Ti, Cu, Sr, Ag, As, Sb and W analyzed by ICP-AES. Y, Zr, Nb, REE, Hf, Ta, Th and U analyzed by ICP-MS. Analyses of major elements (including Zn and Pb) in per cent; analyses of minor elements in p.p.m (continued).

"High-temperature" hydrothermal Mn deposits	Y	Zr	Nb	La	Ce	Pr	Nd	Sm	Eu	Gd	Tb	Dy	Ho	Er	Tm	Yb	Lu	Hf	Ta	Th	U
CV1a	25.0	40.4	3.7	24.6	32.6	4.4	16.6	3.4	4.5	4.7	0.3	3.6	0.6	2.3	0.1	2.0	0.3	1.5	4.8	5.7	6.9
CV1b	24.6	36.2	4.4	25.1	36.6	4.6	17.0	3.5	5.7	4.8	0.3	3.8	0.6	2.3	0.1	2.1	0.3	1.2	4.2	5.9	8.0
CV2	18.8	60.6	12.5	22.0	39.7	4.2	14.9	3.1	4.1	3.8	0.2	2.9	0.4	1.8	0.1	1.6	0.2	2.2	5.5	10.1	4.5
CV3	15.0	16.5	1.7	18.8	22.3	4.1	15.9	3.4	4.8	3.7	0.2	2.6	0.3	1.4	n.d.	1.0	0.1	0.5	3.9	3.4	6.0
CV49	26.4	39.4	11.0	23.9	33.7	4.9	18.1	3.7	5.5	4.3	0.7	3.5	0.8	2.2	0.4	2.0	0.4	1.1	10.5	4.9	5.6
CV50	10.9	n.d.	6.5	16.2	17.2	3.0	11.4	2.5	4.5	3.0	0.4	1.8	0.4	1.1	0.2	0.9	0.2	n.d.	9.1	0.3	4.5
CV51	19.0	38.2	8.8	22.4	30.5	4.1	14.8	3.2	6.7	5.1	0.5	2.9	0.6	1.9	0.3	1.7	0.4	0.7	9.3	5.4	4.7
CV52	10.9	52.2	7.5	13.1	16.7	2.4	8.5	2.0	5.3	3.0	0.4	1.9	0.4	1.2	0.2	1.2	0.3	0.9	8.9	3.7	4.1
CV53	7.9	23.2	7.3	7.4	11.1	1.4	5.5	1.3	2.1	1.8	0.3	1.3	0.3	0.9	0.2	0.9	0.2	0.1	12.3	3.0	2.4
CV54	18.7	70.3	12.5	24.5	30.2	4.7	17.2	3.5	5.9	4.3	0.3	3.4	0.5	2.0	0.1	1.8	0.2	2.3	8.8	6.5	8.1
CV55	17.4	21.5	8.3	20.3	23.7	3.7	13.8	3.0	5.3	4.6	0.5	3.0	0.6	1.9	0.3	1.7	0.4	0.1	8.9	2.0	6.5
CV56	22.9	69.4	9.5	30.7	20.4	6.4	26.3	5.7	7.0	6.7	0.6	5.1	0.7	2.7	0.1	2.1	0.3	2.2	8.4	6.2	19.0
CV57	21.4	59.5	9.1	19.6	14.7	3.6	15.6	3.5	4.5	4.8	0.4	4.5	0.7	2.5	0.1	1.8	0.2	1.9	6.4	4.8	5.5
CV58	16.3	43.1	16.0	18.2	20.8	3.8	15.1	3.4	6.4	5.4	0.2	3.0	0.4	1.8	0.1	1.6	0.2	1.7	13.1	7.5	7.0
CV59	20.3	20.7	6.7	21.6	19.8	4.8	19.3	4.1	6.5	5.5	0.6	3.3	0.7	1.9	0.3	1.5	0.3	n.d.	8.9	3.5	6.1
CV60	17.3	97.8	8.2	21.3	29.6	4.6	17.9	4.0	5.8	4.5	0.7	3.5	0.7	2.2	0.4	2.1	0.4	2.4	8.8	4.5	7.4
CV61	21.5	21.9	7.0	19.7	18.7	3.8	15.3	3.2	5.9	5.1	0.6	3.3	0.7	2.1	0.3	1.6	0.4	0.0	9.4	3.7	6.3
CV62	19.5	25.1	6.5	23.7	24.4	4.2	16.0	3.4	6.5	5.2	0.4	3.6	0.5	2.0	0.1	1.8	0.2	0.7	4.5	3.7	8.0
CV63	29.0	25.4	8.9	26.6	35.6	4.8	18.4	4.0	6.0	5.4	0.4	4.5	0.7	2.7	0.2	2.4	0.3	0.8	10.7	2.9	5.1
CV64	34.5	9.0	4.1	17.1	28.2	3.7	15.5	3.3	6.7	4.3	0.3	3.7	0.7	2.5	0.2	2.4	0.4	1.4	7.8	1.6	2.4
CV65	17.8	27.7	10.8	13.3	21.8	2.6	10.2	2.3	4.9	4.6	0.1	2.7	0.4	1.8	0.1	1.8	0.2	1.5	11.7	4.4	2.6
CV66	11.9	23.3	5.9	10.9	17.2	1.8	7.1	1.5	1.3	1.9	n.d.	1.7	0.2	1.2	n.d.	1.0	0.1	0.8	5.9	3.7	1.9
CV67	13.5	24.5	5.3	13.9	18.4	2.5	9.3	2.1	2.4	2.6	0.4	2.1	0.5	1.5	0.3	1.5	0.3	0.0	7.8	1.6	4.9
CV68	15.9	18.2	6.4	19.9	27.5	3.1	11.1	2.2	3.9	3.1	0.1	2.3	0.3	1.6	0.1	1.7	0.2	0.5	5.4	2.8	2.7
CV69	12.6	26.1	8.2	18.3	28.7	2.6	9.0	1.9	2.6	2.2	0.4	2.0	0.5	1.5	0.3	1.6	0.4	0.1	8.7	1.5	2.9
CV70	9.1	17.2	6.8	21.0	25.8	3.1	10.5	2.1	4.6	3.3	0.3	1.4	0.3	1.0	0.2	1.0	0.2	n.d.	9.1	2.3	5.0
CV71	15.8	22.4	8.0	24.7	34.0	3.5	12.0	2.2	3.4	2.6	0.1	2.3	0.3	1.4	n.d.	1.3	0.1	0.7	5.5	3.2	5.2

Table 3. Major elements including Zn and Pb but excluding Ti analyzed by AAS. Ti, Cu, Sr, Ag, As, Sb and W analyzed by ICP-AES. Y, Zr, Nb, REE, Hf, Ta, Th and U analyzed by ICP-MS. Analyses of major elements (including Zn and Pb) in per cent; analyses of minor elements in p.p.m (continued).

"High-temperature" hydrothermal Mn deposits	Y	Zr	Nb	La	Ce	Pr	Nd	Sm	Eu	Gd	Tb	Dy	Ho	Er	Tm	Yb	Lu	Hf	Ta	Th	U
CV1a	25.0	40.4	3.7	24.6	32.6	4.4	16.6	3.4	4.5	4.7	0.3	3.6	0.6	2.3	0.1	2.0	0.3	1.5	4.8	5.7	6.9
CV1b	24.6	36.2	4.4	25.1	36.6	4.6	17.0	3.5	5.7	4.8	0.3	3.8	0.6	2.3	0.1	2.1	0.3	1.2	4.2	5.9	8.0
CV2	18.8	60.6	12.5	22.0	39.7	4.2	14.9	3.1	4.1	3.8	0.2	2.9	0.4	1.8	0.1	1.6	0.2	2.2	5.5	10.1	4.5
CV3	15.0	16.5	1.7	18.8	22.3	4.1	15.9	3.4	4.8	3.7	0.2	2.6	0.3	1.4	n.d.	1.0	0.1	0.5	3.9	3.4	6.0
CV49	26.4	39.4	11.0	23.9	33.7	4.9	18.1	3.7	5.5	4.3	0.7	3.5	0.8	2.2	0.4	2.0	0.4	1.1	10.5	4.9	5.6
CV50	10.9	n.d.	6.5	16.2	17.2	3.0	11.4	2.5	4.5	3.0	0.4	1.8	0.4	1.1	0.2	0.9	0.2	n.d.	9.1	0.3	4.5
CV51	19.0	38.2	8.8	22.4	30.5	4.1	14.8	3.2	6.7	5.1	0.5	2.9	0.6	1.9	0.3	1.7	0.4	0.7	9.3	5.4	4.7
CV52	10.9	52.2	7.5	13.1	16.7	2.4	8.5	2.0	5.3	3.0	0.4	1.9	0.4	1.2	0.2	1.2	0.3	0.9	8.9	3.7	4.1
CV53	7.9	23.2	7.3	7.4	11.1	1.4	5.5	1.3	2.1	1.8	0.3	1.3	0.3	0.9	0.2	0.9	0.2	0.1	12.3	3.0	2.4
CV54	18.7	70.3	12.5	24.5	30.2	4.7	17.2	3.5	5.9	4.3	0.3	3.4	0.5	2.0	0.1	1.8	0.2	2.3	8.8	6.5	8.1
CV55	17.4	21.5	8.3	20.3	23.7	3.7	13.8	3.0	5.3	4.6	0.5	3.0	0.6	1.9	0.3	1.7	0.4	0.1	8.9	2.0	6.5
CV56	22.9	69.4	9.5	30.7	20.4	6.4	26.3	5.7	7.0	6.7	0.6	5.1	0.7	2.7	0.1	2.1	0.3	2.2	8.4	6.2	19.0
CV57	21.4	59.5	9.1	19.6	14.7	3.6	15.6	3.5	4.5	4.8	0.4	4.5	0.7	2.5	0.1	1.8	0.2	1.9	6.4	4.8	5.5
CV58	16.3	43.1	16.0	18.2	20.8	3.8	15.1	3.4	6.4	5.4	0.2	3.0	0.4	1.8	0.1	1.6	0.2	1.7	13.1	7.5	7.0
CV59	20.3	20.7	6.7	21.6	19.8	4.8	19.3	4.1	6.5	5.5	0.6	3.3	0.7	1.9	0.3	1.5	0.3	n.d.	8.9	3.5	6.1
CV60	17.3	97.8	8.2	21.3	29.6	4.6	17.9	4.0	5.8	4.5	0.7	3.5	0.7	2.2	0.4	2.1	0.4	2.4	8.8	4.5	7.4
CV61	21.5	21.9	7.0	19.7	18.7	3.8	15.3	3.2	5.9	5.1	0.6	3.3	0.7	2.1	0.3	1.6	0.4	0.0	9.4	3.7	6.3
CV62	19.5	25.1	6.5	23.7	24.4	4.2	16.0	3.4	6.5	5.2	0.4	3.6	0.5	2.0	0.1	1.8	0.2	0.7	4.5	3.7	8.0
CV63	29.0	25.4	8.9	26.6	35.6	4.8	18.4	4.0	6.0	5.4	0.4	4.5	0.7	2.7	0.2	2.4	0.3	0.8	10.7	2.9	5.1
CV64	34.5	9.0	4.1	17.1	28.2	3.7	15.5	3.3	6.7	4.3	0.3	3.7	0.7	2.5	0.2	2.4	0.4	1.4	7.8	1.6	2.4
CV65	17.8	27.7	10.8	13.3	21.8	2.6	10.2	2.3	4.9	4.6	0.1	2.7	0.4	1.8	0.1	1.8	0.2	1.5	11.7	4.4	2.6
CV66	11.9	23.3	5.9	10.9	17.2	1.8	7.1	1.5	1.3	1.9	n.d.	1.7	0.2	1.2	n.d.	1.0	0.1	0.8	5.9	3.7	1.9
CV67	13.5	24.5	5.3	13.9	18.4	2.5	9.3	2.1	2.4	2.6	0.4	2.1	0.5	1.5	0.3	1.5	0.3	0.0	7.8	1.6	4.9
CV68	15.9	18.2	6.4	19.9	27.5	3.1	11.1	2.2	3.9	3.1	0.1	2.3	0.3	1.6	0.1	1.7	0.2	0.5	5.4	2.8	2.7
CV69	12.6	26.1	8.2	18.3	28.7	2.6	9.0	1.9	2.6	2.2	0.4	2.0	0.5	1.5	0.3	1.6	0.4	0.1	8.7	1.5	2.9
CV70	9.1	17.2	6.8	21.0	25.8	3.1	10.5	2.1	4.6	3.3	0.3	1.4	0.3	1.0	0.2	1.0	0.2	n.d.	9.1	2.3	5.0
CV71	15.8	22.4	8.0	24.7	34.0	3.5	12.0	2.2	3.4	2.6	0.1	2.3	0.3	1.4	n.d.	1.3	0.1	0.7	5.5	3.2	5.2

Table 3. Major elements including Zn and Pb but excluding Ti analyzed by AAS. Ti, Cu, Sr, Ag, As, Sb and W analyzed by ICP-AES. Y, Zr, Nb, REE, Hf, Ta, Th and U analyzed by ICP-MS. Analyses of major elements (including Zn and Pb) in per cent; analyses of minor elements in p.p.m (continued).

"High-temperature" hydrothermal Mn deposits	Y	Zr	Nb	La	Ce	Pr	Nd	Sm	Eu	Gd	Tb	Dy	Ho	Er	Tm	Yb	Lu	Hf	Ta	Th	U
Av.	18.3	35.8	7.8	20.0	25.2	3.7	14.2	3.0	4.9	4.1	0.4	3.0	0.5	1.8	0.2	1.6	0.3	1.0	8.1	4.0	5.7
CV72	24.3	32.1	5.9	21.9	19.7	3.3	12.6	2.7	5.5	3.8	0.6	3.4	0.8	2.4	0.4	2.4	0.5	0.3	8.7	2.5	3.5
CV73	13.2	25.4	6.4	22.6	28.8	3.4	11.8	2.2	5.1	3.2	0.4	2.1	0.5	1.4	0.3	1.4	0.3	0.2	8.3	3.3	4.2
CV74	13.8	14.0	7.6	16.2	22.5	2.5	8.9	2.0	2.9	2.5	0.4	2.3	0.5	1.7	0.3	1.9	0.4	n.d.	8.2	1.7	4.5
Av.	17.1	23.8	6.6	20.2	23.7	3.1	11.1	2.3	4.5	3.2	0.5	2.6	0.6	1.8	0.3	1.9	0.4	0.2	8.4	2.5	4.1
CV90	19.5	26.2	9.1	20.7	36.6	3.4	13.2	2.8	3.8	4.3	0.2	3.0	0.4	2.0	0.1	1.8	0.2	0.8	5.3	4.1	4.1
CV91	12.3	20.0	5.4	13.8	20.6	2.3	8.7	1.8	2.1	2.3	0.1	2.2	0.3	1.5	n.d.	1.3	0.1	0.6	4.9	2.7	5.9
CV92	12.3	21.0	6.3	16.8	25.5	2.7	10.6	2.1	2.4	2.7	0.1	2.4	0.3	1.6	n.d.	1.5	0.2	0.6	4.9	2.9	5.7
CV93	13.6	17.5	5.2	39.5	54.4	4.5	15.5	2.9	5.6	3.8	0.5	2.8	0.6	1.8	0.3	1.8	0.4	n.d.	8.7	1.3	6.2
CV94	11.2	21.8	6.1	18.7	24.5	3.0	10.2	2.1	2.8	2.8	0.4	2.2	0.5	1.5	0.3	1.7	0.4	n.d.	8.0	3.4	6.6
Av.	13.7	21.3	6.4	21.9	32.3	3.2	11.7	2.3	3.3	3.2	0.3	2.5	0.4	1.7	0.1	1.6	0.3	0.4	6.4	2.9	5.7
CV96	25.5	14.2	4.4	9.3	16.5	2.3	10.0	2.6	0.8	3.2	0.4	4.7	0.9	4.0	0.5	5.2	0.9	0.4	5.0	2.2	1.7
CV97	24.0	13.6	5.5	9.8	16.5	2.4	10.0	2.4	1.0	3.1	0.6	4.0	1.0	3.2	0.6	3.7	0.7	n.d.	8.2	1.8	1.4
CV98	26.6	16.4	4.8	12.0	25.5	3.0	12.4	3.1	1.5	4.0	0.5	5.3	1.1	4.1	0.5	4.5	0.7	0.4	5.6	2.8	1.9
CV99	15.2	14.4	5.5	8.3	14.5	2.0	7.9	2.0	0.8	2.4	0.5	2.9	0.7	2.2	0.4	2.3	0.5	n.d.	8.1	3.7	2.2
CV100	21.0	13.8	4.5	11.1	20.9	2.6	10.9	2.7	1.3	3.3	0.3	4.6	0.9	3.7	0.5	5.1	0.8	0.3	7.8	2.0	4.9
CV101	17.0	14.0	3.9	9.8	19.7	2.4	9.8	2.5	1.8	3.2	0.3	4.2	0.8	3.6	0.5	5.3	0.9	0.4	5.4	2.2	3.0
CV102	20.7	n.d.	5.0	9.6	16.3	2.2	8.5	2.1	1.2	2.8	0.5	3.4	0.8	2.9	0.5	3.6	0.7	n.d.	10.1	1.3	3.9
CV103	21.5	n.d.	5.1	11.2	22.7	2.7	10.6	2.6	1.3	3.3	0.6	4.2	1.0	3.5	0.6	4.7	0.9	n.d.	9.6	1.2	2.8
CV104	24.9	267	5.9	12.6	19.5	3.2	13.3	3.4	1.4	4.0	0.8	4.8	1.1	3.9	0.7	4.5	0.9	8.1	9.2	2.7	5.7
CV105	18.7	18.2	4.3	18.1	20.1	3.0	12.0	2.5	1.8	3.4	0.3	3.8	0.7	3.0	0.3	3.6	0.5	0.4	5.2	2.2	2.2
CV106	8.5	14.1	3.9	10.8	12.7	1.5	5.9	1.2	0.8	1.3	n.d.	1.6	0.2	1.3	n.d.	1.3	0.2	0.3	4.8	1.9	2.1
Av.	20.3	35.1	4.8	11.1	18.6	2.5	10.1	2.5	1.2	3.1	0.4	4.0	0.8	3.2	0.6	4.9	0.7	0.9	7.1	2.2	2.9
CV109	5.7	20.1	7.2	17.1	16.6	1.6	5.3	1.0	5.4	2.4	n.d.	0.7	n.d.	0.5	n.d.	0.4	n.d.	0.7	7.9	3.8	4.2
Av.	5.7	20.1	7.2	17.1	16.6	1.6	5.3	1.0	5.4	2.4	n.d.	0.7	n.d.	0.5	n.d.	0.4	n.d.	0.7	7.9	3.8	4.2

Table 3. Major elements including Zn and Pb but excluding Ti analyzed by AAS. Ti, Cu, Sr, Ag, As, Sb and W analyzed by ICP-AES. Y, Zr, Nb, REE, Hf, Ta, Th and U analyzed by ICP-MS. Analyses of major elements (including Zn and Pb) in per cent; analyses of minor elements in p.p.m (continued).

Bedded hydrothermal Mn deposits	Na_2O	K_2O	MgO	CaO	Al_2O_3	TiO_2	MnO	Fe_2O_3	Zn	Pb	Cu	Sr	Ag	As	Sb	W
Phi 5																
CV30	2.4	7.0	0.7	0.4	8.9	0.1	30.1	2.9	0.47	1.14	610	1291	n.d.	1217	355	497
CV31	1.4	6.0	0.2	0.2	6.9	0.2	29.9	1.4	0.30	0.82	383	1533	20	803	120	672
CV32	1.2	8.9	0.3	0.2	9.1	0.1	31.8	0.2	0.37	0.54	769	627	28	590	102	1594
CV34	1.8	6.1	0.3	0.2	7.1	0.1	26.4	1.4	0.22	0.34	318	1472	n.d.	469	94	418
CV35	0.9	7.0	0.1	0.2	7.7	0.2	35.0	3.1	0.34	1.63	474	652	n.d.	1475	324	522
Av.	1.5	7.0	0.3	0.2	7.9	0.1	30.6	1.8	0.30	0.90	511	1115	10	911	199	741
Phi 8																
CV36	0.5	5.2	0.4	0.3	7.3	0.2	34.1	2.8	0.36	1.74	1393	774	n.d.	2206	184	237
CV37	1.9	8.8	0.7	0.3	9.7	0.1	23.6	1.8	0.32	1.00	1137	739	n.d.	1077	n.d.	146
CV38	4.2	9.2	1.0	0.5	10.4	0.1	19.6	1.9	0.43	0.09	555	1617	n.d.	999	64	319
CV39	2.0	7.0	0.8	0.4	8.7	0.1	26.6	2.8	0.44	0.02	108	1396	n.d.	1396	72	126
CV40	2.8	6.5	1.5	0.5	7.3	0.1	29.2	4.7	0.71	0.08	121	3092	n.d.	1328	67	138
CV41	2.6	9.2	0.6	0.2	9.5	0.2	21.6	4.8	0.87	0.43	365	250	n.d.	990	157	428
CV42	2.2	5.4	1.1	0.4	6.9	0.1	31.5	4.6	0.54	0.02	83	882	n.d.	661	56	159
CV45	1.9	5.1	0.6	0.3	6.8	0.1	14.3	1.7	0.39	0.63	428	1258	32	434	n.d.	267
CV46	3.2	7.1	0.5	0.3	7.8	0.2	28.4	2.6	0.29	0.02	51	1235	n.d.	803	59	172
CV47	3.5	8.9	0.8	0.4	9.3	0.2	23.9	3.6	0.32	0.13	89	795	n.d.	981	138	433
CV48	2.6	6.4	0.9	0.3	7.7	0.1	29.4	3.2	0.35	0.02	16	678	n.d.	857	61	268
Av.	2.5	7.1	0.8	0.4	8.3	0.1	25.6	3.1	0.38	0.40	395	1156	3	1067	78	245
Phi 7																
CV75	4.7	1.7	0.9	3.8	16.3	0.8	9.6	4.3	0.38	0.37	374	644	n.d.	676	n.d.	308
CV76	3.1	2.1	1.5	2.3	14.6	0.6	10.3	3.5	0.89	0.56	727	775	n.d.	378	n.d.	136
CV77	5.1	2.1	2.1	1.9	15.9	0.2	18.2	2.3	0.94	0.07	1482	744	n.d.	598	65	71
CV78	4.7	2.4	0.7	3.2	17.2	0.8	19.9	3.2	0.35	0.02	678	1347	40	378	n.d.	1194

Table 3. Major elements including Zn and Pb but excluding Ti analyzed by AAS. Ti, Cu, Sr, Ag, As, Sb and W analyzed by ICP-AES. Y, Zr, Nb, REE, Hf, Ta, Th and U analyzed by ICP-MS. Analyses of major elements (including Zn and Pb) in per cent; analyses of minor elements in p.p.m (continued).

Bedded hydrothermal Mn deposits	Na_2O	K_2O	MgO	CaO	Al_2O_3	TiO_2	MnO	Fe_2O_3	Zn	Pb	Cu	Sr	Ag	As	Sb	W
CV79	4.9	2.4	0.9	2.7	14.5	0.1	17.6	1.7	0.48	0.01	624	668	n.d.	631	n.d.	492
CV80	5.5	2.8	1.2	3.1	13.1	0.3	18.9	2.0	0.40	0.02	344	921	n.d.	586	50	1102
CV82	6.5	1.9	1.5	2.4	9.1	0.1	28.6	1.2	0.38	0.98	655	622	n.d.	981	335	618
CV84	4.3	1.9	1.5	1.7	9.1	0.3	24.1	3.0	0.49	1.76	748	943	n.d.	1229	549	158
CV87	1.2	4.6	0.4	0.3	1.3	0.0	69.0	0.9	0.32	4.69	602	835	n.d.	2070	2289	723
CV89	3.2	4.5	0.9	0.8	7.6	0.1	42.1	1.1	0.36	1.32	1422	1456	114	2664	375	504
Av.	4.3	2.6	1.2	2.2	11.8	0.3	25.8	2.3	0.45	1.00	766	895	15	1019	366	530
Grand Av.	3.0	5.4	0.8	1.0	9.5	0.2	26.7	2.6	0.50	0.71	560	1048		1018	218	450

Bedded hydrothermal Mn deposits	Y	Zr	Nb	La	Ce	Pr	Nd	Sm	Eu	Gd	Tb	Dy	Ho	Er	Tm	Yb	Lu	Hf	Ta	Th	U
CV30	13.4	38.0	6.8	20.9	31.5	3.4	12.5	2.4	4.4	2.8	0.1	2.6	0.3	1.7	0.1	1.8	0.3	1.2	6.4	3.8	4.0
CV31	16.8	35.4	6.3	18.5	27.9	3.5	13.6	2.9	3.9	3.3	0.2	3.0	0.4	1.9	0.1	1.9	0.3	1.3	6.0	4.0	6.8
CV32b	6.7	38.8	6.3	23.6	38.7	3.6	12.0	2.1	3.8	2.4	n.d.	1.6	0.1	0.9	n.d.	0.8	0.1	1.6	6.5	5.4	6.2
CV34	13.1	37.3	5.4	15.1	26.1	2.8	10.3	2.0	2.8	2.4	n.d.	2.2	0.2	1.5	n.d.	1.5	0.2	1.2	5.4	3.7	3.8
CV35	16.6	52.8	18.1	31.4	61.9	4.9	17.2	3.5	4.2	3.6	0.3	3.2	0.5	2.0	0.1	1.9	0.2	1.7	8.4	5.6	3.6
Av.	13.3	40.5	8.6	21.9	37.1	3.6	13.1	2.6	3.8	2.9	0.1	2.5	0.3	1.6	0.0	1.6	0.2	1.4	6.5	4.5	4.9
CV36	10.8	44.8	10.0	22.9	28.3	3.5	11.7	2.1	7.7	2.8	n.d.	1.8	0.2	1.3	0.0	1.5	0.2	1.5	6.5	5.5	4.2
CV37	5.6	23.1	5.4	12.0	15.4	1.6	6.5	1.4	3.9	1.8	n.d.	1.2	n.d.	0.7	n.d.	0.7	n.d.	0.7	4.9	2.6	2.4
CV38	8.6	30.4	6.5	12.6	17.3	1.9	7.2	1.4	2.8	1.7	n.d.	1.6	0.1	1.2	n.d.	1.2	0.1	0.9	4.9	3.9	2.6
CV39	16.9	28.8	3.8	54.3	20.5	2.8	10.7	2.1	5.3	3.5	0.1	2.7	0.4	1.9	0.1	1.8	0.2	0.8	4.2	3.8	3.0
CV40	13.5	38.1	8.2	13.9	22.6	2.5	9.6	2.1	3.8	2.5	0.1	2.3	0.3	1.5	n.d.	1.5	0.2	1.2	4.7	5.8	3.5
CV41	16.2	44.9	7.5	17.8	24.6	2.8	10.8	2.3	4.0	2.8	0.1	2.5	0.3	1.5	n.d.	1.3	0.2	1.5	4.9	4.3	2.6
CV42	26.3	29.4	7.7	14.7	22.0	2.7	10.4	2.3	4.4	3.6	0.2	2.9	0.5	2.1	0.1	2.4	0.3	0.8	4.7	4.2	2.8
CV45	9.5	15.5	4.8	4.5	7.7	0.8	3.5	0.8	2.2	1.4	n.d.	0.9	0.0	0.6	n.d.	0.5	n.d.	0.3	4.8	2.1	1.9
CV46	12.9	34.2	9.5	13.5	23.2	2.5	9.0	1.8	2.8	2.9	n.d.	2.0	0.2	1.3	n.d.	1.3	0.1	1.0	5.2	4.9	2.8

Table 3. Major elements including Zn and Pb but excluding Ti analyzed by AAS. Ti, Cu, Sr, Ag, As, Sb and W analyzed by ICP-AES. Y, Zr, Nb, REE, Hf, Ta, Th and U analyzed by ICP-MS. Analyses of major elements (including Zn and Pb) in per cent; analyses of minor elements in p.p.m (continued).

Bedded hydrothermal Mn deposits	Y	Zr	Nb	La	Ce	Pr	Nd	Sm	Eu	Gd	Tb	Dy	Ho	Er	Tm	Yb	Lu	Hf	Ta	Th	U
CV46	12.9	34.2	9.5	13.5	23.2	2.5	9.0	1.8	2.8	2.9	n.d.	2.0	0.2	1.3	n.d.	1.3	0.1	1.0	5.2	4.9	2.8
CV47	13.0	45.0	9.4	13.1	18.3	2.1	8.0	1.7	2.8	2.6	n.d.	2.0	0.2	1.4	n.d.	1.3	0.1	1.4	5.3	4.9	2.7
CV48	11.5	37.5	7.7	10.8	17.1	2.0	7.8	1.7	3.0	2.8	n.d.	1.8	0.2	1.3	n.d.	1.1	0.1	1.1	4.9	5.1	2.5
Av.	13.3	33.8	7.3	17.3	19.7	2.3	8.7	1.8	3.9	2.6	n.d.	2.0	0.2	1.3	n.d.	1.3	0.1	1.0	5.0	4.3	2.8
CV75	11.5	34.1	11.7	21.1	27.1	3.2	11.5	2.4	2.2	3.1	0.5	2.4	0.5	1.6	0.3	1.5	0.3	0.5	8.7	3.5	1.2
CV76	12.6	52.6	11.1	16.5	29.3	3.3	11.9	2.5	2.1	2.8	0.4	2.4	0.5	1.6	0.3	1.5	0.3	1.4	8.6	6.9	1.5
CV77	17.5	57.9	13.8	22.7	35.4	4.3	15.5	3.0	2.3	4.3	0.2	3.3	0.5	2.1	0.1	1.9	0.3	2.3	7.9	10.4	2.0
CV78	21.2	61.7	18.6	16.8	29.3	3.0	10.6	2.0	2.3	2.3	0.1	2.3	0.4	1.8	n.d.	1.5	0.2	2.2	13.6	4.5	2.4
CV79	18.8	55.6	9.5	18.1	24.2	3.2	11.7	2.2	1.0	2.3	0.1	2.6	0.4	1.8	0.1	1.7	0.2	2.0	7.2	6.8	3.4
CV80	10.0	43.3	9.0	14.0	22.6	2.4	9.1	1.8	1.4	1.9	0.0	1.9	0.2	1.3	n.d.	1.2	0.1	1.5	7.0	4.3	2.7
CV82	10.5	27.7	5.6	14.2	22.6	2.8	11.2	2.3	3.0	3.0	0.1	2.5	0.4	1.8	0.1	1.6	0.2	0.8	5.5	2.6	6.3
CV84	11.1	30.2	7.6	12.2	20.5	2.3	8.2	1.8	2.5	2.5	0.4	1.9	0.5	1.4	0.2	1.3	0.3	0.3	8.1	3.0	3.7
CV87	10.9	8.4	2.9	12.4	22.8	1.8	6.9	1.4	2.1	1.8	n.d.	2.0	0.3	1.5	n.d.	1.4	0.2	0.0	4.5	1.0	7.5
CV89	34.4	29.7	6.1	52.9	91.0	9.5	32.9	6.4	6.3	8.0	0.8	7.2	1.3	5.0	0.6	6.0	0.9	0.9	5.3	3.3	7.9
Av.	15.8	40.1	9.6	20.1	32.5	3.6	13.0	2.6	2.5	3.2	0.3	2.9	0.5	2.0	0.2	2.0	0.3	1.2	7.6	4.6	3.8
	14.2	37.5	8.4	19.2	28.0	3.0	11.2	2.2	3.3	2.9	0.2	2.4	0.3	1.6	0.1	1.6	0.2	1.2	6.3	4.5	3.6

In a previous study, Liakopoulos (1987) has identified two generations of manganese oxides within the Vani manganese deposit based on X-ray diffraction analysis, optical mineralogy and electron microprobe analysis of individual mineral grains; the first generation consisted of pyrolusite and ramsdellite; the second generation consisted of oxides of the isostructural series cryptomelane-hollandite-coronadite and hydrohetaerolite which are characterized by high contents of K, Ba, Pb and Zn, respectively (Liakopoulos et al., 2001). This sequence was thought to be the result of a two-stage process of formation of the manganese oxide minerals. In the initial phase, boiling of the hydrothermal fluid were thought to have taken place during its ascent to the surface with the formation of stockwork (sulphide) mineralization in the crust such that the hydrothermal fluid was depleted in elements such as Ba, Pb and Zn. As a result, the hydrothermal fluid deposited pyrolusite and ramsdellite, which are characterized by low contents of these elements within the volcaniclastic sandstone. The subsequent phase probably took place following tectonic uplift. In this case, boiling and phase separation probably took place within the geothermal reservoir (cf. Christanis and Seymour, 1995). The high-salinity hydrothermal fluid became enriched in Ba, Pb and Zn during its ascent to the surface as a result of dissolution of barite and sulphide minerals and remineralized the original manganese oxide assemblage to cryptomelane, hollandite, coronadite and hydrohetaerolite. This model presupposes a significant time interval between the two phases of mineralization. Significantly, the Vani deposit is located only 4 km from Galana, the ancient mining area of Triades, where both barite and sulphides are located (Vavelidis and Melfos, 1997; Liakopoulos et al., 2001). This two-stage process was thought to be mainly responsible for the unique characteristics of the Vani manganese deposit.

However, the X-ray diffraction data presented here were not able to confirm this hypothesis possibly because the mineralogical transformations took place within individual mineral grains within the deposit and were therefore not amenable to study using bulk samples for analysis.

6. COMPOSITION

The chemical compositions of the 75 samples from the Vani manganese deposit are presented in Table 3. These data show that the Vani manganese deposit is very variable in composition and anomalously enriched in a number of elements. On average, the "high-temperature" deposits have higher concentrations of Mn, Pb, Cu, Sr, As, Sb, W, Y, Pr-Lu, Ta and U and lower concentrations of Na, K, Mg, Ca, Al, Ti, Fe, Zn, Ag, Zr, Nb, La, Ce, Hf and Th than the bedded deposits (Fig. 4). However, there is a very significant overlap in concentrations of all these elements between the two types of ore. The deposit is also characterized by high Mn/Fe ratios in both the "high-temperature" and bedded samples.

In order to test these observations statistically, the 75×36 array of compositional data values was first transformed into logratios (Aitchison, 1986). The expected logratio for each element was assumed to be of the form $\mu \pm \alpha$, where μ is the overall mean

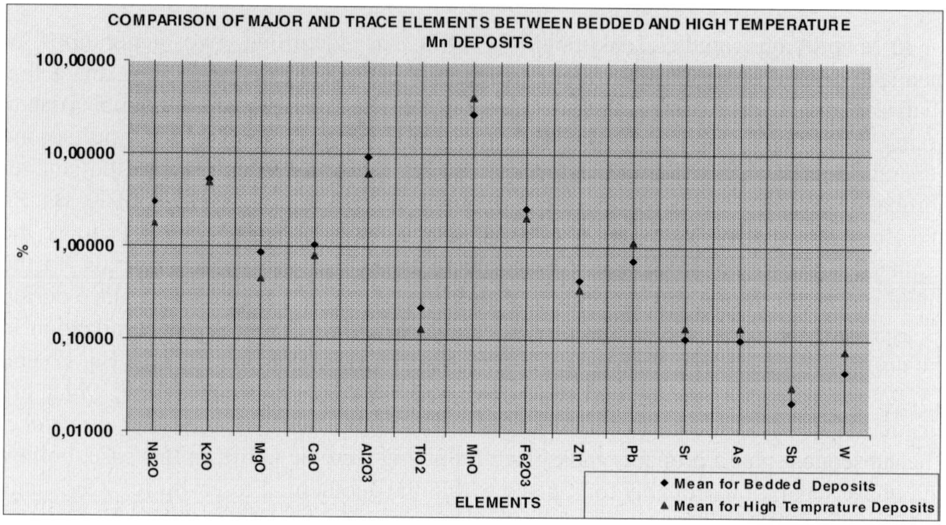

(a)

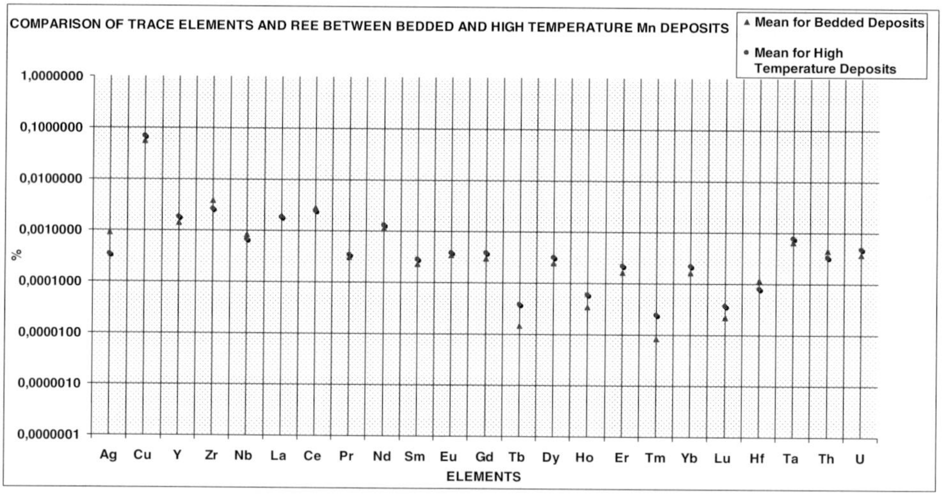

(b)

Fig. 4. Plots showing the average concentrations of major and trace elements, including REE, in "high-temperature" deposits and bedded hydrothermal manganese deposits from Vani. Concentrations in %.

logratio

for the whole Vani deposit, $+\alpha$ is the contribution to this for the "high-temperature" deposits, and $-\alpha$ is the contribution (negative) for the bedded deposits. Least squares regression was then undertaken to estimate μ and α, assuming a full model (i.e., the maximum number of free parameters, in this case, two; Dobson, 2002). The absolute values of the t-statistic associated with the estimates for α were extremely significant for

Table 4. "High-temperature" deposits: Elements with significantly high log perturbations, the associated *t*-statistics, and *p*-values in percentages. When the trace element concentration in a given sample was too low to be detected, the imputed value in this analysis was taken to be one half of the measured minimum value for that element.

Mn	Tb	Ho	Tm	Lu
6.56	4.88	3.40	3.82	2.63
0.01	0.01	0.11	0.03	1.05

Table 5. Bedded deposits: Elements with significantly high log perturbations, the associated *t*-statistics, and *p*-values in percentages. When the trace element concentration in a given sample was too low to be detected, the imputed value in this analysis was taken to be one half of the measured minimum value for that element.

Na	Mg	Ca	Al	Ti	Fe	Zn	Zr	Nb	Ce	Hf	Th
4.31	4.47	2.43	4.98	3.31	2.98	3.76	3.44	3.16	2.72	3.77	3.55
0.01	0.01	1.75	0.01	0.15	0.40	0.03	0.10	0.23	0.82	0.03	0.07

17 of the 36 elements. These elements, their *t*-statistics and associated *p*-values (significance values) are set out in Tables 4 and 5 where the first row lists the elements, the second row lists the *t*-statistics, and the third lists the *p*-values as percentages in each table (Note that in these tables, 0.01 is to be interpreted as $\leq$ 0.01%). Table 4 displays the elements for which the contribution α was significantly high in the "high-temperature" deposits (and therefore, significantly low (negative) in the bedded deposits). Table 5 displays the elements for which the contribution $-\alpha$ was significantly high (positive) in the bedded deposits (and therefore, significantly low (negative) in the "high-temperature" deposits). For the "high-temperature" deposits, the highly significant enrichment of Mn is not unexpected but the apparent enrichment of Tb, Ho, Tm and Lu, odd atomic number heavy rare earth elements (REE), in these deposits almost certainly represents an artefact in the data resulting from their very low concentrations in the samples.

These data therefore demonstrate that, of the ore-forming elements (Pb, Zn, As, Sb and W), only Mn is significantly enriched in the "high-temperature" deposits. On the other hand, most of the elements associated with the lithogenous fraction of the deposit (i.e. Na, Al, Ti, Fe, Zr, Nb) are significantly enriched in the bedded deposits.

In a previous publication, Hein et al. (2000) have shown that the Vani manganese deposit is enriched in Ba, Pb, Zn and As as well as in Sb, Cu, Cd, Zr, W, Sr and Ag in some samples. The maximum concentrations of these elements recorded in this study are MnO 69.0%, Pb 6.6%, Zn 1.2%, As 0.35%, Sb 0.37%, Cu 0.19%, W 0.29%, Sr 0.35% and Ag 0.004%, all displaying significantly higher concentrations than the maxima recorded by Hein et al., 2000 (MnO 58.9%, Pb 3.3%, Zn 0.81%, As 0.31%, Sb 0.22%, Cu 0.09%, W 0.01%, Sr 0.59% and Ag 0.001%). Interestingly, the maximum concentrations of all these elements (with the exception of Ag) are observed in the "high-temperature" deposits. However, the wide range in concentrations can be seen from the minimum concentrations of these elements in these "high-temperature" deposits (MnO 20,3%, Pb 0.01%, Zn 0.07%, As 0.037%, Sb n.d., Cu 0.0006%, W 0.0023%, Sr 0.03% and Ag n.d.). Hein et al. (2000) have also estimated the concentration of Ba in the Vani manganese deposit to be in the range 1.1-2.8% in those samples which do not include barite veins. In light of these data, the Vani manganese deposit may therefore be classified as a Mn-Ba-Pb-Zn-As-Sb-W deposit.

The average Mn/Fe ratio of the "high-temperature" deposits is higher (21.3) than that of the bedded deposits (12.3). This reflects the fact that these deposits precipitated first when the Mn content of the hydrothermal solutions was higher. However, the Mn/Fe ratio of the dump sample is only 17.1 (less than the average for the "high-temperature" deposits). This is somewhat surprising since this type of deposit consisting of botryoidal Mn oxides with lustrous steel grey appearance would have been expected to have precipitated first from the hydrothermal solutions. However, in absolute terms, the dump sample has a higher Mn content (45.5%) than the average for the "high-temperature" deposits (31.3%) and an even higher Mn content than the average for the bedded deposits (20.6%). Differences in the Mn contents of the Vani manganese deposit were well known to the miners. The manganese ore was known to be in two grades, one with an average Mn content of 20% which was minable and and the other with an average Mn content of 16% which was considered uneconomic (Glasby et al., 2001).

Within the Vani deposit, element concentrations range from 1.2-17.2% Al_2O_3, 0.2-11.7% Fe_2O_3, 9.6-69.0% MnO, 0.07-1.2% Zn, 0.02-6.58% Pb, 367-3515 ppm As, 50-3679 ppm Sb and 23-2420 ppm W. Some of this variability is undoubtedly related to the variability in the abundances of manganese oxides and the lithogenous fraction in the samples. Significantly, samples with Mn concentrations >40% (30 samples) are restricted to typical "high-temperature" deposits (including 2 samples CV87 and CV89 taken from a bedded sequence Phi7) whereas those with Mn concentrations <20% (8 samples) are restricted to bedded deposits. On the other hand, samples with the highest Al_2O_3 contents (>10%) (8 samples) are, with one exception, taken from bedded deposits whereas the samples with the lowest Al_2O_3 contents (<4%) (4 samples) are typical "high-temperature" deposits including CV87. The higher average concentrations of Mn, Pb, As, Sb, W in the "high-temperature" hydrothermal Mn deposits relative to the bedded hydrothermal Mn deposits therefore reflects the fact that the "high-temperature" deposits were formed initially when the hydrothermal fluids penetrated the faults and fissures within the volcaniclastic sandstone leading to higher abundances of the hydrothermal manganese minerals. The bedded deposits formed subsequently as the cooling hydrothermal fluids migrated along the bedding plane of the porous sandstone. These deposits are characterized by higher concentrations of Na_2O, MgO, Al_2O_3, TiO_2, Fe_2O_3, Zr and Nb derived from the associated volcaniclastic sandstone.

Despite this, several of the ore-forming elements show a lack of any clear association with sample type, sample location or concentrations of other elements. For example, the highest concentrations of Pb (>2.0%) were observed in sections Phi 4 (5 samples), Phi 9 (1 samples) and Phi 10 (4 samples) in the "high-temperature" hydrothermal Mn deposits. These samples could variously be described as hydrothermal vent tubes and samples having a lustrous steel grey appearance. However, samples with the lowest Pb contents (Pb <0.1%) were located in both the "high-temperature" deposits in sections Phi 4 (4 samples), Phi 11 (9 samples), Phi 13 (1 sample) and Dump sample (1 sample) and in the bedded deposits in sections Phi 5 (1 sample), Phi 8 (6 samples) and Phi 7 (4 samples). The "high-temperature" samples with the lowest Pb contents were quite similar in appearance to the "high-temperature" deposits with the highest Pb contents whereas the bedded deposits with the lowest Pb contents were quite different in character being tuffaceous material almost completely replaced by Mn oxides. The Pb concentrations did not therefore appear to be controlled by sample type. In addition, there appeared to be no clear association of the Pb concentrations with location. Whilst the "high-temperature" deposits from sections Phi 4 and Phi 9 display consistently high Pb contents (av. 1.2 and 1.7%, respectively) and samples from Phi 11 a uniformly low Pb content (av. 0.04%), the "high-temperature" samples from section Phi 4 display a wide range of Pb concentrations (0.02-5.14%) and bedded samples from Phi 7 a range of 0.02-4.69% Pb. As a consequence of this, Pb shows no well-defined association with any of the other ore-forming elements.

From the section on mineralogy, it is apparent that samples from the Vani manganese deposit consist of volcaniclastic material enriched in Na_2O, MgO, Al_2O_3, TiO_2 and Fe_2O_3, manganese minerals (pyrolusite, ramsdellite, cryptomelane, hollandite, coronadite, romanechite and possibly hydrohetaerolite) enriched in K, Ba, Pb and possibly Zn and barite. Although the concentration of K_2O is higher on average in the bedded deposits than in the "high-temperature" deposits, the average concentration of K_2O in the Vani manganese deposit (5.1%) is much higher than that in andesites which is typically about 1.4% K_2O (Taylor and McClellan, 1985, Table 1). This confirms that K_2O is present in the Vani manganese deposit mainly as cryptomelane. From the compositional data, it can also be calculated that the abundance of coronadite in the Vani deposit varies from about 0.3 to 20%. The association of Pb with coronadite explains why Pb varies independently of the other elements.

NASC-REE patterns are very useful in elucidating the origin of mineral deposits. The main features of these patterns are the minor negative Ce anomalies and very high positive Eu anomalies observed in most samples (Fig. 5). Thermodynamic studies have shown that divalent Eu dominates in hydrothermal fluids with temperatures greater than 250°C and trivalent Eu at temperatures lower than 250°C (Sverjensky, 1984; Glasby et al., 1997; Douville et al., 1999; Mills et al., 2001). The high Eu anomalies in the Vani manganese deposit therefore indicate leaching of divalent Eu^{2+} from the host rocks at temperatures greater than 250°C, possibly related to the release of Eu^{2+} during the sericitization of K-feldspar as occurs at Vani (Sverjensky, 1984; Liakopoulos et al., 2001). According to Liakopoulos (1987) and Liakopoulos et al. (1991), the geothermal waters of the modern deep geothermal reservoir located on the Zephyria Plain in the central-eastern part of Milos are characterized by a maximum temperature of 318°C and chloride concentration of 87 ppt at a depth of 1150-1220 m. This temperature is well above that necessary for the leaching of the Eu^{2+} ions. Although no temperature data are available for the formation of the manganese deposits themselves, they are clearly late-stage, low-temperature deposits (Cronan, 1979; Liakopoulos et al., 2001). Interestingly, the samples with the highest average Eu anomaly (from section ϕ12) have the lowest

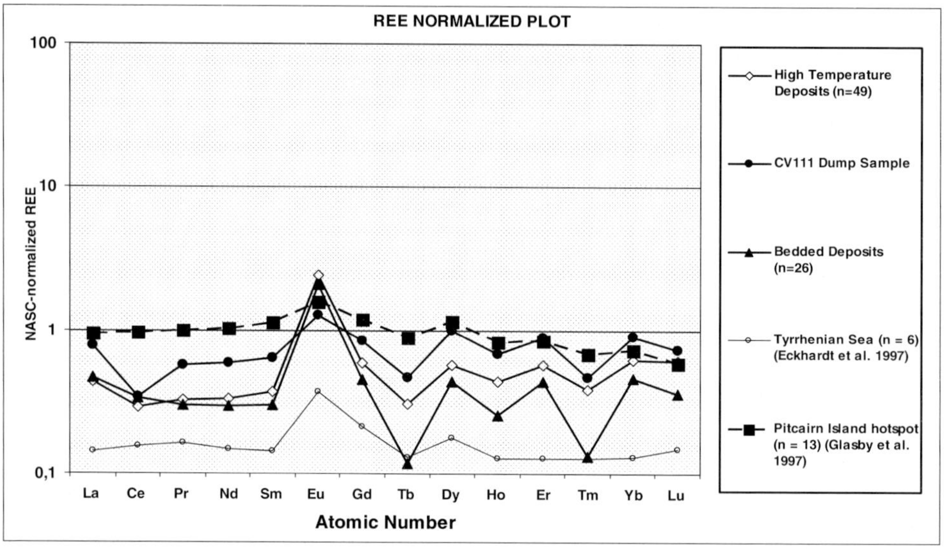

Fig. 5. North American Shale Composite (NASC)-normalized REE distribution patterns for the average REE patterns from the six sections of the "high-temperature" hydrothermal Mn deposits and three sections of the bedded hydrothermal Mn deposits shown in Fig. 2. The NASC REE data used to normalize the Vani REE data is taken from Piper (1974).

concentrations of most other REE, although there does not appear to be any consistent trend between the average Eu anomaly and the concentrations of the other REE in each section (Fig. 5). By contrast, the Vani manganese deposit does not display the pronounced negative Ce anomalies typical of submarine hydrothermal Mn deposits (Glasby, 2000). This probably reflects the mixing of the hydrothermal fluids with seawater prior to deposition of the Mn oxides (Fleet, 1984; Mills et al., 2001).

With the exception of La and Ce, the REE concentrations are slightly higher on average in the "high-temperature" hydrothermal Mn deposits than in the bedded hydrothermal Mn deposits (Fig. 4). This difference probably reflects the fact that the REE are associated mainly with the manganese oxide fraction of the deposit. However, a plot of the NASC-normalized REE distribution patterns for the average REE patterns from the each section of the "high-temperature" and bedded hydrothermal Mn deposits shows that this enrichment sequence is somewhat erratic (Fig. 5). For comparison, it should be noted that the average REE concentrations in island arc andesites are similar to those in the Vani manganese deposit but they are distinguished from them by the absence of a positive Eu anomaly (Taylor and McLennan, 1985, Table 3). This supports the idea that the bulk of the REE is associated with the manganese oxide phase. The Vani manganese deposit is closely associated with the K-feldspar zone described by Liakopoulos et al. (2001) and Eu is known to be enriched in K-feldspars relative to its neighbours in the periodic table (Taylor and McLennan, 1985, Table 2). However, the average REE concentrations in K-feldspars are very much lower than those of the Vani manganese deposit (Taylor and McLennan 1985, Table 2) indicating that K-feldspars can not be the main source of the positive Eu anomaly in that the Vani manganese deposit.

Cd is also known to be enriched in the Vani manganese deposit with a maximum concentration of 0.0065% (Hein et al., 2000). Randall et al. (1998) have proposed a model for the uptake of Cd within the tunnels of cryptomelane but it is more likely that Cd would be taken up by coronadite in the Vani manganese deposit in view of the greater similarity in the effective ionic radii of Cd^{2+} (0.95Å) and Pb^{2+} (1.19 Å) compared to those of K^{+} (1.38 Å), Ba^{2+} (1.35 Å) and Zn^{2+} (0.68 Å) (Shannon, 1976).

7. DISCUSSION

Arc-backarc settings are considered to be particularly favourable for the formation of hydrothermal mineral deposits. During the past 20 years, submarine hydrothermal manganese deposits have been discovered on the Tonga-Kermadec Arc (Cronan et al., 1982), the Aeolian Arc (Eckhardt et al., 1997) and the Izu-Bonin and Mariana Arcs (Usui and Glasby, 1998). These deposits are generally characterized by high Mn/Fe ratios, low contents of Cu, Ni, Zn, Co, Pb and detrital silicate minerals, pronounced negative Ce anomalies and positive Eu anomalies, the presence of 10 Å manganate as the principal manganese oxide phase and their occurrence as dense crusts, up to 50 mm thick; the surfaces of some of these crusts are dark grey with a submetallic sheen. In addition, the crusts are characterized by high growth rates exceeding 1000 mm/10^6yrs in some cases (Hein et al., 1997; Glasby, 2000). A key question is whether submarine hydrothermal manganese deposits may be considered to be a possible analogue for the Vani manganese deposit or whether some other model is more appropriate. As will be shown later, the Vani manganese deposit is quite different in thickness, mineralogy and composition from known submarine hydrothermal manganese deposits. Even though the Vani manganese deposit was formed under shallow submarine conditions, it can not therefore be considered an analogue of these deposits.

More realistically, the Vani manganese deposit can be compared to the epithermal vein deposits of the southwestern United States and elsewhere which have been described by Hewett and Fleischer (1960), Hewett et al. (1963) and Hewett (1964, 1971). Hewett (1964) recognized psilomelane (romanechite), hollandite, cryptomelane, coronadite, pyrolusite and ramsdellite as the principal manganese oxide minerals in these deposits. Jacobsite and franklinite have also been identified in these deposits (Hewett and Fleischer, 1960). Chemical analysis showed that these deposits contained higher concentrations of W, Ba, Sr, Be, As, Sb, Tl and Ge than supergene oxides (Hewett et al., 1963). Based on several criteria, Hewett (1964) concluded that these deposits were hypogene. The deposits are characterized by persistent veins of barite and fluorite and, in many cases, occur in close proximity to explored deposits of these minerals. The host rocks are mainly layered volcanic rocks of Tertiary age with persistent alteration of plagioclase in the host rock to K-feldspar (adularia). These deposits also tend to be associated with epithermal gold and epithermal base-metal sulphide deposits. However, these deposits are fault-controlled and occur in the form of veins which may be up to 1.5 m thick, extend horizontally up to 500 m and can be traced to depths of up to 150 m. Both Hewett (1964) and Roy (1968, 1981) concluded that these vein deposits formed in the upper reaches of epithermal systems as a result of mixing of ascending hydrothermal fluids with descending meteoric water enriched in oxygen. Indeed, Roy (1968) attributed the formation of these higher oxides of manganese to the high concentrations of oxygen and falling temperatures. Although these deposits are similar in mineralogy and composition to the Vani manganese deposit, their mode of emplacement is different. They can not therefore be considered exact analogues of the Vani manganese deposit (Liakopoulos et al., 2001). Although epithermal vein deposits containing pyrolusite, ramsdellite, cryptomelane, hollandite, coronadite and romanechite are well documented, bedded deposits of manganese oxides containing this mineral assemblage are extremely rare. Possible examples are the deposits from Mellila and Imini, Morocco, and Alcerreca, Chile (Hewett and Olivares, 1968; Hewett, 1971; Roy, 1981). Nonetheless, it appears probable that the Vani manganese deposit is unique in combining this mineral assemblage in a massive economic ore deposit.

From the work of Hewett et al. (1963), Roy (1981) and others, it appears that a number of elements are characteristic of hydrothermal manganese oxides. Nicholson (1992) has proposed that the element association As-Ba-Cu-Li-Mo-Pb-Sb-Sr-V-Zn in manganese oxides is a diagnostic feature of hydrothermal manganese minerals.

The data presented here suggest that Mn, Ba, Pb, Zn, As, Sb and W are sigificantly enriched in the Vani manganese deposit. This pattern of element association has also been reported in a Långban-type Fe-Mn deposit from Central Sweden (Holtsam and Mansfield, 2001) and in Fe-Mn deposits from the Swiss Alps (Brugger and Gieré, 2000). Submarine hydrothermal sulphides from the JADE site in the Okinawa Trough have also been shown to be enriched in in Pb, Zn, As and Sb relative to hydrothermal sulphides from mid-ocean ridges (Herzig and Hannington, 1995, 2000; Glasby and Notsu, 2003). This is a function of the source rock. In the Okinawa Trough, back arc rifting has taken place in continental crust and is characterized by a complex volcanic suite of basalt, andesite, dacite and rhyolite (Herzig and Hannington, 1995). Fouquet et al. (2000) considered this pattern of element association to be typical of immature felsic environments. In this regard, it is significant that the Aegean Arc may also be considered to be an intracontinental arc characterized by felsic volcanism (Fytikas et al., 1986).

The forms of As, Sb and W are transported in hydrothermal fluids in continental settings have been summarized by Yang and Blum (1999). These authors considered that As may be transported as $H_3AsO_3^o$ or $HaAsS_2^o$ (depending on the sulphur content), Sb as $Sb(OH)_3$ or SbS_2^-, $HSb_2S_4^-$ and $H_2Sb_2S_4^0$ (depending on the sulphur content) and HWO_4^-, and W as WO_4^{2-} and $NaHWO_4^o$. Other authors have also discounted the role of chloro complexes of As and Sb in natural hydrothermal systems (Tossell, 1997; Sherman et al., 2000). However, based on a study of submarine hydrothermal fluids from the Juan de Fuca Ridge, Trefry et al. (1994) concluded that elements such as Pb,

Zn and Sb are enriched in deeply-penetrating Cl^--rich hydrothermal fluids as a result of the formation of chloro complexes following phase separation. Lüders et al. (2001) proposed a similar mechanism for the leaching of Pb, Zn, As and Sb from felsic rocks at the JADE site (cf. Halbach et al., 2002). Oelkers et al. (1998) have also presented experimental evidence which shows that Sb speciation is dominated by chloro complexes in chloride-rich solutions under acidic conditions. As previously noted, the geothermal waters of the modern deep geothermal reservoir located on the Zephyria Plain are characterized by high chloride ion concentrations. Rubin (1997), on the other hand, has presented evidence that a significant proportion of the As, Sb and W in submarine hydrothermal systems may be magmatic in origin (cf. Schwarz-Schampera et al., 2001).

In comparing the Vani manganese deposit to submarine hydrothermal manganese deposits and the epithermal vein deposits of the southwestern United States and elsewhere, it is worthwhile to speculate on the origin of these two types of deposits. The high Mn/Fe ratios and low contents of Cu, Ni, Zn, Co and Pb in submarine hydrothermal manganese deposits is thought to reflect the fractionation of Mn from Fe and the associated trace metals as a result of the deposition of these elements as stockwork mineralization within the underlying oceanic crust (Glasby, 2000). 10 Å manganate is the dominant manganese mineral in these deposits. These characteristics clearly differentiate submarine hydrothermal manganese deposits from the Vani deposit. In addition, the Vani manganese deposit is much thicker than any known submarine hydrothermal manganese deposit. This reflects the fact that the Vani deposit was formed by the replacement of the volcaniclastic sandstone which effectively trapped most of the hydrothermal manganese in its matrix in contrast to most modern submarine hydrothermal systems where the bulk of the manganese is discharged directly into sea water (Lavelle et al., 1992). However, both deposits are characterized by pronounced positive Eu anomalies and the surfaces of some submarine hydothermal manganese crusts have been described as a hard, smooth layer of manganese oxides with a characteristic metallic sheen (Eckhardt et al., 1997; Usui and Glasby, 1998) somewhat similar to the lustrous, steel grey manganese oxides in the Vani deposit.

The epithermal vein deposits, on the other hand, are significantly enriched in Mn, Ba, Pb, Zn, As, Sb and W. It seems probable that this element enrichment pattern reflects the leaching of these elements from proximal sulphide deposits by high-temperature hydrothermal solutions in accordance with the paragenetic sequence proposed by Hewett (1964). The high concentrations of K, Ba, and Pb in the hydrothermal solutions resulted in the formation of cryptomelane, hollandite, coronadite, romanechite and barite in these deposits.

Submarine hydrothermal and epithermal vein manganese deposits can therefore be considered to be at the opposite end of the spectrum with respect to their uptake of metals from hydrothermal solutions. Clearly, the Vani hydrothermal manganese deposit is more closely related to the epithermal vein deposits of the southwestern United States and elsewhere. The difference between these two types of deposit lies in structural control. The Vani deposit is a stratabound hydrothermal deposit formed as a result of the penetration of hydrothermal fluids into fissures within the volcaniclastic sandstone which effectively trapped most of the hydrothermal manganese in its matrix, thus forming an economic ore deposit. The epithermal vein deposits, on the other hand, are fault-controlled and occur in the form of veins which may be up to 1.5 m thick. Both deposits are characterized by the alteration of plagioclase in the host rock to K-feldspar (adularia) reflecting similar conditions of hydrothermal mineralization in the two types of deposit.

The Vani deposit is the only large manganese deposit found within the Aegean Arc. It owes its formation to the presence of a geothermal reservoir to supply the high-temperature, high-salinity hydrothermal fluids necessary to mineralize the area, the proximity of large faults such as the Vromolimni-Kondaro fault to transport the hydrothermal fluids to the surface and the deposition and compaction of a large volume of pyroclastic material within the Vani volcano-sedimentary basin to act as host to the hydrothermal manganese deposit. By contrast, submarine hydrothermal mineralization in Santorini is limited in extent and consists mainly of either hydrothermal pyritiferous diatomaceous ooze or amorphous iron oxyhydroxides and goethite depending on location (Cronan et al., 2000). The reason for this is that, at Santorini, the hydrothermal fluid discharges directly into seawater within embayments on Palaea and Nea Kameni. Only the Fe oxyhydroxides settle out locally. Manganese is rapidly dispersed within the hydrothermal plume in the sometimes rough conditions encountered in the area and is deposited mainly within the caldera with the rest transported into the Aegean Sea (Boström et al., 1990; Papavassiliou et al., 1990).

8. CONCLUSIONS

This study emphasizes the unique characteristics of the Vani manganese deposit. In particular, this deposit displays a characteristic mineral assemblage of pyrolusite, ramsdellite, cryptomelane, hollandite, coronadite and romancheite associated with high concentrations of Mn, Ba, Pb, Zn, As, Sb and W. As such, it can be considered to be an analogue of the epithermal vein deposits of the southwestern United States. Differences in the nature of these deposits appear to be, in part, structurally controlled. Future research needs to be directed towards a better understanding of the mechanism of uptake of these ore-forming elements in the Vani manganese deposit.

Acknowledgements

We thank Professor N. Skarpelis for his active participation in the field work and preparation of the sample location map.

REFERENCES

Aitchison, J., 1986. The Statistical Analysis of Compositional Data. Chapman & Hall,

London and New York.
Boström, K., Perissoratis, C., Galanopoulos, V., Papavassiliou, C., Boström, B., Ingri, J. and Kalogeropoulos, S., 1990. Geochemistry and structural control of hydrothermal sediments and new hot springs in the caldera of Santorini, Greece. In: D.A. Hardy (Editor), Thera and the Aegean World III Vol. 2, Thera Foundation, London, pp. 325-336.
Brugger, J. and Gieré, R., 2000. Origin and distribution of some trace elements in metamorphosed Fe-Mn deposits, Val Ferrera, eastern Swiss Alps. Can. Mineral., 38: 1075-1101.
Christanis, K. and Seymour, K.St., 1995. A study of scale deposition: an analogue of meso- to epithermal ore formation in the volcano of Milos, Aegean Arc, Greece. Geothermics, 24: 541-552.
Cronan, D.S., 1979. Metallogensis at oceanic spreading centres. J. Geol. Soc. London, 126: 621-626.
Cronan, D.S., Glasby, G.P., Moorby, S.A., Thomson, J., Knedler, K.E. and McDougall, J.C., 1982. A submarine hydrothermal manganese deposit from the south-west Pacific island arc. Nature, 298: 456-458.
Cronan, D.S., Varnavas, S.P. and Hodkinson, R., 2000. Hydrothermal mineralizing processes and associated sedimentation in Santorini hydrothermal embayments. Mar. Georesourc. Geotechnol., 18: 77-118.
Dobson, A.J., 2002. An Introduction to Generalized Linear Models, 2nd Ed. Chapman & Hall/CRC, Boca Raton.
Douville, E., Bienvenue, P., Charlou, J.L., Donval, J.P., Fouquet, Y., Appriou, P. and Gamo, T., 1999. Yttrium and rare earth elements in fluids from various deep-sea hydrothermal fluids. Geochim. Cosmochim. Acta, 63: 627-643.
Eckhardt, J.-D., Glasby, G.P., Puchelt, H. and Berner, Z., 1997. Hydrothermal manganese crusts from Enareta and Palinuro Seamounts in the Tyrrhenian Sea. Mar. Georesourc. Geotechnol., 15: 175-208.
Fleet, A.J., 1984. Aqueous and sedimentary geochemistry of the rare earths. In: P.Henderson (Editor) Rare Earth Element Geochemistry. Elsevier, Amsterdam, pp. 342-373.
Fouquet, Y., Henry, K., Cambon, P., Auzende, J.M., Charlou, J.L., Urabe, T. and Ishibashi, J., 2000. Hydrothermal sulfide deposits mineralogical and chemical composition in relation to the maturity of the back-arc systems in the SW Pacific. EOS Trans. Am. Geophys. Un., 81(22): WP90 (Abstr.)
Fytikas, M., Innocenti, F., Kolios, N., Manetti, P., Mazzuoli, R., Poli, G., Rita, F. and Villari, L., 1986. Volcanology and petrology of volcanic products from Milos and neighbouring islets. J. Volcanol. Geothermal Res., 28: 297-317.
Glasby, G.P., 2000: Manganese: Predominant Role of Nodules and Crusts. In: H.D. Schulz and M. Zabel (Editors) Marine Geochemistry. Springer-Verlag, Heidelberg, pp. 335-372.
Glasby, G.P. and Notsu, K., 2003. Submarine hydrothermal mineralization in the Okinawa Trough, S.W. of Japan: An overview. Ore Geol. Revs, 23: 299-339.
Glasby, G.P., Papavassiliou, C.T., Liakopoulos, A. and Galanopoulos, V., 2001. Past

mining of the Vani manganese deposit, Milos Island, Greece. Proceedings of the 5th International Mining History Congress, which was held 12-15 September 2000 in Milos Island, Greece, pp. 93-105.
Glasby, G.P., Stüben, D., Jeschke, G., Stoffers, P. and Garbe-Schönberg, C.-D., 1997. A model for the formation of hydrothermal manganese crusts from the Pitcairn Island hotspot. Geochim. Cosmochim. Acta, 61: 4583-4597.
Halbach, P., Nayak, B., Pracejus, B. and Fujimoto, H., 2002. Uncommon Au- and Ag-enriched sulfidic mineral assemblages from a basaltic ultraslow spreading ridge environment: Mt. Jourdanne, southwest Indian Ridge. In 'Minerals from the Ocean'. International Conference St. Petersburg 20-23 April (Abstr.)
Hein, J.R., Koschinsky, A., Halbach, P., Manheim. F.T., Bau, M., Kang, J-K. and Lubrich, N., 1997. Iron and manganese oxide mineralization in the Pacific. In: K. Nicholson, J.R. Hein, B. Bühn and S. Dasgupta (Editors), Manganese Mineralization: Geochemistry and Mineralogy of Terrestrial and Marine Deposits. Geol. Soc. Spec. Publ. No. 119: 123-138.
Hein, J.R., Stamatakis, M.G. and Dowling, J.S., 2000. Trace metal-rich Quaternary hydrothermal manganese oxide and barite deposit, Milos Island, Greece. Trans. Instn Min. Metall. (Appl. Earth Sci.), 109B: 67-76.
Herzig, P.M. and Hannington, M., 1995. Polymetallic massive sulfides at the modern seafloor: A review. Ore Geol. Revs, 10: 95-115.
Herzig, P.M. and Hannington, M., 2000. Polymetallic massive sulfides and gold mineralization at mid-ocean ridges and in subduction-related environments, In: D.S. Cronan (Editor), Handbook of Marine Mineral Deposits. CRC Press, Boca Raton, pp. 347-268.
Hewett, D.F., 1964. Veins of hypogene manganese oxide minerals in the southwestern United States. Econ. Geol., 59: 1429-1472.
Hewett, D.F., 1971. Coronadite-modes of occurrence and origin. Econ. Geol., 66: 164-177.
Hewett, D.F. and Fleischer, M., 1960. Deposits of the manganese oxides. Econ. Geol., 55: 1-55.
Hewett, D.F., Fleischer, M. and Conklin, N., 1963. Deposits of the manganese oxides: supplement. Econ. Geol., 58: 1-51.
Hewett, D.F. and Olivares, R.S., 1968. High potassium cryptomelane from Tarapaca Province, Chile. Am. Mineral., 53P: 1551-1557.
Holtsam, D. and Mansfield, J., 2001. Origin of a carbonate-hosted Fe-Mn-(Ba-As-Pb-Sb-W) deposit of Långban-type in Central Sweden. Mineral. Deposita, 36: 641-657.
Lavelle, J.W., Cowen, J.P. and Massoth, G.J., 1992. A model for the deposition of hydrothermal manganese near mid-ocean ridge crests. J. Geophys Res., 97: 7413-7427.
Liakopoulos, A., 1987. Hydrothermalisme et mineralisations metallifères de l'île de Milos (Cyclades, Grèce). Mem. Sc.Terre Univ. Curie, Paris, no 87-36. 276 pp. + 3 Annexes.
Liakopoulos, A., Katerinopoulos, A., Markopoulios, T. and Boulegue, J., 1991. A

mineralogical petrographic and geochemical study of samples from wells in the geothermal field of Milos island (Greece). Geothermics, 20: 237-256.
Liakopoulos, A., Glasby, G.P., Papavassiliou, C.T. and Boulegue, J., 2001. Nature and origin of the Vani manganese deposit, Milos, Greece: an overview. Ore Geol. Revs, 18: 181-209.
Lüders, V., Pracejus, B. and Halbach, P., 2001. Fluid inclusion and sulfur isotope studies in probable modern analogue Kuroko-type ores from the JADE hydrothermal field (Central Okinawa Trough, Japan). Chem. Geol., 173: 45-58.
Mills, R., Wells, D.M. and Roberts, S., 2001. Genesis of ferromanganese crusts from the TAG hydrothermal field. Chem. Geol., 176: 283-293.
Nicholson, K., 1992. Contrasting mineralogical-geochemical signatures of manganese oxides: guides to metallogenesis. Econ. Geol., 87: 1253-1264.
Oelkers, E.H., Sherman, D.M., Ragnarsdottir, K.V. and Collins, C., 1998. An EXAFS spectroscopic study of aqueous antimony(III)-chloride complexation at temperatures from 25-250°C. Chem. Geol., 151: 21-27.
Papanikolaou, D., Lekkas, E., Syskakis, D., Adamopolou, E., 1993. Correlation on neotectonic structures with the geodynamic activity in Milos during the earthquakes of March 1992. Bull. Geol. Soc. Greece, 28 (3): 413-428.
Papavassiliou, C., Boström, K., Paritsis, S., Galanopoulos, V., Arvanitides, N. and Kalogeropoulos, S., 1990. Drilling in an ore-forming shallow hydrothermal system, Santorini Volcano, Greece. In: D.A. Hardy (Editor), Thera and the Aegean World III Vol. 2, Thera Foundation, London, pp. 250-256.
Piper, D.Z., 1974. Rare earth elements in ferromanganese nodules and other marine phases. Geochim. Cosmochim. Acta, 38: 1007-1022.
Post, J.E., 1999. Manganese oxide minerals: Crystal structures and economic and environmental significance. Proc. Natl. Acad. Sci. USA, 96: 3447-3454.
Randall, S.R., Sherman, D.M. and Ragnarsdottir, K.V., 1998. An extended X-ray absorption fine-structure spectroscopy investigation of cadmium sorption on cryptomelane [KMn_8O_{16}]. Chem. Geol., 151: 95-106.
Roser, B.P., 1983. Comparative studies of Cu and Mn in mineralisation Torlesse, Waipapa and Haaast Schist Terranes, New Zealand. Unpubl. Ph.D. thesis (Victoria University of Wellington). 491 pp.
Roy, S., 1968. Mineralogy of the different genetic types of manganese oxides. Econ. Geol., 63: 760-786.
Roy, S., 1981. Manganese Deposits. Academic Press, London.
Rubin, K., 1997. Degassing of metals and metalloids from erupting seamount and mid-ocean ridge volcanoes: Observations and predictions. Geochim. Cosmochim. Acta, 61: 3525-3542.
Schwarz-Schampera, U., Herzig. P., Hannington, M. and Stoffers, P., 2001. Shallow submarine epithermal style As-Sb-Hg-Au mineralization. In: Piestrzynski et al. (Editors), Mineral Deposits at the beginning of the 21st Century. Swets and Zeitlinger Publishers, Lisse, pp. 333-335.
Shannon, R.D., 1976. Revised effective ionic radii and systematic studies of interatomic

distances in halides and chalcogenides. Acta Cryst., A32: 751-767.

Sherman, D.M., Ragnarsdottir, K.V. and Oelkers, E.H., 2000. Antimony transport in hydrothermal solutions: an EXAFS study if antimony(V) complexation in alkaline sulfide and sulfide-chloride brines art temperatures from 25°C to 300°C at P_{sat}. Chem. Geol., 167: 161-167.

Sverjensky, D.A., 1984. Europium redox equilibria in aqueous solution. Earth Planet. Sci. Letts, 67: 70-78.

Taylor, S.R. and McLennan, S.M., 1985. The Continental Crust: its Composition and Evolution. Blackwell Scientific Publications, Oxford.

Tossell, J., 1997.Theoretical studies on arsenic oxide and hydroxide species in minerals and in aqueous solutions. Geochim.Cosmochim. Acta, 61:1613-1623.

Trefry, J.H., Butterfield, D.B., Metz, S., Massoth, G.J., Trocine, R.P. and Feely, R.A., 1994. Trace metals in hydrothermal solutions from Cleft segment on the southern Juan de Fuca Ridge. J. Geophys. Res., 99: 4925-4935.

Usui, A. and Glasby, G.P. 1998. Submarine hydrothermal manganese deposits in the Izu-Bonin-Mariana Arc: An overview. The Island Arc, 7: 422-431.

Usui, A. and Mita, M., 1995. Geochemistry and mineralogy of a modern buserite deposit from a hot spring in Hokkaido, Japan. Clays Clay Mins, 43: 116-127.

Vavelidis, M. and Melfos, V., 1997. Two plumbian tetrahedrite-tennantite occurrences from Maronia area (Thrace) and Milos island (Aegean Sea), Greece. Eur. J. Mineral., 9: 653-657.

Yang, S.X. and Blum, N., 1999. A fossil hydrothermal system or a source-bed in the Madiyi Formation near the Ziangxi Au-Sb-W deposit, NW Hunan, PR China? Chem Geol., 155: 151-169.

APPENDIX

"High-temperature" hydrothermal Mn deposits

ϕ4 (CV1a-CV3, CV49-CV71; 27 samples)

Upper section (CV1a-CV3, CV49-CV64; 20 samples) 2.9 m thick section of Mn oxides within a volcaniclastic matrix 10 m wide; 5 vertical profiles sampled down this section

Central upper part of section consists of almost vertical hydrothermal vent tubes. The largest tube is 0.16 m in diameter with walls 5 mm thick; the tube has been infilled by a mass of Mn botryoids 3-10 mm in diameter; the hydrothermal vent tube consists of very dense black material with numerous small botryoids on the outer face of the tube.

Above this lies a whole mass of hollow hydrothermal vent tubes leaning up to 10° from the vertical; these tubes are typically 20-30 mm in diameter with walls typically 3 mm thick; some of the tubes are infilled with lustrous steel grey Mn oxides.

"Fissure filling" occurs where dense hydrothermal Mn deposits have almost completely infilled a fissure. Three types of fissure filling were observed in this section:

i almost vertical fissures in which Mn has deposited on both walls until the fissure was filled with Mn oxides; these fissures are 8-30 mm wide; this is the predominant type of fissure at this location.

ii in some cases, the permeability of the host rock did not permit the fissure to extend further; Mn then penetrated into the host rock as a mass of massive botryoids 5-20 mm in diameter with lustrous submetallic sheen.

iii in some cases, Mn interpenetrated the fissure as a massive lustrous submetallic layer up to 5 mm thick; Mn-rich fluids then interpenetrated into the pore spaces in the adjacent host rock to form thin dendrites of Mn oxides up to 10 mm in diameter.

This rock face consists of a series of almost vertical open fissures 8-30 mm wide which have been infilled by Mn mineralization; barite has intruded into some fissures later. Many of the samples in this section were almost pure Mn oxide material with lustrous steel grey appearance. Exceptionally, this section consists of hydrothermal vent tubes where hydrothermal fluids have been able to penetrate into open channels within the host rock. These features are relatively rare within the Vani manganese deposit. The host rock is a thickly bedded volcaniclastic sandstone.

Middle section (CV65-CV67; 3 samples) 9.5 m section of of thinly bedded to laminated volcaniclastic sandstone containing relatively pure, banded Mn deposits with variable amounts of barite; at the base of the section, the volcaniclastic sandstone is highly replaced by Mn and Fe oxides directly overlying greenish glauconitic material.

Bottom section (CV68-CV71; 4 samples) 3.2 m thick section of thickly bedded volcaniclastic sandstone replaced by botryoidal Mn ore with lustrous steel grey colour in the interior or crosscut by fissures filled with Mn ore. The lowest part of this section is massively replaced by Mn ore.

ϕ9 (CV72-CV74; 3 samples) section immediately to the left of section **ϕ4;** separated by a fault from section **ϕ4**; section up to 1.2 m thick of volcaniclastic sandstone strongly replaced by high purity Mn oxides; one Fe-rich Mn oxide deposit; no stratigraphic control on this section.

ϕ10 Upper section (CV90-CV93; 4 samples) 1.45 m of highly weathered bedded volcaniclastic sandstone interpenetrated to strongly replaced by Mn oxides with steel grey, lustrous texture in places; contains variable, minor amounts of barite; underhanging layer of very dense, almost pure lustrous steel grey Mn oxides with irregular botryoidal surface texture at base of section.
Lower section (CV94; one sample) 3.8 m thick finely laminated volcaniclastic sandstone containing irregularly spaced layers of massive (dense) layers of lustrous, steel grey Mn oxides; 5-20 mm thick.

ϕ11 (CV96-CV108; 11 samples) six vertical sections taken over a distance of 82 m on an exposed face of volcaniclastic sandstone

ϕ11(1) Left hand section of deposit (CV96-CV99; 4 samples) thick bedded volcaniclastic sandstone with intercalations of conglomerates; 2.6 m section of "high-temperature" hydrothermal Mn oxides. These include vertical fissure 0.15 m in diameter (hollow) from which hydrothermal Mn oxides have interpenetrated into the adjacent sandstone along fissures and stratiform planes, hydrothermal vent tubes (15 mm in diameter) containg numerous hanging Mn dendrites and hydrothermal vent tube walls 2 mm thick and horizontally bedded Mn oxides containing disseminated barite; inner walls of vent tubes characterized by lustrous, steel grey appearance.

ϕ11(2) 10 m to right of ϕ11(1) (CV100-CV102; 3 samples) 2.35 m high section; hydrothermal vent tubes infilled Mn dendrites formed within the volcaniclastic sandstone matrix plus bedded hydrothermal Mn oxides with steel grey, lustrous appearance in part; inner walls of vent tubes characterized by lustrous, steel grey appearance.

ϕ11(3) 2 m to right of ϕ11(2) (CV103; one sample) hydrothermal vent tube 0.25 x 0.13 m in diameter; vent tube walls 1-2 mm thick; vent tube infilled with volcaniclastic sandstone and Mn dendrites

ϕ11(4) 20 m to right of ϕ11(3) (CV104; one sample) 1.6 m thick bed of volcaniclastic sandstone with pronounced reddish coloration due to presence of Fe oxyhydroxides; section displays numerous horizontal bands of Mn oxides 3-15 mm thick which are lustrous steel grey in appearance; occasional large vent tube up to 0.14 m in diameter; abundant Mn dendrites in part of section.

ϕ11(5) 50 m to right of ϕ11(4) (CV105-CV106; 2 samples) 2.05 m thick section of thinly bedded volcaniclastic sandstone massively replaced by Mn oxides; band of lustrous steel grey Mn oxides underlain by botryoidal Mn oxides coated with reddish volcaniclastic sandstone.

ϕ12 (CV109; one sample) botryoidal Mn deposit with lustrous, steel grey appearance; one of the purest samples of Mn oxides recovered at Vani.

ϕ13 (CV110; one sample) botryoidal Mn deposit with lustrous, steel grey appearance.

Dump sample (CV111; one sample) botryoidal Mn deposit with lustrous, steel grey appearance; one of the purest samples of Mn oxides recovered at Vani; sample was taken from a mine dump; its stratigraphy is therefore unknown.

Bedded hydrothermal Mn deposits

ϕ5 (CV30-CV35; 5 samples) 2 thick section of volcaniclastic sandstone with dull black Mn oxides with botryoidal layers (5-20 mm) or Mn oxides (1-8 mm) with granulated surface texture on exposed surfaces.

ϕ8 (CV36-CV48; 11 samples) 2.8 m, 0.6 m and 1.0 m thick sections of bedded manganese deposits separated by beds of volcaniclastic sandstone 0.3 m and 0.45 m thick respectively; bedded manganese deposits consist of altered to highly altered volcaniclastic sandstone showing variable replacement by Mn oxides, from lightly replaced to strongly interpenetrated to almost completely replaced by Mn oxides.

ϕ7 (CV75-CV89; 10 samples) approx. 8m thick section of thinly bedded to laminated volcaniclastic sandstone with minor incalations of conglomerate and impregnated by Mn oxides at various intervals; sample CV87 included hydrothermal vent tube and displayed steel grey, lustrous texture typical of "high-temperature" hydrothermal activity in part.

An FT-Raman, Raman and FTIR study of hydrothermally altered volcanic rocks from Kos Island (Southeastern Aegean, Greece)

D. Papoulis, P. Tsolis-Katagas, B. Tsikouras* and C. Katagas

University of Patras, Department of Geology, Section of Earth Materials, GR-265 00 Patra, Greece.

ABSTRACT

Both Tertiary and Quaternary volcanic rocks occur in Kos Island. The Quaternary volcanism produced a successive series of rhyolitic-dacitic volcanic rocks both lavas and pyroclastics. Hydrothermal alteration of rhyolitic rocks at Kefalos peninsula (at southern tip of the island) and Asfendiou (NE Kos) resulted in kaolin formations. At both places, alteration is clearly associated with faults. The mineralogical study of the kaolin samples was carried out using various vibrational spectroscopic techniques, covering a broad wavenumber range (Raman, FT-Raman and FTIR) along with XRD and SEM analytical methods.

The study of the kaolin samples showed that mixed-layer kaolinite-smectite developed from the alteration of K-feldspar, while the pure kaolinite from both Kefalos and Asfendiou areas, is the result of Na-plagioclase alteration. Book-type dickite forms after kaolinite and represents peak conditions of hydrothermal alteration. The presence of zunyite, a rare mineral that occurs in hydrothermal alteration regimes, has been verified in the kaolin samples from both areas. The assemblage kaolinite + dickite + zunyite ± pyrophyllite, in the most altered samples, is stable at a temperature range between 250°C and 290°C, at low pH (<3.5). The occurrence of zunyite and the presence of alunite in trace amounts suggest the involvement of hydrothermal fluids rich in F- and Cl- and poor in S-2, thus indicating contamination by seawater.

Keywords: Kaolinite; dickite; zunyite; Raman; FT-Raman; FTIR.

1. INTRODUCTION

FTIR, FT-Raman and Raman spectroscopy studies of kaolins are of considerable interest due to their applications to structural studies of kaolinite group minerals (Frost

* Corresponding author: e-mail: v.tsikouras@upatras.gr

and Kloprogge, 2001). The hydroxyl bands are considered to be very sensitive probes for distinguishing between kaolinite minerals and determining their structure (Frost, 1997, 1998). The assignment of the hydroxyl stretching bands remains under constant review, and differing viewpoints concerning the position and origin of both the inner and inner-surface hydroxyl groups still exist (Giese, 1988; Shoval et al., 2001, 2002).

Previous Raman and FT-Raman studies of clay minerals were mainly focused in the determination and analysis of the bands of standard phases (e.g. Frost et al., 1993; Frost, 1997; Frost and Kloprogge, 2001). FTIR, Raman and FT-Raman spectroscopic techniques were proved to be useful tools for the identification of particular clay and rare minerals in natural samples from Kos. Our purpose was 1) to compare a conventional analytical method like XRD to FTIR, Raman and FT-Raman techniques to distinguish kaolinite polymorphs and to determine minerals present in low amounts and 2) to obtain the required data for the determination of conditions of hydrothermal alteration of rhyolites and formation of kaolins in Kos Island.

2. GEOLOGICAL SETTING

Kos Island lies at the southeastern part of the Aegean Sea, and it is a part of the SAAVA (Fig. 1). The lowermost unit comprises a metamorphic series of

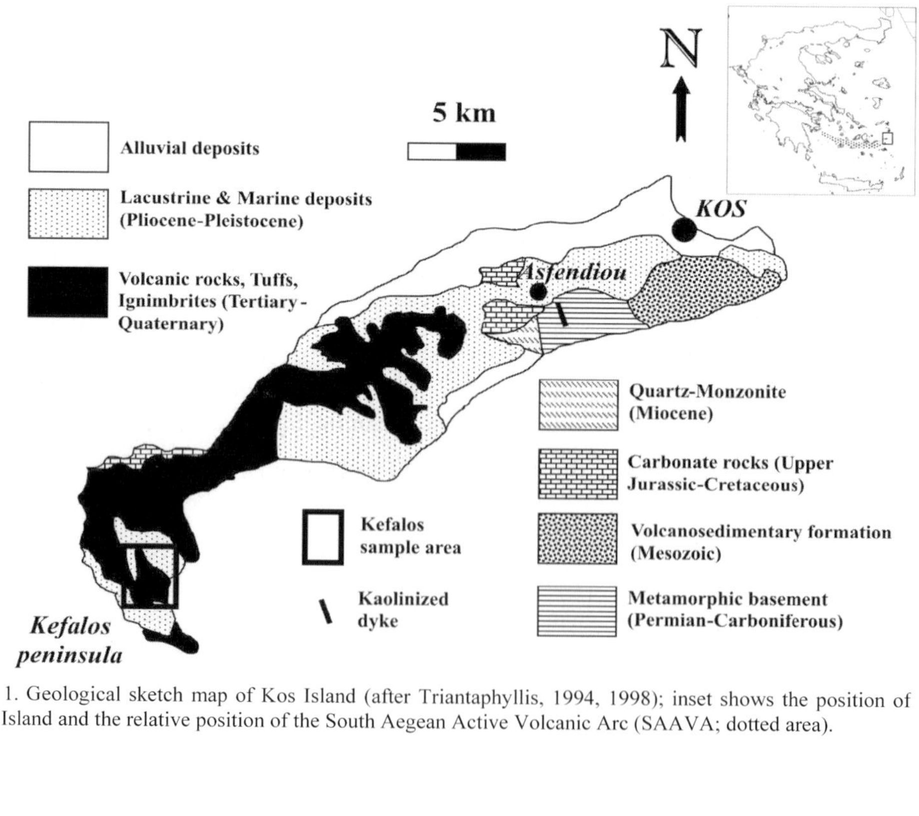

Fig. 1. Geological sketch map of Kos Island (after Triantaphyllis, 1994, 1998); inset shows the position of Kos Island and the relative position of the South Aegean Active Volcanic Arc (SAAVA; dotted area).

Permocarboniferous marls, impure limestones and sandstones, phyllites and rare mafic intercalations (Altherr et al., 1976). At 12 Ma, this series was partly contact metamorphosed by the intrusion of a large quartz monzonite (Fig. 1; Altherr et al., 1982; Henjes-Kunst et al., 1988; Kalt et al., 1998). Jurassic to Eocene limestones and dolomitic limestones tectonically overlie the above metamorphic basement. A successive series of volcanic rocks, mainly of rhyolitic-dacitic composition and pyroclastic formations covered the island of Kos (Fig. 1), while Pliocene and Pleistocene sediments occur mainly at the peripheral parts of the island (Keller, 1969; Dürr and Jacobshagen, 1986; Allen et al., 1999; Allen, 2001). The volcanic products of Kos derive from two periods of intense volcanic activity. The first began 10.4 Ma ago and continued until about 7.3 Ma ago, with effusions of magma belonging to the K–alkaline series and characterised by ignimbrite deposits. The second period began 3.4 Ma ago, with effusions of magma belonging to the Ca–alkaline series and continued up to the most recent volcanic activity on the island, some 140 ka ago (Boven et al., 1987; Dalabakis, 1987; Keller et al., 1990; Dalabakis and Vougioukalakis, 1993; Allen et al., 1999).

Rhyolite occurrences of Pliocene age (Boven et al., 1987) at Kefalos display variable degrees of kaolinization. The kaolinized rocks are generally white in colour, however they are frequently stained reddish by iron oxides; they extend over an area of about 0.5 km^2. Rhyolitic dykes at Asfendiou penetrate the metamorphic basement. The studied rocks were sampled from a completely kaolinized rhyolitic dyke (Papoulis, 2003).

3. ANALYTICAL TECHNIQUES

Mineralogical compositions of bulk rocks and of the clay fractions ($<2\mu m$), extracted by sedimentation, were determined by X-ray diffraction, using a Philips PW1050/25 diffractometer, with Ni-filtered, CuK_α radiation. Oriented clay powders were prepared by the dropper method and were scanned at $1^0 2\theta$/min from 3 to $60^o 2\theta$. For each $<2\mu m$ specimen, clay minerals were identified from three XRD patterns (air-dried at 25°C, ethylene-glycol solvated, and heated at 490°C for 2h).

Mineral textures and morphology were determined using a JEOL 6300 Scanning Electron Microscope (SEM) equipped with ED and WD spectrometers.

The Raman spectra were excited with linearly polarized light; an air-cooled Ar^+ laser was used with wavelength 514.5 nm (Spectra-Physics model 163-A42). A small-band-pass interference filter was used for the elimination of the laser plasma lines. The excitation beam was directed to a properly modulated sample compartment of a metallurgical microscope (Olympus BHSM-BH2). The focusing objective lens was a long working distance (8 mm) 50x/0.55 Olympus lens.

The spectra were obtained with a power of 3mW on the specimen for a total integration time of 12s; a viewing screen connected to the microscope offered good sample positioning, good laser beam focusing, and direct surface inspection. The scattered light was filtered (HNF-514-1.0 from Kaiser Optical Systems), to remove the elastic Rayleigh scattering. The T-64000 (Jobin Yvon) Raman system, equipped with a Spectraview-2D liquid N_2-cooled CCD detector, was used, in the single spectrograph configuration, to disperse and detect the Raman signal. The spectral resolution for the Raman spectra used for molecular orientation was ~5cm^{-1}. The spectral window centered at 3600cm^{-1} (in the Stokes region). The Raman spectrum has been recorded with the FT-Raman FRA-106-S module of an Equinox 55 FTIR spectrometer of Bruker.

Infrared spectra were measured in the specular reflectance mode on a Bruker vacuum spectrometer (IFS 113v) equipped with a near-normal incidence (11°) reflectance accessory. A KRS-5 wire grid polarizer was positioned before the sample to polarize the infrared radiation perpendicular to the plane of incidence. The polarizer remained fixed and the sample was rotated by 90° to obtain reflectance spectra with different polarization directions. All spectra were measured at room temperature and represent the average of 200 scans at 2cm^{-1} resolution.

4. RESULTS

Representative FTIR, Raman and FT-Raman spectra along with XRD patterns are illustrated in Figs. 2-13. Two parageneses occur at Kefalos, at moderately and highly altered samples, respectively:

1. Kaolinite ± dickite ± mixed-layer kaolinite/smectite + quartz + cristobalite
2. Dickite + kaolinite + quartz + zunyite ± pyrophyllite ± calcite

At Asfendiou, only highly altered samples were collected, consisting of:

1. Kaolinite + dickite + quartz + zunyite ± pyrophyllite ± calcite

Traces of alunite occur too, sporadically in both areas.

Evidently, FTIR, Raman and FT-Raman techniques indicate that kaolinite is the most abundant kaolinite-group clay mineral in moderately altered samples from Kefalos, while in the highly altered ones dickite predominates. The FTIR, Raman and FT-Raman spectra of the Asfendiou kaolinites suggest that kaolinite prevails. This is particularly evident from the comparison of FT-Raman spectra of highly altered samples from both areas of Kos to the San Juanito standard dickite, illustrated in Fig. 10.

SEM observations and microanalyses assisted the present study revealing the additional existence of smectite, illite, barite and halite. The mixed-layer kaolinite-smectite developed from the alteration of K-feldspar (Fig. 14) while Na-plagioclase alters to kaolinite. Book-type dickite forms after kaolinite and represents peak conditions of the hydrothermal alteration (Fig. 15).

5. DISCUSSION

5.1 Comparison of FTIR, Raman and FT-Raman spectroscopic techniques with XRD

Clearly, the FTIR, Raman and FT-Raman spectroscopic techniques provide significant advantages over XRD analysis. The occurrence of kaolinite and cristobalite

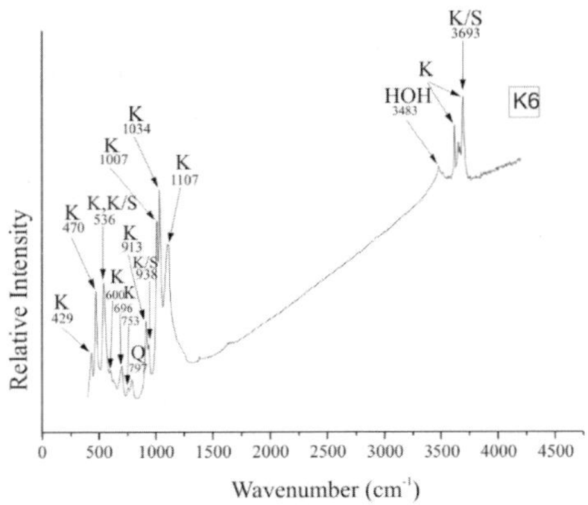

Fig. 2. FTIR spectrum of sample K6 (moderately altered) from Kefalos, in the region 400-4200 cm-1. K=kaolinite, Q=quartz, K/S=mixed-layer kaolinite-smectite. The presence of mixed-layer kaolinite-smectite is indicated by the band appearing at 536cm-1, instead of 540cm-1, as well as by the band at 938cm-1 (Madejova et al., 2002).

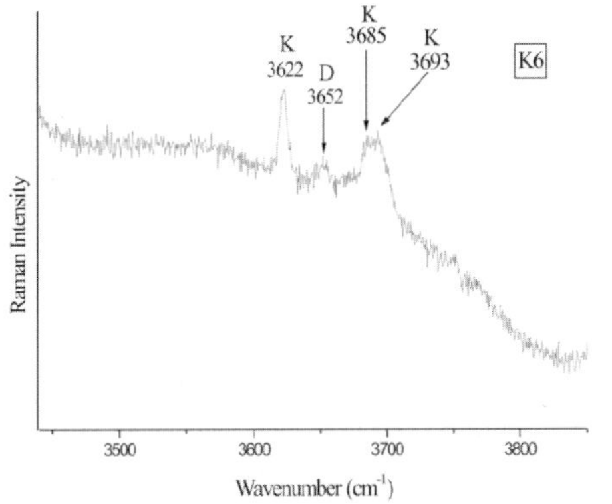

Fig. 3. Raman spectrum of sample K6 (moderately altered) from Kefalos, in the hydroxyl stretching region 3550-3750cm^{-1}. K=kaolinite, D=dickite. Note the presence of traces of dickite.

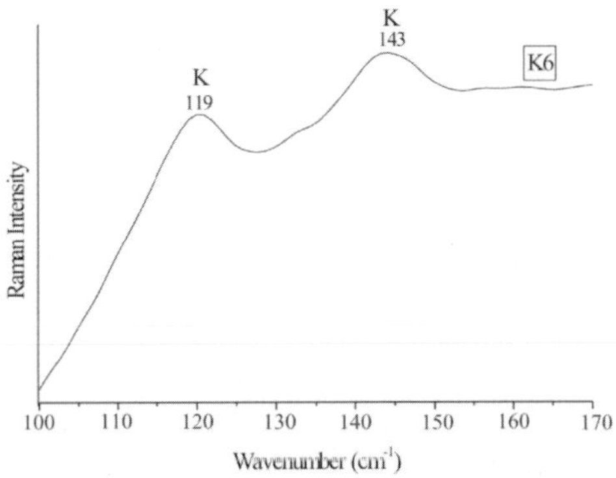

Fig. 4. FT-Raman spectrum of sample K6 (moderately altered) from Kefalos, in the region 100-170 cm-1. K=kaolinite. The high background is probably due to the fluorescence caused by the presence of Fe-oxides.

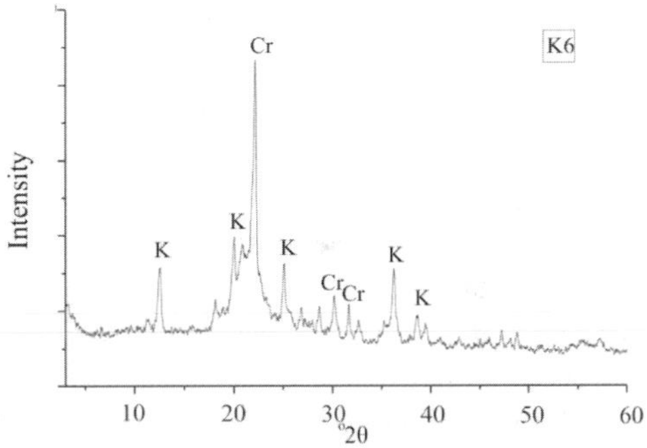

Fig. 5. Representative X-ray powder diffraction pattern of sample K6 (moderately altered) from Kefalos. Cr=cristobalite, K=kaolinite.

can be easily verified in moderately altered samples, from X-Ray Diffraction patterns, however, certain peaks of kaolinite are weak compared to the peaks of cristobalite and partially overlap with each other (Fig. 5). Generally, the kaolinite-group of minerals can be easier distinguished using FTIR, Raman and FT-Raman spectroscopic techniques. The existence of dickite in moderately altered samples cannot be ascertained from the X-Ray diffractograms (Fig. 5), probably due to its restricted amount; in the highly altered samples, it is not possible to determine the relative abundance of dickite and kaolinite (Figs. 12, 13). The occurrence of dickite and the relative abundance of dickite

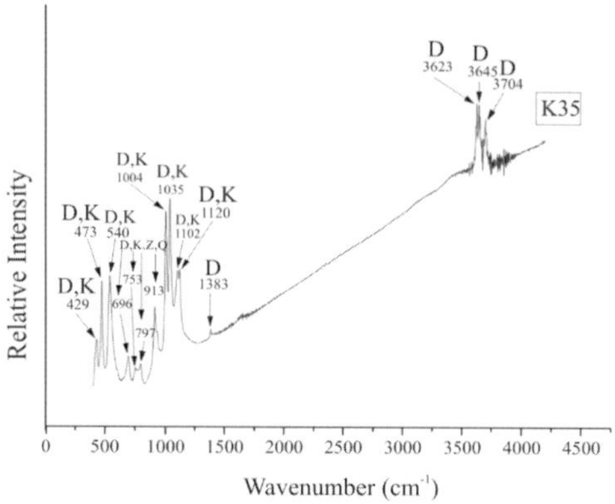

Fig. 6. FTIR spectrum of sample K35 (highly altered) from Kefalos, in the region 400-4200cm^{-1}. K=kaolinite, Q=quartz, D=dickite and Z=zunyite.

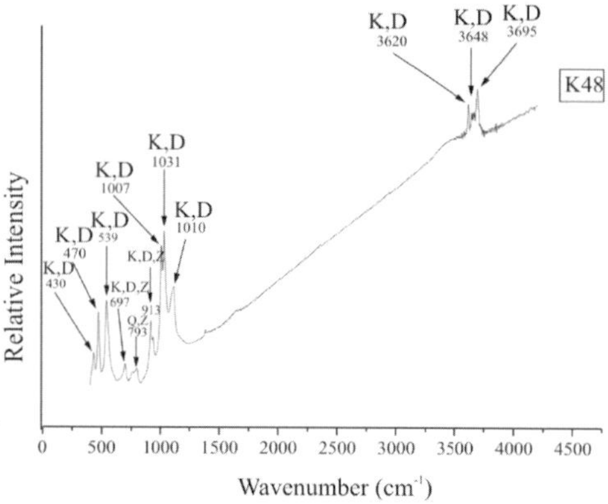

Fig. 7. FTIR spectrum of sample K48 (highly altered) from Asfendiou, in the region 400-4200 cm^{-1}. K=kaolinite, Q=quartz, D=dickite and Z=zunyite.

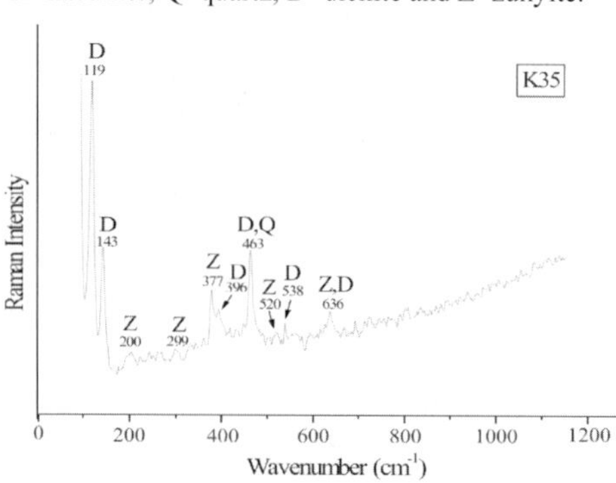

Fig. 8. FT-Raman spectrum of sample K35 (highly altered) from Kefalos, in the region 100-1200 cm^{-1}. D=dickite, Q=quartz and Z=zunyite. Dickite predominates over kaolinite.

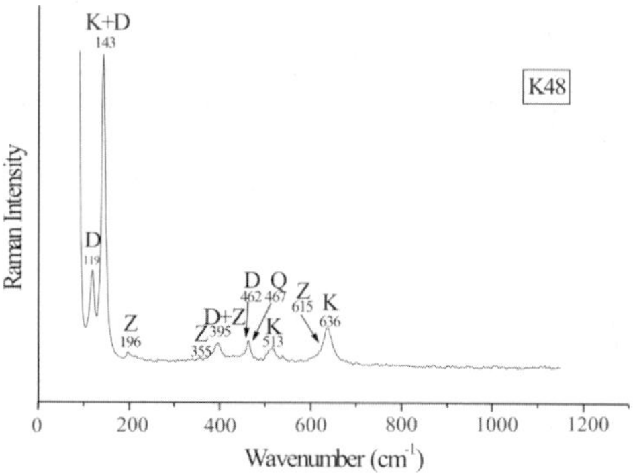

Fig. 9. FT-Raman spectrum of sample K48 (highly altered) from Asfendiou, in the region 100-1200 cm^{-1}. K=kaolinite, D=dickite, Q=quartz and Z=zunyite. Kaolinite predominates over dickite.

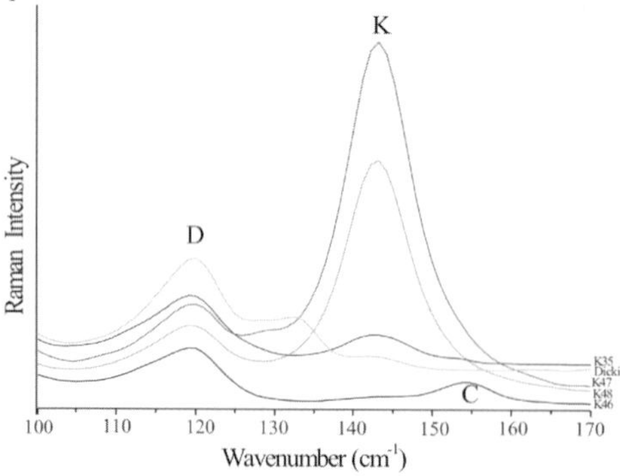

Fig. 10. Comparative FT-Raman spectra of moderately (K47, K48), highly altered (K35, K46) samples and the San Juanito dickite standard (Frost, 1997) in the region 100-170cm-1. D=dickite, K=kaolinite, C=calcite.

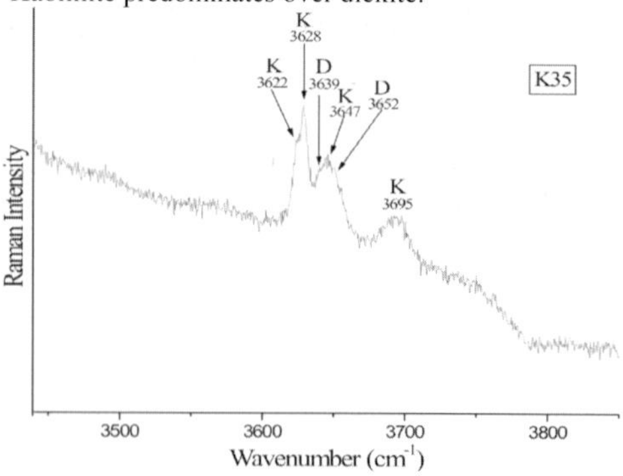

Fig. 11. Raman spectrum of sample K35 (highly altered) from Kefalos, in the region 3550-3750 cm^{-1}. D=dickite, K=kaolinite. Dickite predominates over kaolinite.

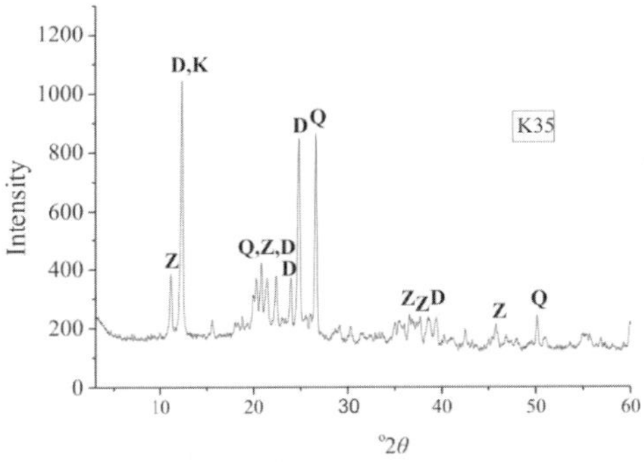

Fig. 12. Representative X-ray powder diffraction pattern of sample K35 (highly altered) from Kefalos. Q=quartz, K=kaolinite, D=dickite, Z=zunyite.

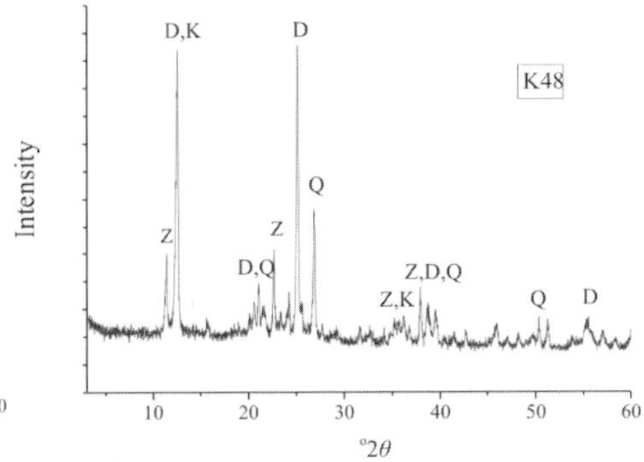

Fig. 13. Representative X-ray powder diffraction pattern of sample K48 (highly altered) from Asfendiou. Q=quartz, K=kaolinite, D=dickite, Z=zunyite.

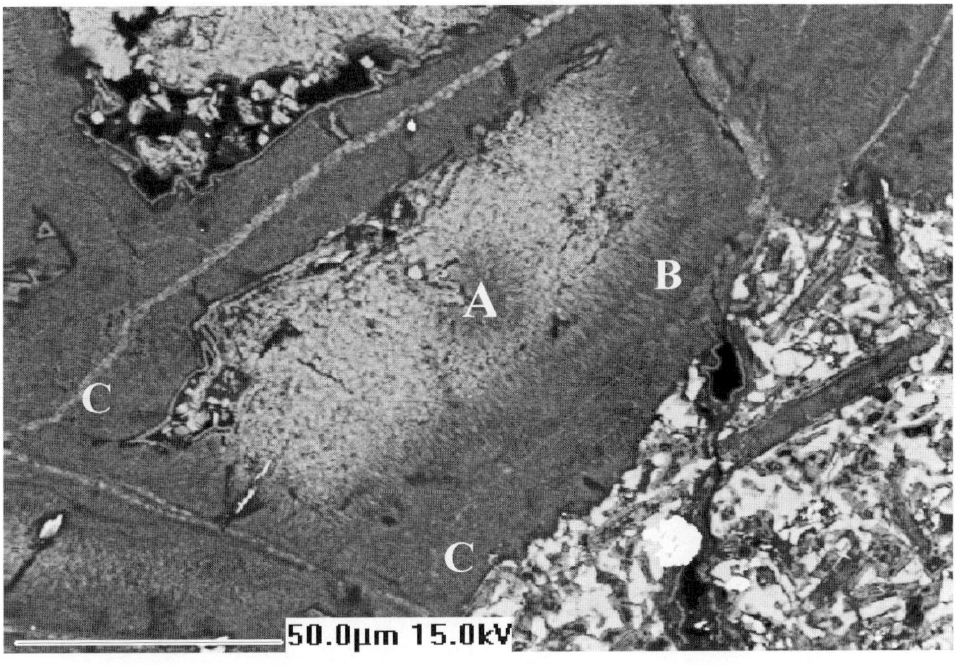

Fig. 14. Back-scattered image of a partially altered K-feldspar (A) to: a mixed-layer kaolinite-smectite zone (B) and a kaolinite-rich zone (C) (sample K6, moderately altered).

and kaolinite can be easily detected in the FT-Raman and Raman spectra (Figs. 3, 8, 9, 10, 11). Unlike the FTIR spectra (Fig. 5), no clear evidence for the presence of mixed-layer kaolinite-smectite can be obtained from XRD data (Fig. 5). This is probably due to

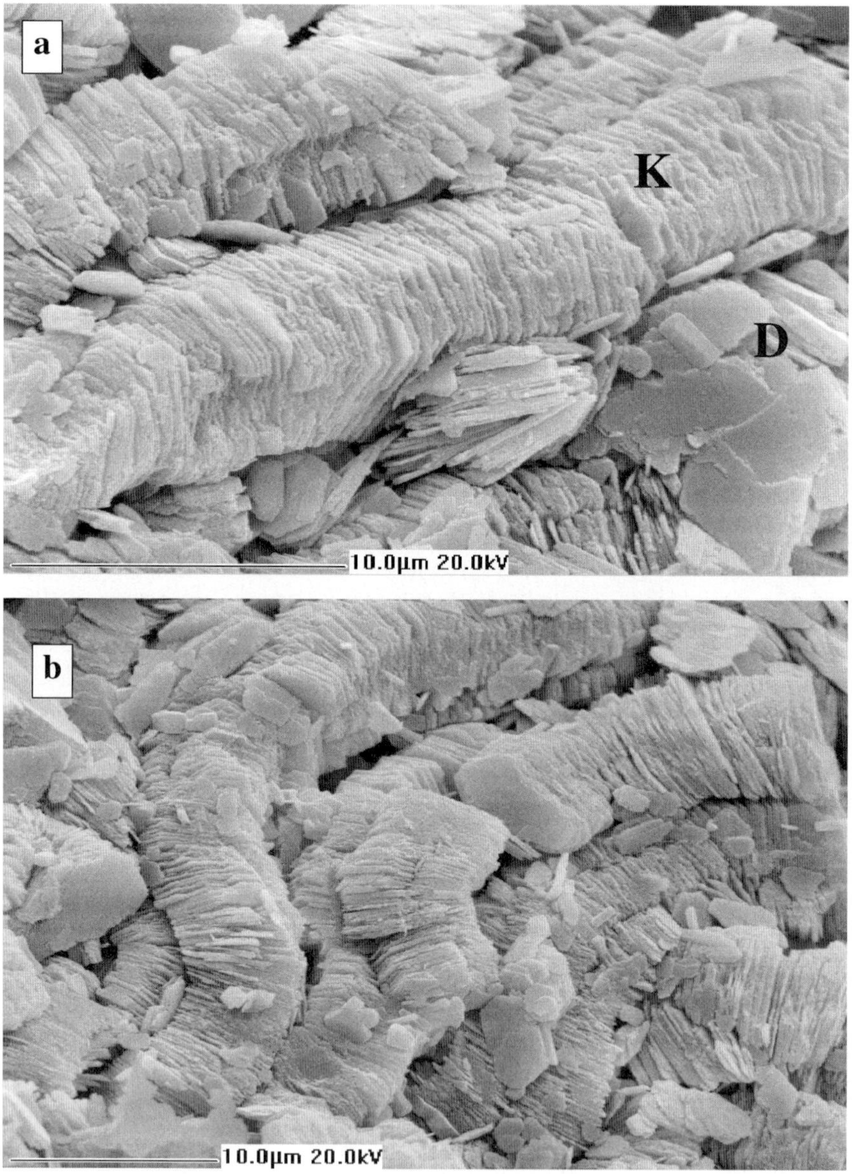

Fig. 15. a) SEM image of kaolinite (K) coexisting with dickite (D) (sample K6, moderately altered); b) book-type dickite (sample K35, highly altered).

the small amount of mixed-layer kaolinite-smectite present. The occurrence of zunyite, which is a rare mineral, is verified from the data obtained from the spectroscopic methods (Figs. 6, 7, 8, 9) whereas the XRD patterns of the samples commonly display very weak or unclear peaks (Figs. 12, 13), due to the restricted appearance of this mineral.

5.2 Conditions of hydrothermal alteration of rhyolites in Kos Island

FTIR, FT-Raman and Raman spectroscopic study of hydrothermally altered rocks of rhyolitic composition from Kos Island, showed that kaolinite is the most abundant clay mineral present in the moderately altered samples from Kefalos and in the highly altered samples from Asfendiou. Dickite, on the other hand, dominates the highly altered samples from Kefalos.

The coexistence of kaolinite and mixed-layer kaolinite-smectite, in some samples, is thought to represent an intermediate stage in the kaolinitization process of K-feldspars. The presence of mixed-layer kaolinite-smectite is common in pedogenesis; it has been rarely found in hydrothermally altered rocks (Wiewiora, 1971, 1973; Thomas, 1989). It has been shown that such abiotic environments could likely produce the necessary rate of removal of SiO_2 or addition of Al_2O_3 to form this phase. The sporadic formation of mixed-layer kaolinite-smectite, in Kefalos kaolins, is probably related to local fluctuations of the fluid phase.

With advancing alteration, kaolinite progressively transforms to dickite. Kaolinite is replaced by dickite in the highly altered samples from Kefalos and Asfendiou, therefore dickite is considered to be the final product of the kaolinitization process, instead of kaolinite. The transformation of kaolinite to dickite in Asfendiou was not so extensive as in Kefalos. The kaolinitization process in Kos rhyolites follows two main alteration paths: (i) K-feldspar→mixed-layer kaolinite-smectite→kaolinite→dickite (Fig. 14), (ii) Na-plagioclase→kaolinite→dickite. Nevertheless, a minor amount of kaolinite could also originate from alteration of volcanic glass.

To our knowledge, this is the first occurrence of zunyite reported in Greece. Coexistence of zunyite, kaolinite, dickite and traces of pyrophyllite, is not feasible in a wide range of conditions. According to Inoue (1995), Dill et al. (1997) and Willan and Armstrong (2001) the assemblage dickite + kaolinite + quartz + zunyite ± pyrophyllite is stable at a temperature range between 250°C and 290°C, at low pH (<3.5), thus restricting conditions of alteration in Kos Island.

The occurrence of zunyite ($Al_{13}Si_5O_{20}(OH,F)_{18}Cl$) in the kaolins of Kos Island, implies incorporation of hydrothermal fluids rich in F^- and Cl^-, thus indicating contamination by seawater. The formation of halite is consistent with the involvement of hydrothermal fluids rich in Cl^-. Alunite is a common mineral in advanced argilic alteration zones; occurrence of traces of alunite in the Kos assemblages indicates the involvement of S^{-2}-poor hydrothermal fluids. Papoulis and Tsolis-Katagas (2001) argue that the advanced argilic alteration zone at Kefalos extends to a small distance around faults and the hydrothermal fluids were channelled along them.

6. CONCLUSIONS

FTIR, FT-Raman and Raman spectroscopic methods are powerful tools in determining certain clay mineral phases, such as zunyite, mixed-layer kaolinite-smectite, participating in low amounts, and in distinguishing coexisting kaolinite from dickite, in Kos samples. It was also possible to estimate their relative abundance in the studied samples. Discriminating between the latter minerals is very important because dickite can indicate higher temperatures and potentially slightly different pH than kaolinite. FTIR, FT-Raman and Raman are more sensitive methods in determining polymorphs and minerals present in low amounts, relative to X-Ray diffraction analysis.

The moderately altered samples from Kefalos contain kaolinite ± dickite ± mixed-layer kaolinite/smectite + quartz + cristobalite while the assemblage of the highly altered samples from both Kefalos and Asfendiou contains dickite + kaolinite + quartz + zunyite ± pyrophyllite ± calcite, suggesting the involvement of a single hydrothermal alteration event. The kaolinitization process in Kos Island rhyolites follows two main alteration paths: (i) K-feldspar→mixed-layer kaolinite-smectite→kaolinite→dickite and (ii) Na-plagioclase→kaolinite→dickite.

The assemblage kaolinite + dickite + zunyite ± pyrophyllite constrains temperature at a range of 250-290°C, and pH below 3.5. The occurrence of zunyite, coupled with the presence of halite, suggest involvement of hydrothermal fluids rich in F^- and Cl^-, thus indicating contamination by seawater, while presence of alunite in trace amounts, is compatible with hydrothermal fluids poor in S^{-2}.

Acknowledgements

The authors wish to thank Dr. Voyiagis of the Laboratory of molecular spectroscopy, Research Center of High Temperature and Chemical Processes, University of Patras, for his help with FTIR, FT-Raman and Raman spectroscopy and Mr. Kotsopoulos, of the Laboratory of Electron Microscopy and Microanalysis, University of Patras, for his help with the SEM micrographs. D.P. is thankful to the Greek State Scholarship Foundation for financial support during his Ph.D. study. Suggestions by anonymous reviewers are gratefully acknowledged.

REFERENCES

Allen, S.R., 2001. Reconstruction of a major caldera-forming eruption from pyroclastic deposit characteristics: Kos Plateau Tuff, eastern Aegean Sea. J. Volc. Geoth. Res., 105: 141-162.

Allen, S.R., Stadlbauer, E. and Keller, J., 1999. Stratigraphy of the Kos Plateau Tuff: Products of a major Quaternary explosive rhyolitic eruption in the eastern Aegean Sea. Int. J. Earth Sci., 88: 132–156.

Altherr, R., Keller, J. and Kott, K., 1976. Der jungtertiäre Monzonit von Kos und sein Kontakthof (Ägäis, Griechenland). Bull. Soc. Géol. France 7(XVIII): 403-412.

Altherr, R., Kreuzer, H., Wendt, I., Lenz, H., Wagner, G.A., Keller, J., Harre, W. and Höhndorf, A., 1982. A late Oligocene/early Miocene high temperature belt in the Attic-Cycladic Crystalline Complex (SE Pelagonian, Greece). N. Jb. Geol., E23: 97-164.

Boven, A., Brousse, R., Dalabakis, P. and Pasteels, P., 1987. Geological and geochronological evidences on the evolution of Kos-Jali-Nisyros eruptive centres, Aegean arc, Greece. Terra Cognita, 7: 328-329.

Dalabakis, P., 1987. Le volcanisme récent de l'île de Kos. Ph.D. Thesis, Paris-Sud Orsay, pp. 266.

Dalabakis, P. and Vougioukalakis, G., 1993. The Kefalos tuff ring (W. Kos): Depositional mechanisms, vent position, and model of the evolution of the eruptive activity. Bull. Geol. Soc. Greece, 28/2: 259-273 (in Greek with English Abstr.).

Dill, H.G., Bosse, R., Henning, K.H. and Fricke, A., 1997. Mineralogical and chemical variations in hypogene and supergene kaolin deposits in a mobile fold belt the Central Andes of northwestern Peru. Miner. Deposita, 32: 149-163.

Dürr, St. and Jacobshagen, V., 1986. Ostägäische Inseln. In: V. Jacobshagen (Editor), Geologie von Griechenland. Gebrüder Borntraeger, Berlin, pp. 169-187.

Frost, R.L., 1997. The structure of the kaolinite minerals – a FT-Raman study. Clay Minerals, 32: 65-77.

Frost, R.L., 1998. Hydroxyl deformation in kaolins. Clays and Clay Minerals, 46: 280-289.

Frost, R.L., Barlett, J.R. and Fredericks, P.M., 1993. Fourier transform Raman spectra of kandite clays. Spectr. Acta, Part A, 49: 667-674.

Frost, R.L. and Kloprogge, J.T., 2001. Towards a single crystal Raman spectrum of kaolinite at 77K. Spectr. Acta, Part A, 57: 163-175.

Giese, R.F., 1998. Kaolin minerals: Structures and stabilities. In: S.W..Bailey (Editor), Hydrous phyllosilicates. Rev. Mineral., 19. Mineral. Soc. Am. Chelsea, MI: BookCrafters, pp. 29-66.

Henjes-Kunst, F., Altherr, R., Kreuzer, H. and Hansen, B.T., 1988. Disturbed U-Th-Pb systematics of young zircons and uranothorites: the case of the Miocene Aegean granitoids (Greece). Chem. Geol., 73: 125-145.

Inoue, A., 1995. Formation of clay minerals in hydrothermal environments. In: B. Velde (Editor), Origin and Mineralogy of Clays, Clays and the Environment, pp. 268-329.

Kalt, A., Altherr, R. and Ludwig, T., 1998. Contact metamorphism in pelitic rocks on the Island of Kos (Greece, Eastern Aegean Sea): a test for the Na-in-cordierite thermometer. J. Petrol., 39: 663-688.

Keller, J., 1969. Origin of rhyolites by anatectic melting of granite and crustal rocks. The example of rhyolitic pumice from the island of Kos (Aegean Sea). Bull. Volcanol., 33: 942–959.

Keller, J., Rehren, T.H. and Stadlbauer, E., 1990. Explosive volcanism in the Hellenic Arc: a summary and review. In: D.A. Hardy, J. Keller, V.P. Galanopoulos, N.C. Flemming and T.H. Druitt (Editors), Thera and the Aegean World III.

Proc. of the Third Intern. Congr. on the Volcano of Thera. Santorini, pp. 13–26.

Madejová, J., Kečkés., J., Pálková, H. and Komadel, P., 2002. Identification of components in smectite/kaolinite mixtures. Clay Minerals, 37: 377-388.

Papoulis, D., 2003. Mineralogical study, kaolinitization processes and properties of kaolins from Leucogia, Drama and Kos Island. Ph.D. Thesis, Univ. Patras (in Greek with English Abstr.).

Papoulis, D. and Tsolis-Katagas, P., 2001. Kaolinization process in the rhyolitic rocks of Kefalos, Kos island, Aegean sea, Greece. Bull. Geol. Soc. Greece, 34/3: 867-874.

Shoval, S., Boudeulle, M., Yariv, S., Lapides, I. and Panczer, G., 2001. Micro-Raman and FT-IR spectroscopy study of the thermal transformations of St. Claire dickite. Optical Materials, 16: 319-327.

Shoval, S., Yariv, S., Michaelian, K.H., Boudeulle, M. and Panczer, G., 2002. Hydroxyl-stretching bands in polarized micro-Raman spectra of oriented single-crystal Keokuk kaolinite. Clays and Clay Minerals, 50: 56-62.

Thomas, A.R., 1989. A new mixed layer clay mineral – Regular 1:1 mixed layer kaolinite/smectite. Abstracts, 26th Annual Meeting Clay Minerals Society, Sacramento, California, 69.

Triantaphyllis, M., 1994. Geological map of Greece, Western Kos (Kefalos) sheet, 1:50000, publ. I.G.M.E., Athens

Triantaphyllis, M., 1998. Geological map of Greece, Eastern Kos sheet, 1:50000, publ. I.G.M.E., Athens

Wiewiora, A., 1971. A mixed layer kaolinite-smectite from Lower Silesia, Poland. Clays and Clay Minerals, 19: 415-416.

Wiewiora, A., 1973. Mixed layer kaolinite-smectite from Lower Silesia, Poland. Final Report: Proc. Int. Clay Conference, Madrid, pp. 75-78.

Willan, R.C.R. and Armstrong, D.C., 2001. Successive geothermal, volcanic-hydrothermal and contact-metasomatic events in islands-arc basalts, South Shetland Islands. Durham meeting abstracts, pp. 57-58.

The South Aegean Active Volcanic Arc
M. Fytikas and G.E. Vougioukalakis

Rhyolitic dykes of Paros Island, Cyclades

A. Hannappel* and T. Reischmann

Institut für Geowissenschaften, Johannes Gutenberg-Universität, Becherweg 21, D-55099 Mainz, Germany

ABSTRACT

The discovery of rhyolitic dykes from the NE part of the island of Paros is described here for the first time. The dykes that can be mapped for a length of ca. 1.1 km are striking about 6° and 38° NE. The width reaches up to 11 m maximum. The rhyolitic rocks are porphyric with a fine-grained matrix of mainly feldspar, quartz, some biotites and opaques. K-feldspar, biotite and plagioclase occur as phenocrysts, and mafic-intermediate enclaves are common. At the chilled margin in contact to the gneissic country rocks the dykes are vitric. There, the enclaves have been preserved from contact reaction with the melt and alteration effects and show primary igneous minerals such as clinopyroxene and plagioclase. This implies that the enclaves are no xenoliths from the basement gneisses, but cogenetic with the dykes, probably from less differentiated parts of the magma.

The dykes are leucocratic with 68.20-74.03 wt. % SiO_2. They display fractionation trends in variation diagrams implying that the compositional variation is largely based on crystal fractionation from a parental magma. Since no other Neogene volcanics are known from Paros, we compare the dykes with rhyolites from Antiparos. The dykes are slightly less silicic than the Antiparos rhyolites. This corresponds to higher concentrations of Al_2O_3, Fe_2O_3(t), MgO, CaO and TiO_2. However, because of higher concentrations of HFSE such as Zr, Nb, Y, and LILE such as Ba and LREE the dykes are unlikely to be comagmatic with the Antiparos rhyolites.

In commonly used discrimination diagrams for granites (Pearce et al., 1984) the Paros rhyolithic dykes are classified as WPG (within-plate granites). According to this discrimination the relation to the South Aegean Volcanic arc appears to be more complex and might involve melting or assimilation of continental material in the back-arc region.

Keywords: Rhyolites, dykes, Cyclades, Hellenic arc, enclaves.

* Corresponding author: e-mail: cybernetic@t-online.de

1. INTRODUCTION

The South Aegean Volcanic Arc is a zone of Pliocene to present volcanism that is related to the subduction of the African plate beneath Eurasia (Nicholls, 1971; Pe and Piper, 1972). In historical time volcanic centers were active in Methana, Milos, Santorini and Nisyros. Volcanic rocks of Pliocene age are found on the areas of Aegina, Crommyonia and Poros. Volcanism began in Pliocene and reached its maximum in the Quaternary. The active volcanoes are located 130-150 km above a seismically defined Wadati-Benioff zone consisting of a shallow branch dipping at low angle (ca. 30°) and a deep branch dipping at high angle (ca. 45°) (Papazachos et al., 2000). The influence of partial melting of the asthenosphere is indicated by high S-wave attenuation beneath Milos and Santorini (Mitropoulos and Tarney, 1992). According to seismic tomography the subducted slab beneath the Aegean Arc can be traced to a depth of at least 1200 km, which implies at least 26 Ma of subduction (Meulenkamp et al., 1988).

The voluminous igneous activity shifted southwards during the Cenozoic from northern Greece and Bulgaria, through the north Aegean islands and the Cyclades to the present South Aegean Volcanic arc. This migration is a result of a rapid extension of the Aegean Sea (Fytikas et al., 1984) and subduction roll-back (Royden, 1993) in the Miocene. The extension has its maximum in the central part of the arc (Makris, 1977). The geochemical character of the rocks of the South Aegean Volcanic arc ranges from basaltic andesite through dacite to rhyolite.

The Central Aegean area behind the arc has geophysical features of a back-arc basin such as high heat flow (Fytikas and Kolios, 1979) and a shallower Moho depth of ca. 30 km (Makris and Röwer, 1986). However, there are some important differences to typical back-arc regions (Berkhemer, 1978) such as the continental character of the crust (Makris, 1977) and the absence of basaltic back-arc volcanism.

A wide range of different Miocene to Quaternary volcanic rocks occurs in the back-arc region further away from the arc in Macedonia, Evia, Skyros, Chios-Psara, Samos, Patmos and Cyclades-Kos-Bodrum (Pe-Piper and Piper, 2002).

In this contribution we describe the new discovery of rhyolitic dykes, which were found by detailed mapping in the NE corner of the island of Paros. We present the composition of the major minerals and the whole-rock chemistry and discuss possible relations to the rhyolites of the neighbor island of Antiparos and the volcanic rocks of the active South Aegean arc.

2. GEOLOGICAL BACKGROUND

2.1 Attic-Cycladic crystalline complex

The Attic-Cycladic crystalline complex is part of the southeastern Pelagonian zone, which extends from western Macedonia trough Thessaly, Euboecos, Attika and the Cyclades to the Menderes Massif in Anatolia.

Its exposed parts that are visible nowadays were still situated in the middle-crust during early Miocene and were exhumed not until Middle-Miocene after neohellenic overprinting. After Dürr et al. (1978a) and Altherr et al. (1979) two major tectonic units can be distinguished. The lower unit is a nappe pile consisting of pre-alpine Permo-Carboniferous basement (Reischmann, 1998; Engel and Reischmann, 1998) and Mesozoic neritic marbles, metapelites, metavolcanic rocks and metaophiolites. The upper unit consists of variegated klippen: Ophiolites that are covered by late Cretaceous neritic limestones and meso-autochthone molasse of Aquitan and Burdigal age (Jansen, 1977; Dürr et al., 1978a; Angelier et al., 1978; Dermitzakis and Papanikolaou, 1979; Papanikolaou, 1980a), Permian and Triassic (meta)-sediments (Dürr and Altherr, 1979; Papanikolaou, 1979, 1980; Marks and Schuiling, 1965) and late Cretaceous high-grade metamorphites, as well as I-type and S-type granites (Dürr et al., 1978b; Reinecke et al., 1982).

In early Eocene, some 40-45 Ma ago, the lower unit was overprinted by high-grade (450-500° C and min. 14 kbars) metamorphism (Dixon, 1976; Altherr et al., 1982). This metamorphic event was followed by a Barrovian metamorphic phase in late Oligocene, 25 Ma ago, which culminated in the formation of thermal domes on the islands of Naxos and Sifnos (Jansen and Schuiling, 1976; Okrusch et al., 1978; Andriessen, 1978; Schliestedt et al., 1987), where it could reach 670-700°C and 2 kbar resulting in the formation of migmatites (Wijbrans and McDougall, 1988; Buick and Holland, 1989).

The subsequent uplift and cooling in the Miocene was accompanied by the emplacement of I- and S-type granitoids that intruded all major tectonic units (Marakis, 1970; Altherr et al., 1982; Henjes-Kunst et al., 1988). Altherr et al. (1982) has shown that the granitoids feature a systematic spatial variation in chemical composition:

- The K_2O contents of the I-type granites increase continuously from WSW to ENE with increasing distance from the former trench (Bateman, 1979).
- The occurrence of S-type granites is limited to areas where I-type granites with high K_2O abundances occur (Miller and Bradfish, 1980).

The initial emplacement of the plutons was associated with extension and subsequent transtensional deformation that was followed by further magma emplacement leading to brittle transtensonial conditions (Boronkay and Doutsos, 1994). The last phase of igneous activity is the active magmatism of the South Aegean arc that began in the Middle Miocene (Fytikas et al., 1976).

2.2 Geology of Paros

Paros has been mapped in detail by Papanikolaou (1977, 1980a) and Robert (1982). According to Papanikolaou (1980a) the following three main tectonic nappes can be distinguished (Fig. 1): The lower Marathi nappe is of amphibolite grade and consists of an orthogneiss group and a group of marbles and amphibolites. These two groups of the Marathi nappe are separated by a major tectonic contact. The occurrence of metabauxites in Paros and their analogy with the Menderes Massif (Dürr, 1975) and Samos (Papanikolaou, 1980b) led Papanikolaou (1980a) to conclude a possible Triassic-Cretaceous age for the amphibolites and for the marbles.

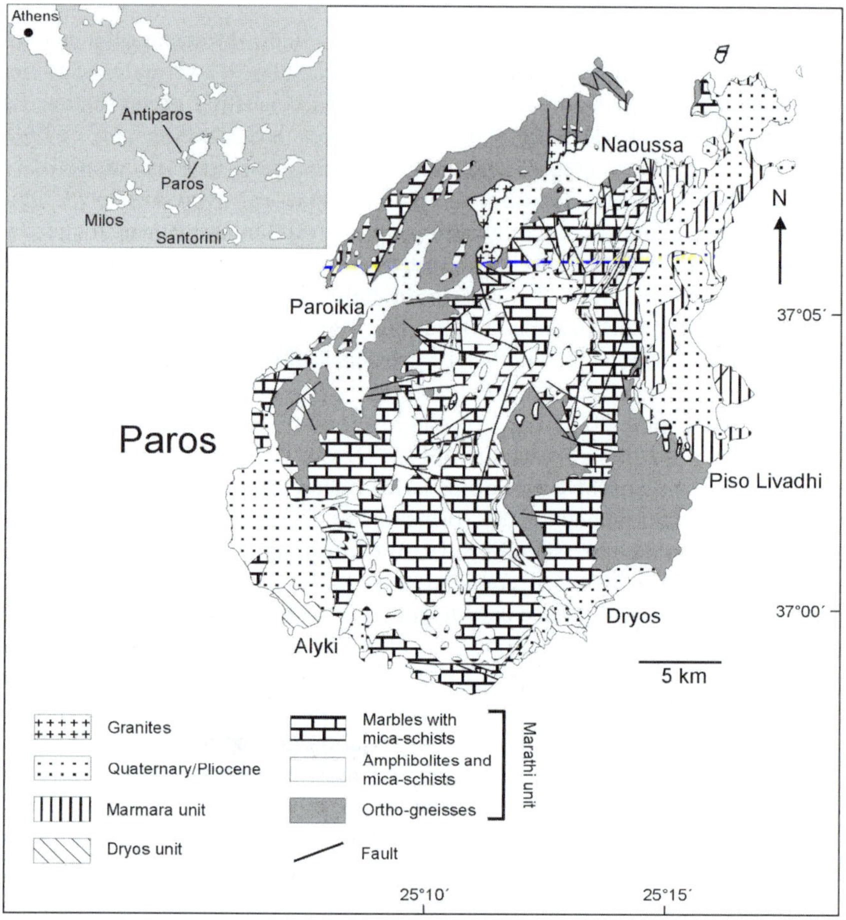

Fig. 1. Simplified geological map of Paros (modified after Papanikolaou, 1977).

In contrast to Papanikolaou (1980a), Robert (1982) and Altherr et al. (1982) interpreted the gneisses of the Marathi nappe as paragneisses. Furthermore, Robert (1982) found no evidence for a major tectonic contact between the two groups of the Marathi nappe.

The intermediate low-grade nappe of Dryos comprises metadiabases, marbles and phyllites that are – at least partly – of Permian age. The upper Marmara nappe is non-metamorphic and comprises ophiolites that are overlain by neritic limestones of Late-Cretaceous age (Papanikolaou, 1980a). The ophiolites and the Cretaceous limestones are covered by a meso-autochtonous molasse of early to middle Miocene (Roesler, 1978; Dermitzakis and Papanikolaou, 1979).

In northeast Paros, near the town of Naoussa, gneisses and metasediments have been intruded by several smaller S-type granites. Biotite from the well-exposed Naoussa granite yielded an age of 12.4 Ma (Altherr et al., 1982), interpreted as cooling age. Apatite fission-track ages of samples from Paros indicate rapid cooling between 12 and 8 Ma ago (Hejl et al., 2003).

2.3 Geology of Antiparos

Antiparos and its two adjacent islands Despotiko and Strongylo comprise mainly gneisses and schists that are intercalated with marbles. The marbles comprise fine-grained fossiliferous limestones that have been interpreted as Permian by Anastopoulos (1963). In the southernmost part of Antiparos Pliocene volcanic rocks are exposed. They consist of rhyolitic domes, lava flows and pyroclastics (Fig. 2). The lava flows are often vesiculated and comprise lenses of obsidian (Innocenti et al., 1982). They are

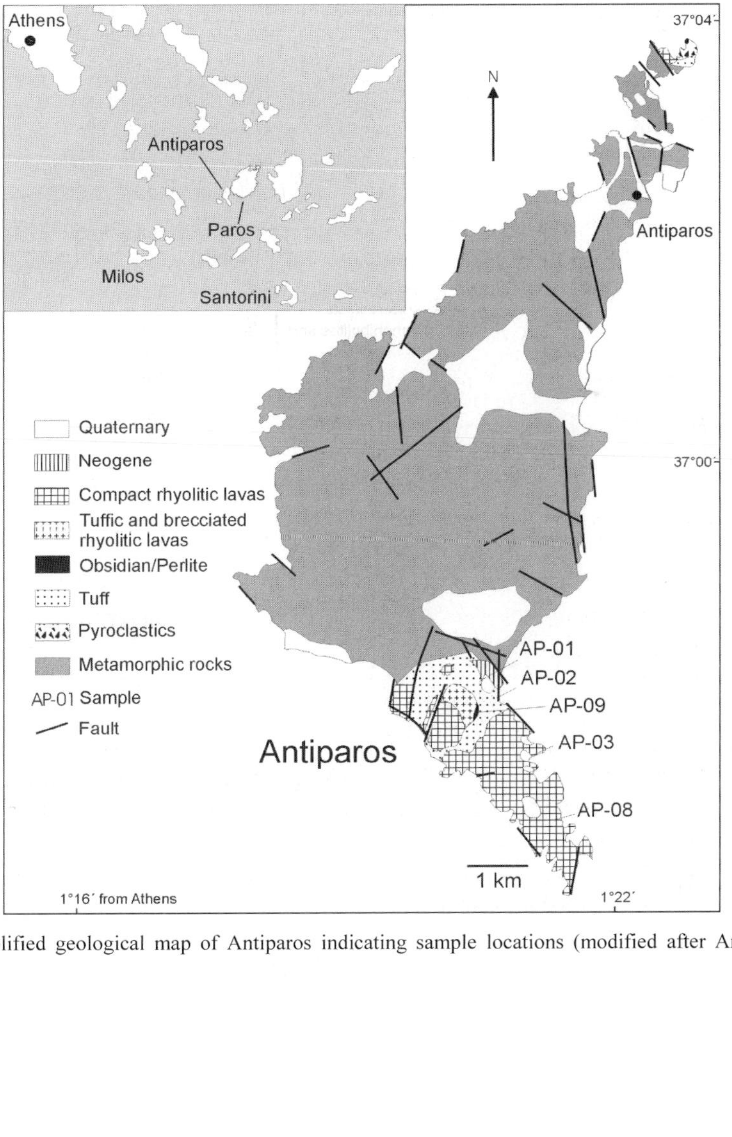

Fig. 2. Simplified geological map of Antiparos indicating sample locations (modified after Anastopoulos, 1963).

geochemically similar to rocks of the South Aegean volcanic arc (Keller, 1982) and yielded a radiometric age between 4.0 and 5.4 Ma (Innocenti et al., 1982).

2.4 Geological and tectonic setting of study area

The study area is located in the NE part of Paros, 3 km northwest of Naoussa (Fig. 3). The dominant rock types are well-foliated orthogneisses with an assemblage of quartz, feldspar, biotite and muscovite. Zircon, apatite and Fe-Ti oxides occur as accessory minerals. The gneisses of this part of the island were dated by Engel and Reischmann (1998) at 314±3 and 319±2 Ma. The gneisses were intruded by the Naoussa granite, mostly along its dominant foliation. Pegmatitic and aplitic dykes generally crosscut foliation of the gneisses. The major constituents of the Pliocene Naoussa granite are quartz, plagioclase, biotite, alkalifeldspar, ± muscovite and Fe-Ti oxides. Only a few remnants of Mesozoic marble occur in the study area. They are separated by

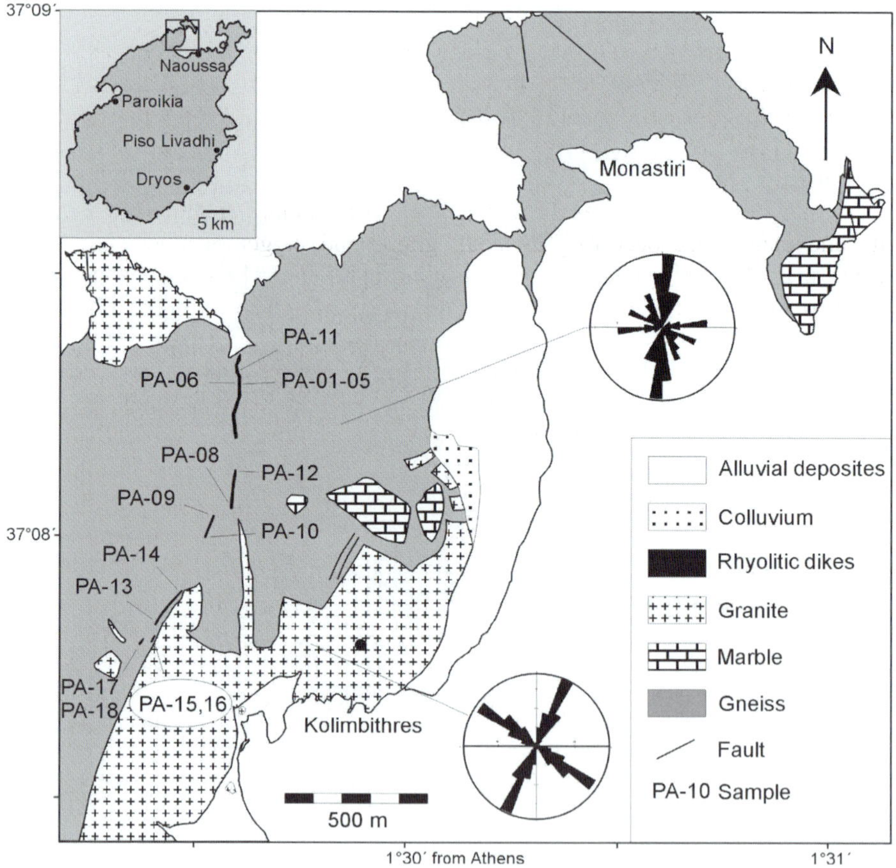

Fig. 3. Geological map of study area in the NE part of Paros, indicating the sample locations. The dykes are orientated according to the joint systems shown as rose diagrams.

a tectonic contact from the underlying gneiss and are obviously tectonic klippen. All metamorphic and plutonic rocks of the study area in northwestern Paros show a single predominant foliation and an associated stretching lineation. This deformation is considered to be related to E-W-extensional conditions (Papanikolaou, 1980), which is confirmed by microkinematic indicators such as feldspar porphyroclasts, domino boudins and shearband cleavage. Along the contact to the underlying gneisses the marble is fragmented and brecciated and truncated by joints are filled with travertine.

3. PETROGRAPHY AND MINERAL CHEMISTRY OF THE RHYOLITIC DYKES

The rhyolitic dykes can be mapped over a length of 1.1 km and their width reaches up to 11 m maximum. They are striking about 6° and 38° NE. Their two main directions are orientated parallel to the prevailing joint directions of the gneiss and the Naoussa granite (Fig. 3).

Fig. 4 shows a cross-section of a well-exposed rhyolitic dyke where several samples have been taken. The leucocratic rhyolitic rocks are porphyric with a fine-grained matrix of mainly feldspar, quartz, ± biotite, and Fe-Ti oxides. Plagioclase, biotite, and K-feldspar occur as phenocrysts and mafic-intermediate enclaves are dispersed throughout the rock.

The mesocratic mafic-intermediate enclaves consist mainly of subhedral to euhedral biotite and plagioclase occurring as small grains and bigger phenocrysts. They are enriched in Fe-Ti oxides and also contain clino-pyroxene. The enclaves are strongly alterated to fine-grained whitish secondary minerals, which due to their high volatile content caused the high loss on ignition. At the margin of the enclaves the minerals display a close intergrowth. A reaction rim could not be observed. The shape of the enclaves is mainly round or oval, but sometimes irregular as well. Plagioclase phenocrysts often show zoning, reaction rims and are strongly corroded in places.

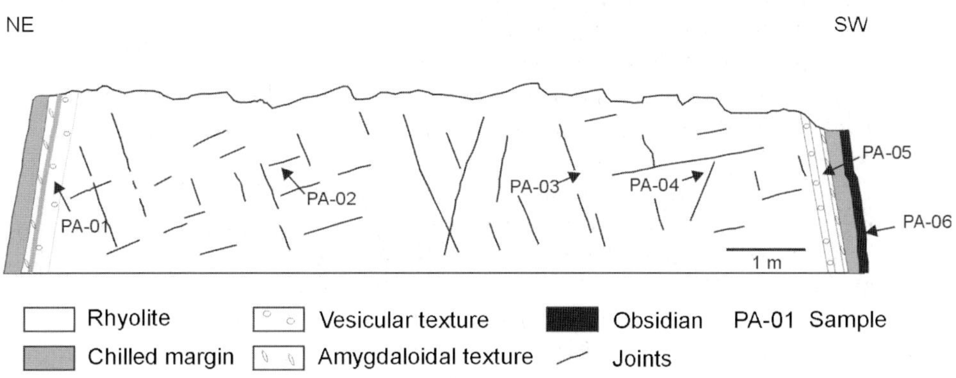

Fig. 4. Cross-section of a rhyolitic dyke with sample site of PA-01 to PA-06.

Feldspars of the rock matrix are mostly K-feldspars and show skeletal growth as a result of rapid cooling.

The rocks show generally a vesicular texture. The vesicles are often elongated to pipe vesicles as a result of movement of bubbles or melt. The marginal zones of the dykes show an amygdaloidal texture. At the chilled margin in contact to the gneissic country rock the dykes are vitric (partly obsidian) and more vesiculated than in the inner parts of the dyke. There, the enclaves have been preserved from contact reaction with the melt and alteration effects because of higher undercooling rates and contain primary igneous minerals such as clinopyroxene and An-rich plagioclase (Sample PA-06).

Since no other Neogene volcanic rocks are known from Paros we sampled rhyolitic lavaflows, pyroclastics and volcanic glasses from the south of Antiparos that have been mapped in detail by Anastoploulos (1963) (Fig. 2).

The lava flows are mostly brecciated, vesiculated and comprise obsidian blocks. All the volcanic rocks are rhyolites and generally glassy with microphenocrysts of sanidine, plagioclase (oligoclase), Fe-Ti oxides and ± biotite. The glass is largely recrystallized (devitrified) causing the formation of spherulites. Spherulites are spherical to ellipsoidal clusters of radiating fibrous quartz and alkali feldspar. Samples of obsidian show concentric fractures resulting in a perlitic texture. This perlitic texture is a result of hydration of the obsidian on fracture surfaces that are exposed to moisture or to meteoric water.

Chemical analysis of minerals (Fig. 5) were obtained using an electron microprobe type JEOL JXA 8900 with an acceleration voltage of 15 kV and a probe current of 12 nA. It shows that the feldspars of the mafic-intermediate enclaves from Paros are plagioclases ranging from bytownite to andesine (An_{40-80}), but the majority has a labradorite composition (Fig. 5B). The feldspar-phenocrysts occur as oligoclases with minor Na sanidin whereas the feldspars of the matrix are generally anorthoclase and Na-sanidine with Ab_{35-72}. Thus two different types of K-feldspar occur. Fig. 5A shows analysis of the feldspars from the Antiparos rhyolites. It is obvious that they are lower in the Ca component than the feldspars of the Paros rhyolites.

Clinopyroxene within the mafic-intermediate enclaves has been identified in sample PA-06. The classification after Morimoto (1988) shows that all clinopyroxenes have high abundances of Ca and Mg component and are of diopsidic composition (Fig. 5C). The microprobe analysis reveals that the core of the zoned clinopyroxene-crystal has higher abundances of Mg in comparison to its rim, which can be related to fractional crystallization.

Most biotites can be considered as members of the four component system phlogopite, annite, eastonite and siderophyllite. Fig. 5D shows the compositions of biotites from the mafic-intermediate enclaves and biotites that occur as phenocrysts and groundmass minerals within the rhyolite. They show Fe/(Fe+Mg) ratios between 0.25 and 0.6. However, there is no clear chemical distinction between the three groups of biotite. The generally high abundances of biotite of the rhyolites from Paros indicate considerable amounts of water in the magma.

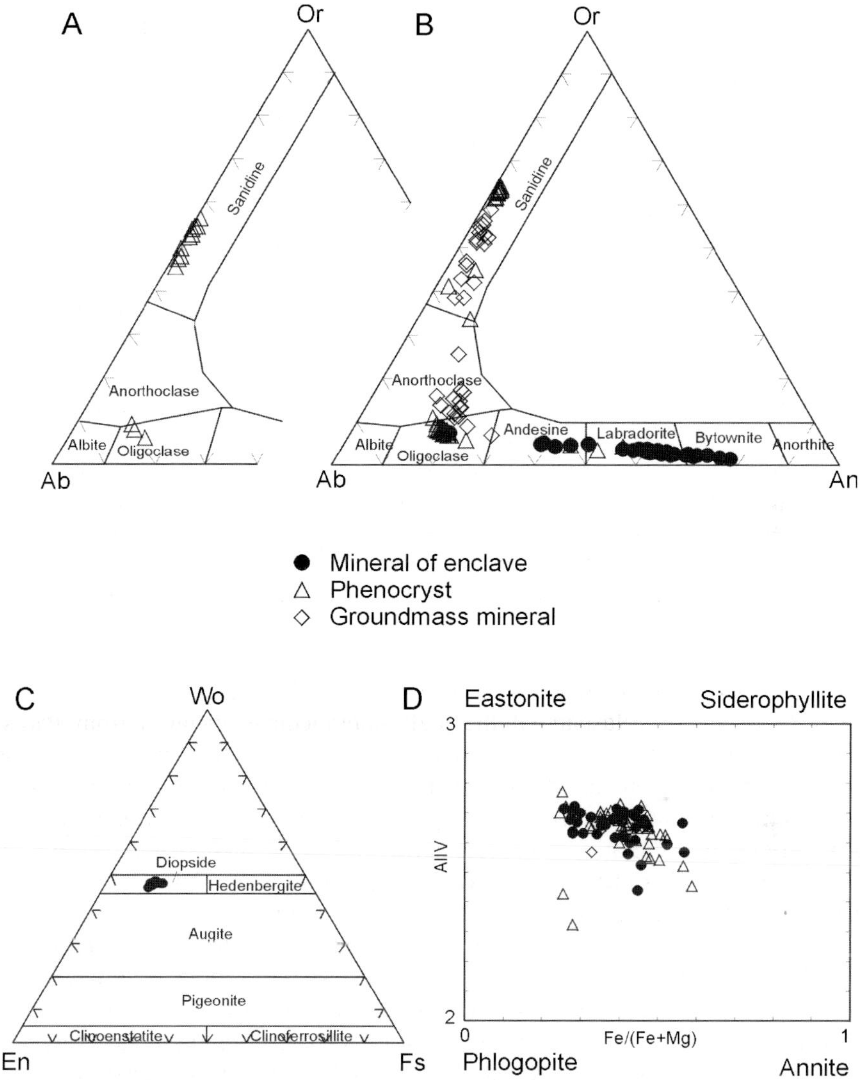

Fig. 5. Mineral composition of rhyolitic rocks and enclaves: Feldspars from Antiparos (A), Feldspars from Paros (B), Pyroxene of mafic enclaves from Paros (C) and Biotite from Paros (D).

4. GEOCHEMISTRY

The major and trace elements were determined by X-Ray Fluorescence (XRF). By using the TAS diagram (total alkalis-silica) after Le Maitre et al. (1989) the sampled rocks are classified as rhyolites that plot within then field of the subalkaline series (Fig. 6). The ternary AFM diagram is used to distinguish between tholeiitic and calc-

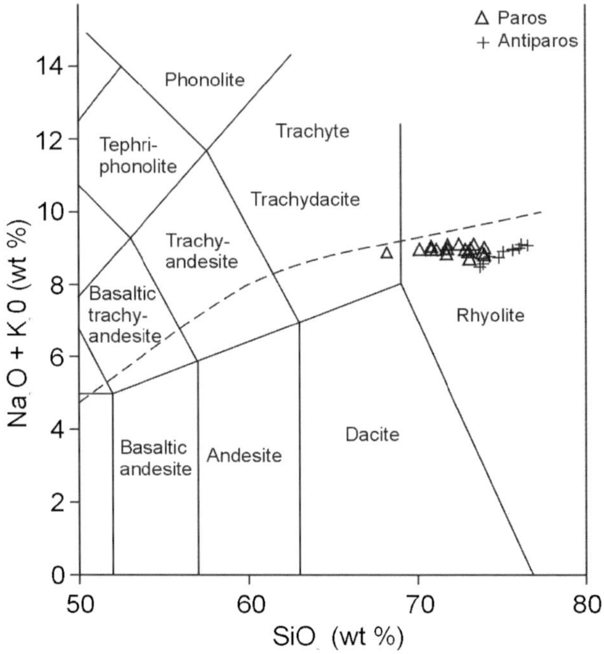

Fig. 6. Classification of the Paros and Antiparos rhyolites (after Le Maitre et al., 1989). The subdivision of alkaline (above) and subalkaline (below) is marked by the dashed line after Irvine and Baragar (1971).

alkaline differentiation trends in the subalkaline magma series. Fig. 7 shows that the rhyolites are members of a calc-alkaline differentiation trend. A further classification on the basis of their concentrations of K_2O and SiO_2 using the subdivisions after Le Maitre et al. (1989) reveals that they belong to the High-K series.

The geochemical analysis (Table 1) shows that the rhyolitic dykes of Paros with 68.20-74.03 wt % SiO_2 are slightly less silicic than the rhyolites from Antiparos with 73.73-76.59 wt % SiO_2. This corresponds to higher concentrations of Al_2O_3, Fe_2O_3 (t), MgO, CaO, TiO_2, and lower Na_2O and K_2O in the Paros dykes.

The major elements of the dykes such as TiO_2, Al_2O_3, Fe_2O_3(t) and MgO show a distinct correlation with SiO_2 whereas Na_2O and K_2O positively correlated (Fig. 8). With exception of the alkalis the Antiparos volcanic rocks display a weak positive trend for the major elements. The rhyolitic dykes have higher abundances of HFSE such as Zr, Nb, Y, Th and some LILE such as Ba, Sr in comparison to the Antiparos rocks. The concentrations of LREE such as La and Ce are higher as well. Only U and to some extend Rb are higher in the Antiparos rhyolites. Consequently, the average Th/U ratio of the Paros dykes of 7.2 is higher than the average ratio of 2.3 for the Antiparos rhyolites.

The Harker variation diagrams show a distinct negative trend of Ba and a slightly negative correlation of Zr and Y with SiO_2 for the Paros dykes. Rb and Th show positive

Table 1. Chemical analyses of rhyolites from Paros and Antiparos.

Sample wt %	PA-1	PA-2	PA-3	PA-4	PA-5	PA-6	PA-7	PA-8	PA-8E	PA-9	PA-10	PA-11	PA-12	PA-13	PA-14
SiO_2	70.16	72.52	71.82	70.85	70.82	68.2	72.88	73.05	49.34	74.03	73.9	71.11	73.13	74	73.21
TiO_2	0.35	0.23	0.25	0.33	0.31	0.27	0.19	0.17	1.4	0.12	0.12	0.28	0.17	0.14	0.14
Al_2O_3	14.52	14.18	14.22	14.69	14.33	13.89	14.15	14.6	23.32	13.78	13.77	14.21	13.99	13.8	13.56
Fe_2O_3	1.58	1.50	1.63	1.58	1.91	0.87	1.43	0.58	4.61	1.13	1.26	1.78	0.98	1.10	1.69
FeO	0.85	0.28	0.26	0.37	0.38	1.5	0.15	0.11	n.d.	0.06	0.07	0.31	0.14	0.04	0.12
MnO	0.03	0.02	0.01	0.01	0.02	0.08	0.02	0.01	0.06	0.06	0.01	0.02	0.01	0.02	0.02
MgO	0.6	0.32	0.36	0.45	0.44	0.58	0.24	0.19	2.68	0.11	0.11	0.41	0.25	0.18	0.13
CaO	0.92	0.72	0.7	0.82	0.83	1.59	0.55	0.51	1.19	0.47	0.48	0.9	0.51	0.56	0.62
Na_2O	3.79	3.86	3.82	3.83	3.81	3.76	3.7	3.43	1.19	3.61	3.65	3.83	3.51	3.78	3.84
K_2O	5.17	5.24	5.18	5.21	5.14	5.12	5.26	5.47	3.79	5.21	5.23	5.14	5.19	5.24	5.16
P_2O_5	0.03	0.05	0.04	0.05	0.05	0.07	0.02	0.03	0.36	0.02	0.02	0.08	0.03	0.02	0.03
LOI	2.03	1.60	1.75	2.12	2.09	3.95	1.65	2.01	12.31	1.36	0.70	1.98	2.10	1.21	1.32
Total	100.02	100.52	100.04	100.31	100.13	99.88	100.24	100.16	100.25	99.97	99.32	100.05	100.02	100.08	99.84
ppm															
V	36	19	24	31	28	27	15	10	146	8	9	27	14	8	15
Zn	42	34	35	34	35	56	32	24	151	32	36	40	43	31	41
Ga	18	17	17	17	17	18	17	18	24	17	16	16	17	17	17
Rb	338	359	354	344	330	332	371	379	170	378	380	345	368	371	370
Sr	135	82	85	113	110	141	54	41	126	31	32	105	45	35	38
Y	28	29	30	33	30	36	24	24	33	36	28	27	25	19	27
Zr	210	203	204	208	206	201	203	203	319	195	193	205	198	189	184
Nb	39	39	39	38	38	38	41	41	41	39	40	40	39	39	37
Ba	220	127	150	206	183	191	97	84	877	63	46	166	78	65	65
Pb	39	33	43	41	37	59	39	46	57	49	45	38	46	40	38
Th	79	80	78	78	77	77	84	90	40	88	86	79	83	85	82
La	91	68	75	73	72	66	78	80	n.d.	75	67	74	67	68	71

Table 1. Chemical analyses of rhyolites from Paros and Antiparos (continued).

Sample wt %	PA-1	PA-2	PA-3	PA-4	PA-5	PA-6	PA-7	PA-8	PA-8E	PA-9	PA-10	PA-11	PA-12	PA-13	PA-14
Ce	156	105	130	130	119	138	138	137	n.d.	148	121	112	131	124	111
Pr	15	13	14	12	12	13	14	17	n.d.	13	14	11	10	15	11
Nd	63	48	50	53	48	49	56	53	n.d.	51	48	50	44	45	44
Sm	12	9	8	10	12	9	10	9	n.d.	9	7	8	7	8	10
U	14	12	12	10	12	15	11	11	6	11	10	11	10	9	10

Sample wt %	PA-15	PA-16	PA-17	PA-18	AP-1	AP-2	AP-3a	AP-4	AP-5	AP-6	AP-7	AP-8	AP-9a	AP-9b
SiO_2	73.4	71.82	71.7	71.75	75.72	76.59	76.05	73.92	75.16	73.73	74.86	76.17	74.29	75.67
TiO_2	0.2	0.25	0.27	0.25	0.1	0.11	0.11	0.1	0.1	0.1	0.1	0.11	0.1	0.1
Al_2O_3	14.04	14.28	14.27	14.14	12.54	12.65	12.63	12.51	12.32	12.18	12.34	12.58	12.31	12.49
Fe_2O_3	1.05	1.35	1.41	1.80	0.37	0.74	0.79	0.75	0.54	0.75	0.76	0.85	0.60	0.48
FeO	0.16	0.26	0.28	0.16	0.72	0.18	0.11	0.1	0.37	0.07	0.15	0.03	0.25	0.57
MnO	0.02	0.02	0.01	0.02	0.08	0.07	0.07	0.08	0.08	0.07	0.08	0.08	0.08	0.08
MgO	0.18	0.35	0.33	0.33	0.01	0.04	0.02	0.02	0	0.09	0.03	0.01	0.01	0.01
CaO	0.64	0.77	0.77	0.74	0.4	0.33	0.28	0.36	0.36	0.39	1.03	0.17	0.39	0.4
Na_2O	3.91	3.79	3.82	3.73	4.37	4.31	4.39	3.62	4.3	3.4	4.3	4.47	3.93	4.35
K_2O	5.18	5.31	5.15	5.12	4.59	4.77	4.6	4.94	4.6	5.08	4.42	4.64	4.84	4.57
P_2O_5	0.03	0.07	0.06	0.06	0	0.01	0	0	0	0	0	0	0	0
LOI	1.06	1.70	1.82	2.05	1.31	0.38	1.13	3.85	1.18	4.44	2.23	1.13	3.47	1.44
Total	99.87	99.97	99.89	100.16	100.21	100.18	100.18	100.25	99.01	100.31	100.30	100.24	100.27	100.15
ppm														
V	17	21	24	24	3	7	5	5	5	6	3	4	2	7
Zn	50	40	49	49	22	14	16	18	17	15	14	11	20	13
Ga	17	17	17	17	18	17	18	18	17	17	17	17	17	17
Rb	363	356	357	356	383	384	349	377	383	388	357	389	362	379
Sr	56	82	87	76	15	18	13	15	11	13	16	13	14	14

Table 1. Chemical analyses of rhyolites from Paros and Antiparos (continued).

Sample wt %	PA-15	PA-16	PA-17	PA-18	AP-1	AP-2	AP-3a	AP-4	AP-5	AP-6	AP-7	AP-8	AP-9a	AP-9b
Y	33	28	33	31	10	13	6	11	11	11	10	10	10	11
Zr	194	205	205	204	146	143	134	144	145	145	143	143	138	145
Nb	37	40	40	40	33	32	33	33	34	34	34	35	32	35
Ba	100	169	156	134	39	46	33	63	35	40	50	27	52	34
Pb	55	39	46	39	47	48	46	45	45	46	47	31	47	52
Th	82	82	82	80	60	61	58	60	64	62	63	62	57	60
La	74	67	73	73	42	57	41	47	40	39	43	50	42	42
Ce	141	116	135	117	64	62	57	70	70	70	63	67	64	66
Pr	13	14	17	13	7	5	2	4	8	4	3	9	5	3
Nd	53	47	52	52	9	13	11	12	10	11	10	15	10	11
Sm	8	8	10	6	3	5	3	3	4	n.d.	3	5	4	5
U	13	13	14	12	29	25	22	28	29	28	26	23	26	28

n.d.= not detected

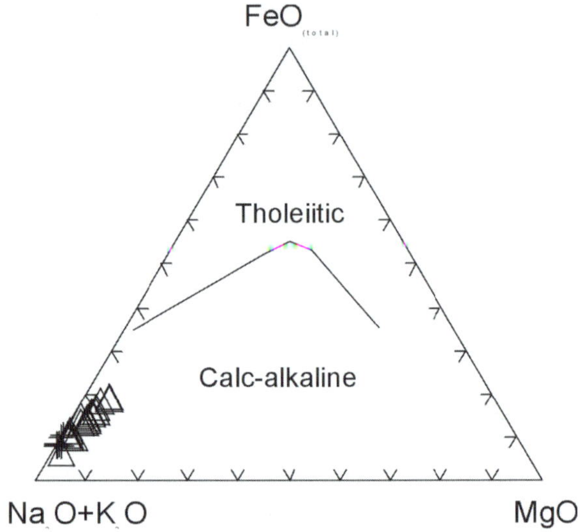

Fig. 7. AFM diagram of the studied rhyolites (after Irvine and Baragar, 1971).

trends whereas Nb is almost constant. The volcanic rocks of Antiparos do not reveal clear correlation trends for the trace elements.

5. DISCUSSION

The majority of major and trace elements of the Paros dykes show clear linear trends that can be explained by crystal fractionation. The geochemical composition of the Antiparos rhyolites does not have such a variation, but appears to be rather constant for major elements. Furthermore, the data of the Antiparos rocks do not correspond to the fractionation trend of the Paros dykes for elements such as the HFSE. This clearly implies that the two rhyolitic sequences are not comagmatic. The trends for TiO_2 and Fe_2O_3(t) are probably a result of fractionation of Fe-Ti oxides. Furthermore, the strong negative anomalies in Ba and Sr in both rhyolitic suites (Fig. 9) indicate that fractional crystallization of feldspar played an important role during magma evolution, since Ba is compatible to alkalifeldspar whereas Sr is retained by fractionation of both alkalifeldspar and plagioclase. Nevertheless, the parental magmas were different for the two rhyolitic occurrences and therefore their petrogenesis will be described separately in this chapter.

5.1 Formation of Antiparos rhyolites

According to their chemical and isotopic characteristics like high $^{87}Sr/^{86}Sr$ ratios of ca. 0.72, Innocenti et al. (1982) interpreted the Antiparos rhyolites as a result of partial melting within the continental crust. They presumed that extensional tectonism within a

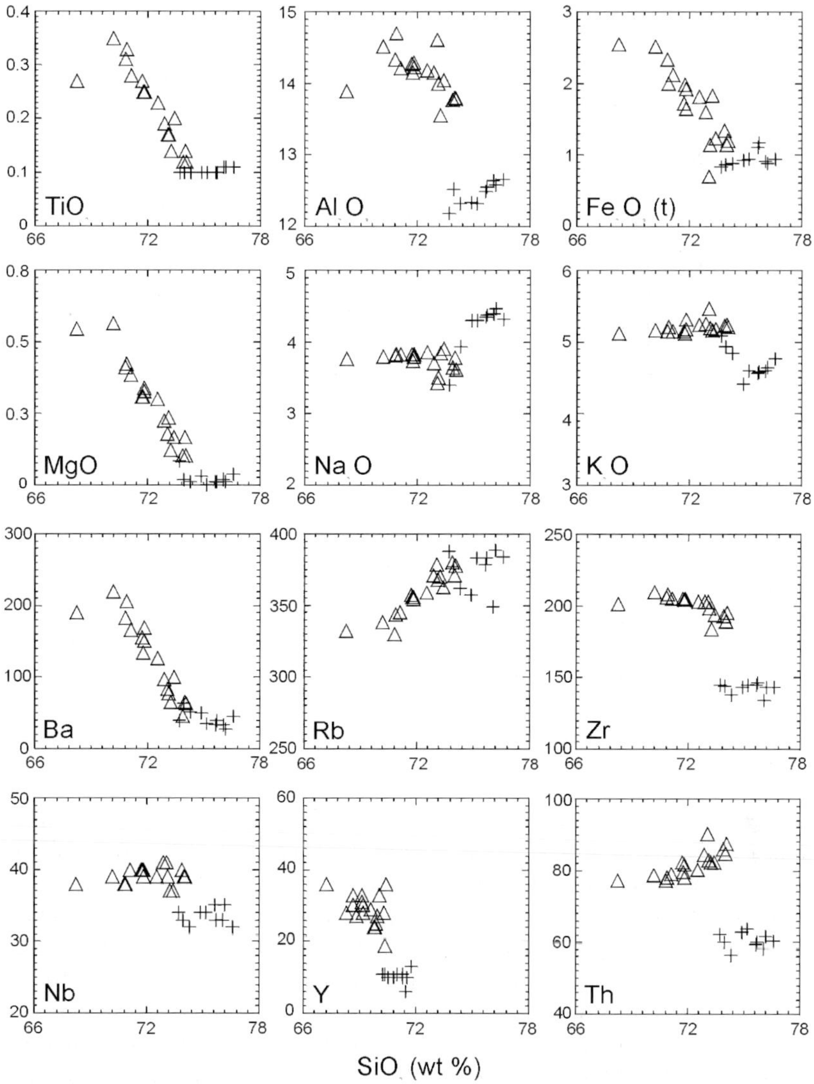

Fig. 8. Plots of selected major and trace elements vs. SiO_2 for the rhyolites of Paros and Antiparos. Major elements as oxides in wt %, trace elements in ppm.

detachment zone allowed the rise of the asthenosphere, creating a thermal anomaly with subsequent melting of shallow parts of continental crust. Since the rhyolites have high abundances of HFSE such as Nb and Zr in comparison to the continental crust the model of Innocenti et al. (1982) seems not to be very likely with respect to magma generation by simple melting of continental crust.

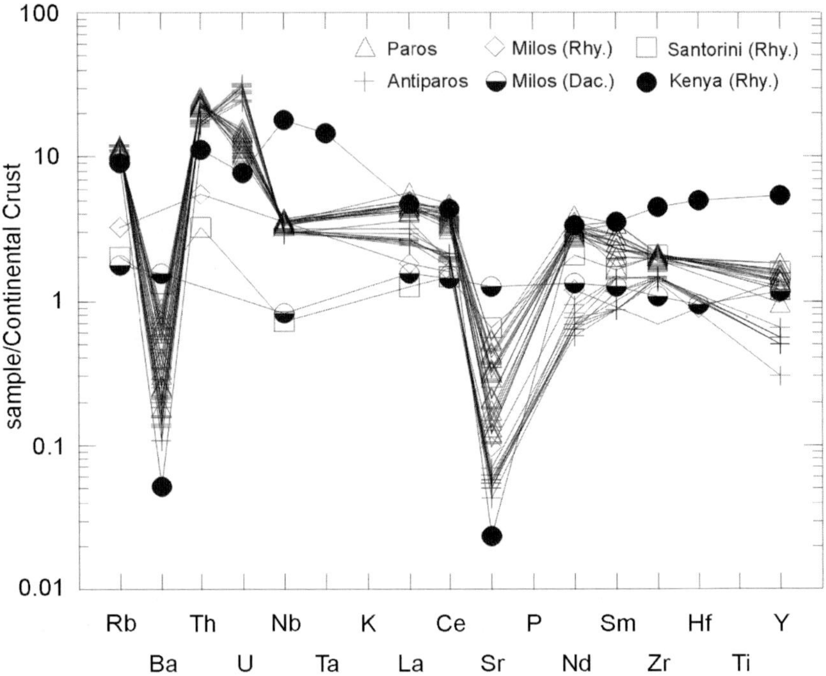

Fig. 9. Trace element abundances normalized to continental crust for the Paros and Antiparos rhyolites (normalization after Taylor and McLennan, 1985). Other volcanic rocks are shown for comparison: Rhyolite (Fytikas et al., 1986) and dacite (Mitropoulos et al., 1987) from Milos, dacite from Santorini (Huijsmans et al., 1988) and a rhyolite from the Kenya rift (Macdonald et al., 1987).

Another approach of Clapsopoulos (1998) suggests that the primary melts of the Antiparos rhyolites were possibly generated by small degrees of partial melting of a felsic granulite source in the middle or lower crust. Basic magmas generated by the subduction of the Aegean arc, which subsequently rose to higher levels provided the heat for the melting of the continental crust. After their generation these primary crustal melts rose to a shallow magma chamber, where they underwent intense fractional crystallisation mainly of alkali-feldspar and plagioclase before their eruption. However, such a crustal melting model is also not without complications since the mid-lower crust is likely to contain garnet, the enrichment of HREE in comparison to MREE presented by Clapsopoulos (1998) is not compatible to partial melts in the presence of garnet.

According to the model of Clapsopoulos (1998) and considering the high abundances of HFSE such as Nb we propose a magma generation under extensional conditions within the back-arc region, taking in account a significant influence of asthenospheric source as follows:

The subduction of the Aegean arc beneath the continental crust generated basic magma, which subsequently rose to higher levels and amplified a convectional motion in the extensional regime in the back-arc region. The subsequent uplift of the

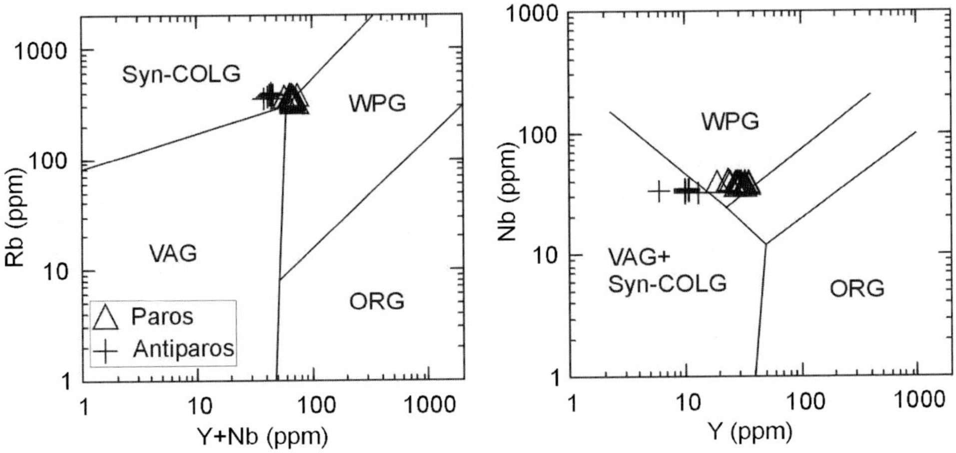

Fig. 10. Rb-(Y+Nb) and Nb-Y discrimination diagrams for granites after Pearce et al. (1984).

asthenosphere led to decompression melting and uprise of the magmas followed by assimilation and fractional crystallisation at the base of the continental crust. The assimilation of continental crust is evident from the high $^{87}Sr/^{86}Sr$ ratios (Innocenti et al., 1982). Such hybrid magmas ascented to shallow magma chambers where they experienced further fractionation, which led to the depletion of Ba, Sr and Eu (Clapsopoulos, 1998).

5.2 Origin of the mafic enclaves of the Paros rhyolites

The widespread occurrence of mafic enclaves within the rhyolitic dykes of Paros might bear information about the origin and evolution of their primary magma. Therefore, it is important to constrain where the enclaves originated from. Following the guiding questions of Grout (1937) it is possible to affirm the igneous and cognate (cogenetic) origin of the enclaves. In contrast to xenoliths the inclusions are distributed fairly regularly throughout the dykes. Higher K_2O than Na_2O and higher MgO than CaO, as well as their rounded shape and mafic minerals such as clino-pyroxene and plagioclase indicate an igneous cognate origin. The inclusions contain also high HFSE concentrations such as Nb, Zr and Th. In discriminaton diagrams for basalts after Pearce and Norry (1979), Pearce (1982) or Meschede (1986) the mafic inclusions plot clearly within the field of within-plate basalts.

Such inclusions or blobs of basaltic/andesitic magma may evolve when new and more mafic magma rises into an evolving silicic magma chamber and is dispersed by convective motion in the chamber due to the heating of the silicic magma by the hotter basaltic magma. The formation of similar mafic enclaves was described e.g. by Flanagan-Brown (1999) as evidence for mingling and mixing of a mafic magma with more silicic magma within the Lucerne Granite of Maine.

5.3 Formation of Paros rhyolites

The melt generation of the rhyolitic dykes of Paros is affected by several different processes. First, mantle melting was caused by extensional tectonism and subsequent uplift of the asthenosphere. The resulting mafic magmas provided the heat for crustal melting by underplating or storage in magma chambers at lower crustal levels and were involved in AFC processes. Such hybrid magmas subsequently rose to higher crustal levels, where they experienced intense fractional crystallization, which is evident from the depletion of Ba, Sr and Eu (Clapsopoulos, 1998). The input of mafic magma is evident from the enclaves, that without much doubt, have developed by mingling and mixing in a replenished silicic magma chamber at shallower crustal depth. This input of mafic magma might have eventually triggered the eruption of the rhyolites.

5.4 Tectonic setting

In the discrimination diagrams for granites (Fig. 10) based upon the elements Rb-Y-Nb after Pearce et al. (1984) the rhyolites from Paros are classified as within-plate granites (WPG) whereas Antiparos rhyolites plot within the field of syn-collisional granites (Syn-COLG). This means that the rhyolites are not related to a simple tectonic setting as implied by such discrimination diagrams.

In order to achieve more information about the origin of the Paros and Antiparos calc-alkaline rhyolites their trace element data were compared to rhyolites from the recently active South Agean arc and a typical rhyolite of a bimodal basaltic-rhyolitic suite from the Kenya Rift as an example for a rhyolite from an intra-continental extensional regime (Fig. 10). As mentioned before the Paros and Antiparos rocks are enriched in incompatible elements such as Rb, Th, U and Nb and show strong negative anomalies in Ba and Sr. Fig. 10 shows that the rhyolites and the dacite from Milos and Santorini, both belonging to the South Agean arc, have generally lower trace element abundances than the rocks from Paros and Antiparos. The rhyolite from the Kenya Rift instead shows some similar characteristics to the rhyolites of this study such as a strong depletion in Ba and Sr as well as the enrichment in Th and U, but the significantly lower contents of HFSE of the Paros and Antiparos samples are not in line with the same setting. Similar HFSE contents are known only from few rhyolitic rocks worldwide e.g. of the Aeolian arc (Renzulli et al., 2001; Gioncada et al., 2003) and the Central Andean Volcanic Zone in Argentina (Siebel et al., 2000). Both regions are complex tectonic settings in a subduction zone environment, which infers that the formation of the Paros and Antiparos rhyolites might as well be a product of such complex processes.

The rhyolites from Paros and Antiparos and the mafic enclaves, present in the Paros rocks, have high abundances of HFSE. This enrichment in HFSE is unlikely to be produced in the crust and confirms the generation of small degrees of partial melting in the mantle by decompression of a rising asthenosphere in an extensional regime within the back-arc region of the Aegean arc. The heat of an underplating mafic magma causes crustal melting followed by AFC processes, magma mixing and mingling. In Paros the mafic component is evident by the occurrence of the mafic enclaves.

The generation of the Antiparos rhyolites might have been caused by crustal melting alone although there are some difficulties to explain all the chemical variations. We therefore suggest AFC processes due to underplating of mafic magma similar to Paros, but in case of the Antiparos rhyolites the part of the magma chamber tapped was free of relicts of mafic magma.

The pronounced differences in the Th and U concentrations of the rhyolitic sequences can be explained by assimilation of crustal material from different crustal levels. The lower crust, which is depleted in U, is likely to be assimilated from the Paros parental magma causing low U concentrations and corresponding high Th/U ratios in the rhyolitic dykes. In contrast, the Antiparos rhyolites have higher U abundances with corresponding lower Th/U ratios. This implies assimilation of or interaction with crustal matetrial from mid-crustal levels, which were not depleted in U.

6. CONCLUSIONS

In contrast to the rhyolites from Antiparos the rhyolitic dykes from Paros show clear fractionation trends, which implies that the rhyolites from both islands are not comagmatic. The occurrence of the mafic enclaves provides strong evidence for mingling and mixing of mafic magma replenishing an evolving silicic magma chamber at shallower depth.

The melt generation of the dykes was related to two major processes of fractional crystallisation and assimilation of mafic magma. High HFSE abundances in both the Paros dykes and the Antiparos rhyolites are unlikely to be produced only by the melting of crust and confirm the generation of small degree partial melts in the mantle by decompression of the rising astenosphere in an extensional regime within the back-arc region of the Aegean arc. The heat of the underplating mafic magma caused melting and AFC processes that finally formed the rhyolitic magmas.

REFERENCES

Altherr, R., Kreuzer, H., Wendt, I., Lenz, H., Wagner, G.A., Keller, J., Harre, W., and Hohndorf, A., 1982. A late Oligocene/early Miocene high temperature belt in the Attic-Cycladic crystalline Complex (S.E. Pelagonian, Greece). Geol. Jb., E23: 97-164.

Altherr, R., Schliestedt, M., Okrusch, M., Seidel, E., Kreuzer, H., Harre, W., Lenz, H., Wendt, I. and Wagner, G.A., 1979. Geochronology of high-pressure rocks on Sifnos (Cyclades, Greece). Contrib. Miner. Petrol., 70: 245-255.

Anastopoulos, J., 1963. Geological study of the Antiparos island group. Geol. Geophys. Res., ISGR, 7: 231-375.

Andriessen, P.A.M., 1978. Isotopic age relations within the polymetamorphic complex of the island of Naxos (Cyclades, Greece). Verhand. Zwo Laboratorium voor Isotopen-Geologie, 3: 71.

Andriessen, P.A.M., Banga, G. and Hebeda, E.H., 1987. Isotopic age study of pre-

Alpine rocks in the basal units on Naxos, Sikinos and Ios, Greek Cyclades. Geologie en Mijnbouw, 66: 3-14.

Angelier, J., Glaçon, G. and Muller, C., 1978. Sur la Présence et la position tectonique du Miocène inferieur marin dans l´archipel de Naxos (Cyclades, Grèce). CR Acad. Sci. Paris, 286: 21-24.

Bateman, P.C., 1979. Generation and Emplacement of the Sierra Nevada Batholith, California. Geol. Soc. Amer., Abstr. With Progr., 11: 385.

Boronkay, K. and Doutsos, T., 1994. Transpression and transtension within different structural levels in the central Aegean region. J. Struct. Geol., 11: 1555-1573.

Buick, I.S. and Holland, T.J.B., 1989. The P-T-t path associated with crustal extension, Naxos, Cyclades, Greece. In: Daly, J.S. (Editor), Evolution of metamorphic belts. Geol. Soc. London, 43: 365-369.

Clapsopoulos, I., 1998. Origin of the Antiparos Island rhyolites, Greece, by subduction-related anatexis of a granulitic source possibly located in the middle to lower continental crust. Bull. Geol. Soc. Greece, 8: 355-366.

Dermitzakis, M. and Papanikolaou, D., 1979. The molasses of Paros Island, Aegean Sea. Ann. Naturhist. Mus. Wien, 83: 59-71.

Dixon, J.E., 1976. Glaucophane schists of Syros, Greece. Bull. Soc. Géol. France 18(7): 280.

Dürr, S., 1975. Über Alter und geotektonische Stellung des Menderes-Kristallins/SW-Anatolien und seiner Äquivalente in der mittleren Ägäis. Habilitations-Schrift, Universität Marburg, 106 p.

Dürr, S. and Altherr, R., 1979. Existence des klippes d´une nappe composite néogène dans l´île de Mykonos/Cyclades (Grèce). Rapp. Comm. Int. Mer Médit., 25/26 (2a): 33-34.

Dürr, S., S.Seidel, E., Kreuzer, H. and Harre, W., 1978b. Témoins d´un métamorphisme d´age crétacé supérieur dans l´Égéide: datations radiométrique de minéraux provenant de l´île de Nikourià (Cyclades, Grèce). Bull. Soc. Géol. France (7), 20: 209-213.

Dürr, S., Altherr, R., Keller, J., Okrusch, M. and Seidel, E., 1978a. The Median Aegean crystalline belt: stratigraphy, structure, metamorphism, magmatism. In: Closs, H., Roeder, D. and Schmidt, K. (Editors), Alps, Apennines, Hellenides. E. Schweizerbart´sche Verlagsbuchhandlung, Stuttgart, 455-477.

Engel, M. and Reischmann, T., 1998. Single zircon geochronology of orthogneiss from Paros, Greece. Bull. Geol. Soc. Greece, 32(3): 91-99.

Flanagan-Brown, R., 1999. Microgranular mafic enclaves as evidence for mingling and mixing of a mafic magma with the Lucerne Granite of central Maine. In: Geological Society of America, 1999 annual meeting, Abstracts with Programs, 31(7): 268.

Fytikas, M.D. and Kolios, N.P., 1979. Preliminary heat flow map of Greece. In: Čermak, V. and Ryback, L. (Editors), Terrestrial Heat Flow in Europe. Springer-Verlag, Berlin, 197-205.

Fytikas, M., Giuliani, O., Innocenti, F., Marinelli, G. and Mazzuoli, R., 1976. Geochronological data on recent magmatism of the Aegean Sea. Tectonophysics, 31: T29-T34.

Fytikas, M., Innocenti, F., Manetti, P., Mazzuoli, R., Peccerillo, A. and Villari, L., 1984. Tertiary to Quaternary evolution of volcanism in the Agean region. In: Dixon, J.E. and Robertson, A.H.F. (Editors), The geological evolution of the eastern Mediterranean. Geol. Soc. Sp. Publ., London, 17: 687-699.

Gioncada, A., Mazzuoli, R., Bisson, M. and Pareschi, M.T., 2003. Petrology of volcanic products younger than 42 ka on the Lipari-Vulcano Complex (Aeolian Islands, Italy): An example of volcanism controlled by tectonics. J. Volcanol. Geotherm. Res., 122: 191-220.

Grout, F.F., 1937. Criteria of origin of inclusions in plutonic rocks. Bull. Geol. Soc. Am., 48: 1521-1572.

Hejl, E., Riedl, H., Soulakellis, N., Van den Haute, P. and Weingartner, H., 2003. Young Neogene tectonics and relief development on the Aegean islands of Naxos, Paros and Ios (Cyclades, Greece). Mitt. Österr. Geol. Ges., 93 (2000): 105-127.

Henjes-Kunst, F., Altherr, R., Kreuzer, H. and Hansen, B.T., 1988. Disturbed U-Th-Pb systematics of young zircons and uranothorites: the case of the Miocene Aegean granitoids (Greece). Chemical Geology (Isotope Geoscience), 73: 125-145.

Innocenti, F., Kolios, N., Manetti, P., Rita, F. and Villari, L., 1982. Acid and basic late Neogene volcanism in central Aegean Sea: its nature and geotectonic significance. Bulletin Volcanologique, 45: 87-97.

Irvine, T.N. and Baragar, W.R.A., 1971. A guide to the chemical classification of the common volcanic rocks. Can. J. Earth Sci., 8: 523-548.

Jansen, J.B.H., 1977. The geology of Naxos. Geol. Geophys. Res., 19: 100 p, IGME, Athens.

Jansen, J.B.H., and Schuiling, R.D., 1976. Metamorphism on Naxos: petrology and geothermal gradients. Amer. J. Sci., 276: 1225-1253.

Keller, J., 1982. Mediterranean Island Arcs. In: Thorpe, R.S. (Editor), Andesites. Wiley, New York, 307-326.

Kreuzer, H., Harre, W., Lenz,H., Wendt, L., Henjes-Kunst, F. and Okrusch, M., 1978. K/Ar and Rb/Sr-Daten von Mineralen aus dem polymetamorphen Kristallin der Kykladen-Insel Ios (Griechenland). Fortschritte der Mineralogie, 56: 69-70.

Le Maitre, R.W., Bateman, P., Dudek, A., Keller, J., Lameyre Le Bas M.J., Sabine, P.A., Schmid, R., Sorensen, H., Streckeisen, A., Woolley, A.R. and Zanettin, B., 1989. A classification of igneous rocks and glossary of terms. Blackwell, Oxford.

Macdonald, R., 1987. Quaternary peralkaline silicic rocks and caldera volcanoes of Kenya. In: Fitton, J.G. and Upton, B.G.J. (Editors), Alkaline igneous rocks. Geol. Soc. Sp. Publ., London, 30: 313-333.

Makris, J., 1977. Geophysical investigations of the Hellenides. Hamburger geophysische Einzelschriften, 34: 124 p.

Makris, J. and Röwer, P., 1986. Struktur und heutige Dynamik der Lithosphäre in der Ägäis. In: Jacobshagen, V. (Editor), Geologie von Griechenland. Gebrüder Borntraeger, Berlin/Stuttgart.

Marakis, G., 1970. Remarks on the age of sulphide mineralization in the Cyclades area

[in Greek, English summary]. Annales géologiques des pays helléniques, 19: 695-700.
Marks, P. and Schuiling, R.D., 1965. Sur la presence du Permien supérieur non-métamorphique à Naxos. Praktika Akadimias Athinon, 40: 96-99.
Meschede, M., 1986. A method of discriminating between different types of mid-ocean ridge basalts and continental tholeiites with the Nb-Zr-Y diagram. Chem. Geol., 56: 207-218.
Meulenkamp, J.E., Wortel, M.J.R., Van Wamel, W.A., Spakman, W. and Hoogerduyn Strating, E., 1988. On the Hellenic subduction zone and the geodynamic evolution of Crete since the late Middle Miocene. Tectonophysics, 146: 203-215.
Miller, C.F. and Bradfish, L.J., 1980. An inner Cordilleran belt of muscovite-bearing plutons. Geology, 5: 685-688.
Mitropoulos, P. and Tarney, J., 1992. Significance of mineral composition variations in the Aegean Island arc. J. volcan. geotherm. Res., 51: 283-303.
Mitropoulos, P., Tarney, J., Saunders, A.D. and Marsh, N.G., 1987. Petrogenesis of Cenozoic volcanic rocks from the Aegean Island Arc. J. volcan. geotherm. Res., 32: 177-193.
Morimoto, N., 1988. Nomenclature of pyroxenes. Fortschritte der Mineralogie, 66: 237-252.
Nicholls, I.A., 1971. Santorini volcano, Cyclades, Greece. J. Petrol., 12: 67-119.
Okrusch, M., Seidel, E. and Davis, E.N., 1978. The Assemblage Jadeite – Quartz in the Glaucophane Rocks of Sifnos (Cyclades Archipelago, Greece). N. Jb. Miner. Abh., 132: 284-308.
Papanikolaou, D.J., 1977. On the structural geology and tectonics of Paros Island (Aegean Sea). Annales géologiques des Pays helléniques, 28: 450-464.
Papanikolaou, D.J., 1979. Unités tectoniques et phases de déformation dans l'île de Samos, Mer Egée, Grèce. Bull. Soc. Géol. France (7), 21: 745-752.
Papanikolaou, D.J., 1980a. Contribution to the geology of the Aegean Sea: The island of Paros. Annales géologique des Pays helléniques, 30 (1): 65-95.
Papanikolaou, D.J., 1980b. Les écailles de l'île de Thymaena; témoins d'un mouvement tectonique Miocène vers l'intérieur de l'arc égéen. C.R. Acad. Sci. Paris, Sér. D, 290: 307-310.
Papazachos, B.C., Karakostas, V.G., Papazachos, C.B. and Scordilis, E.M., 2000. The geometry of the Wadati-Benioff zone and lithospheric kinematics in the Hellenic arc. Tectonophysics, 319: 275-300.
Pe, G.G. and Piper, D.J.W., 1972. Volcanism at subduction zones; The Aegean area. Bull. Geol. Soc. Greece, 9: 133-144.
Pe-Piper, G. and Piper, D.J.W., 2002. The igneous rocks of Greece – The anatomy of an orogen. Gebrüder Borntraeger, Berlin/Stuttgart.
Pearce, J.A., 1982. Trace element characteristics of lavas from destructive plate boundaries. In: Thorpe, R.S. (Editor), Andesites. Wiley, Chichester, 525-548.
Pearce, J.A. and Norry, M.J., 1979. Petrogenetic implications of Ti, Zr, Y and Nb variations in volcanic rocks. Contrib. Mineral. Petrol., 69: 33-47.
Pearce, J.A., Harris, N.B.W. and Tindle, A.G., 1984. Trace element discrimination

diagrams for the tectonic interpretation of granitic rocks. J. Petrol., 25: 956-983.

Reinecke, T., Altherr, R., Hartung, B., Hatzipanagioiou, K., Kreuzer, H., Harre, W., Klein, H., Keller, J., Geenen, E. and Böger, H., 1982: Remnants of a Late Cretaceous High Temperature Belt on the Island of Anáfi (Cyclades, Greece). N. Jb. Mineral. Abh., 145: 157-182.

Reischmann, T., 1998. Pre-alpine origin of tectonic units from the metamorphic complex of Naxos, Greece, identified by single zircon Pb/Pb dating. Bull. Geol. Soc. Greece, 32 (3): 101-111.

Renzulli, A., Serri, G., Santi, P., Mattioli, M. and Holm, P.M., 2001. Origin of high-silica liquids of Stromboli Volcano (Aeolian Islands, Italy) inferred from crustal xenoliths. Bull. Volcanol., 62: 400-419.

Robert, E., 1982. Contribution à l´étude géologique des Cyclades (Grèce): L´île de Paros. Thèse 3eme cycle, Univ. Paris-Sud, France.

Roesler, G., 1978. Relics of non-metamorphic sediments on the central Aegean islands. In: Cloos, H., Roeder, D. and Schmidt, K. (Editors), Alps, Apennines, Hellenides. E. Schweizerbartische Verlagsbuchhandlung, Stuttgart, 480-481.

Royden, L.H., 1993. Evolution of retreating subduction boundaries formed during continental collision. Tectonics, 12: 629-638.

Schliestedt, M., Altherr, R. and Matthews, A., 1987. Evolution of the Cycladic crystalline complex: petrology, isotope geochemistry and geochronology. In: Helgeson, R. (Editor), Chemical transport in metasomatic processes. NATO Advanced Study Institute Series C, Reidel, Dordrecht, 218: 389-428.

Siebel, W., Schnurr, W.B.W., Hahne K., Kraemer, B., Trumbull, R.B., Van den Bogaard, P. and Emmermann, R., 2000. Geochemistry and isotope systematics of small- to medium-volume Neogene-Quaternary ignimbrites in the Southern Central Andes: Evidence for derivation from andesitic magma sources. Chem. Geol., 171: 213-237.

Taylor, S.R. and McLennan, S.M., 1985. The continental crust: its composition and evolution. Blackwell, Oxford.

Wijbrans, J.R. and McDougall, I., 1988. Metamorphic evolution of the Attic Cycladic metamorphic belt on Naxos (Cyclades, Greece) utilizing $^{40}Ar/^{39}Ar$ age spectrum measurements. J. Metamorph. Geol., 6: 571-594.

Investigation of non pristine volcanic structures acting as probable hosts to epithermal gold mineralization in the back arc region of the active Aegean arc, using combined satellite imagery and field data: examples from Lesvos volcanic terrain

C. Vamvoukakis[1], K.St. Seymour[1,2,*], M. Kouli[1], S. Lamera[1], G. Denes[2]

[1] Department of Geology, University of Patras, Rion Patras, Greece.

[2] Department of Chemistry, Concordia University, Canada.

ABSTRACT

The island of Lesvos, located in the northeastern Aegean, just offshore Asia Minor, is mostly covered by Miocene calkalkaline, and predominantly shoshonitic volcanic products. Pyroclastic deposits are voluminous, indicating intense explosive activity. In the eastern part of the island outcrops the Polychnitos Ignimbrite. The western part of Lesvos is covered by the Sigri Pyroclastic formation, the host to the renown lithified forest. The prolific 21.5-17 Ma volcanic activity requires well developed subvolcanic magma chambers and their surfacial expression in the form of sizeable caldera structures, two of which they have been previously inferred from intense alteration zones.

Remote sensing was combined with field geology and geochemical methods to appreciate the actual geometry and number of these volcanic structures, which are severely affected by erosion and fault tectonics. Landsat TM and SPOT satellite images and the Digital Elevation Model (DEM) of the targeted areas were digitally processed in order to reveal specific geological characteristics associated with calderas such as rims and floors, radial and ring faults, areas of hydrothermal alteration, drainage network and lava domes both internal and external to caldera structures. In order to convey as much spectral information as possible and to enhance the outlines of Lesvos volcanic features, we produced a slope map with a resolution of 1^0 derived from DEM of the study area. As a result we have recognized seven caldera structures of which six new ones. The

* Corresponding author: e-mail: kstseymr@upatras.gr

highlights of these investigations are: (a) The Stipsi and Sigri calderas, related with extensive argillic alteration zones and probably high-sulphidation epithermal precious metal mineralization, (b) The source of the Sigri pyroclastics is a (14x10 km) large caldera structure, the Sigri caldera, that encloses the previously recognized Vatoussa structure (nested caldera).

1.INTRODUCTION

The use of Remote Sensing techniques has been applied to the recognition of volcanic features of planetary surfaces with considerable success. Most of these features are volcanic or intrusive in nature and many calderas have been recognized initially on Landsat TM and SPOT images (Smist, 1974; Anguita et al., 2001). Remote Sensing Techniques have met with the same degree of success in their application on epithermal gold deposits research in volcanic terrains since they can locate and characterize the alteration zones that develop around the issues of long established geothermal cells (Gupta, 1991; Sabins, 1997). The validity of such applications for volcanologists and mineral explorationists alike is readily apparent.

The Miocene volcanic field of Lesvos Island presents the ideal pilot case for such Remote Sensing studies. The field comprises a variety of non-folded lavas, domes and pyroclastic rocks resting on exposed basement (see "Lesvos Volcanotectonic Regime" below). However non-complicated by folding, the volcanic structures on Lesvos have been adulterated by erosion and fault tectonics. Fault tectonics enhanced the development of a mature geothermal system providing pathways of hydrothermal water circulation as it is shown by the numerous hot springs on the island and the development of mineralization (Kelepertzis, 1996; Vamvoukakis et al., 2001).

2. LESVOS VOLCANOTECTONIC REGIME

In the island of Lesvos a thick succession of Miocene volcanic rocks overlies a metamorphic Paleozoic basement. Mid-Miocene volcanism on Lesvos was of shoshonitic character both overlain and underlain by products of calkalcaline volcanic affinity. Shoshonitic volcanism lasted only one million years (18.4-17.3 M.a.). Calcalkaline volcanic activity has occurred at 21.5 M.a. and was repeated at 16.5 M.a. (Pe-Piper and Piper, 1993). A chain of stratovolcanoes and submerged calderas (in the Gulf of Kalloni) occupies a long-standing graben orientated NE-SW across the island (Pe-Piper and Piper, 1993) with large caldera structures in the western (Sigri caldera) and the eastern part of the island (Stipsi Caldera) (Kouli, 2004).

The overall stratigraphy as defined by Katsikatsos (1992) consist of:

(i) a pre-alpine Permo-Triassic autochthonous succession of metasediments with interlayers of marbles and dolomitic marbles, all metamorphosed in low grade. The pre-alpine basement outcrops mainly in the central and southeastern part of the island (Fig. 1).

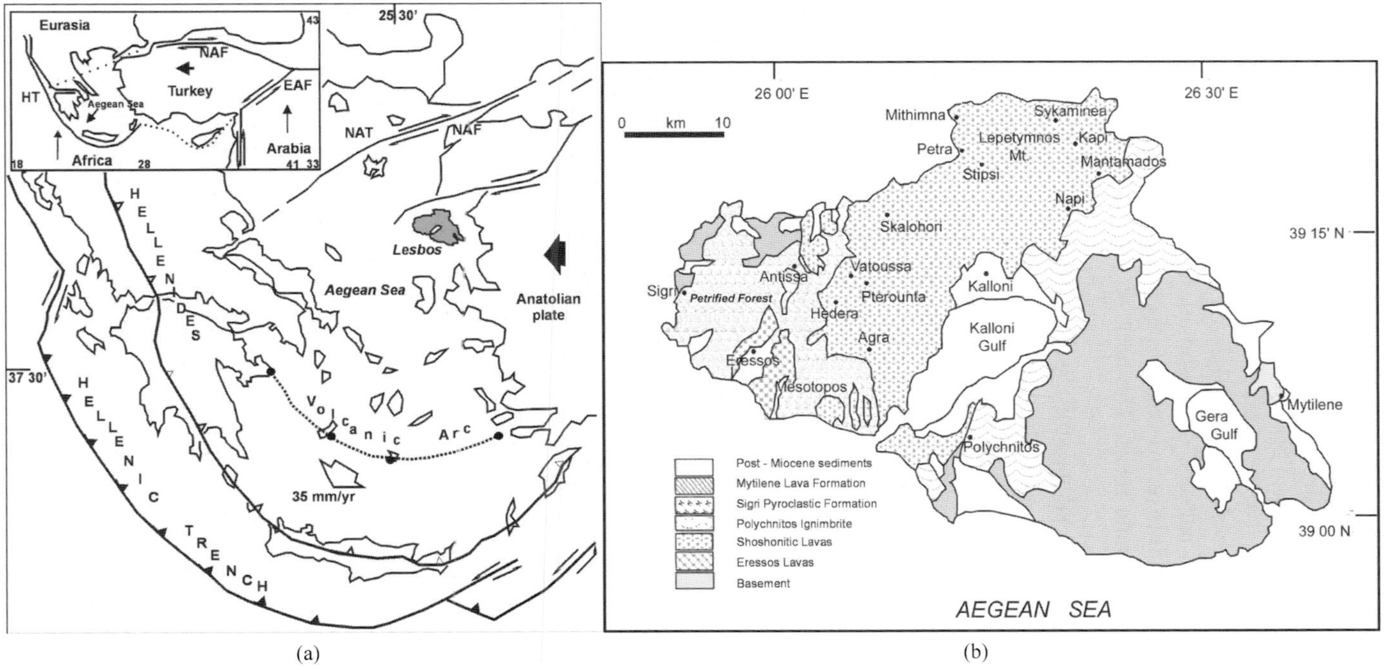

Fig. 1. (a) General map showing the main structures of the Hellenic Arc and Trench system. NAF: Northern Anatolian Fault, NAT: Northern Aegean Trough. Inset: Map summarizing the geodynamic framework around the eastern Mediterranean Sea together with the major plates, which involve in collision process (modified from McKenzie, 1972). EAF: Eastern Anatolian Fault, DSF: Dead Sea Fault, HT: Hellenic Trench. The study area is marked with gray color. (b) Simplified geological map of Lesbos island (modified from Pe-Piper and Piper, 1993).

(ii) Two allochthonus covers: a lower one of metasediments and metamorphosed mafic volcanic rocks which is overlain by the second allocthon made of ophiolites.
(iii) Miocene – Pliocene mostly locally derived volcanic rocks and minor marles and other clastic sedimentary rocks. These formations cover most of the exposed surface of the island.

The Cenozoic post-collision volcanism on Lesvos (iii above) occurred in three cycles which have resulted in a package of shoshonitic domes, lavas and pyroclastic deposits which are both underlain (Eressos lavas, Lower lavas) and overlain (Mytilene formation) by calc-alkaline volcanic rocks. The peak of volcanism occurred between 16-17 M.a.

The area has been mapped by Hecht (1972-1975) on a scale of 1:50.000 and the volcanic stratigraphy, its revisions and absolute age-dating and the petrology and mineralogy of shoshonites was first presented by Pe-Piper (1980a, 1980b, 1992,) and Pe-Piper and Piper (1980, 1993) who divided the Miocene volcanic succession into: the calcalkaline Lower lavas and Eressos formation, the Skoutaros formation (mainly lavas), the Acid member (comprises the Kapi rhyolites, Polychnitos ignimbrite, and Sigri pyroclastics) the Skalohori and Sykaminea lavas and on the top of the volcanic stratigraphy the Mytilene formation.

The Miocene volcanic rocks are deformed by numerous faults, the main directions of which are NNE and NNW with subsidiary N-S and E-W directions (Koukouvelas and Aydin, 2002; Vamvoukakis et al., 2001; Rokos et al., 2000; Pavlides et al., 1990). These directions mimic the tectonic strain in the area being in the proximity to the Northern Anatolian Fault (NAF, Fig. 1a) and to the Izmir-Ankara Suture (Fig. 1a) (Aydar, 1998). Some of these faults are long standing lines of weakness. Pe-Piper and Piper (1993) first indicated that the Kapi rhyolites lie in a northeaster zone running across the island. Our work has indicated that this zone is a graben that has located many important volcanis centers such as the water-submerged calderas in the Gulf of Kalloni (Lamera, 2004) and the large caldera of Stipsi (Vamvoukakis, 2004; Kouli, 2004) from which issues some of the most voluminous ignimbrites in the Polychnitos Ignimbrite Formation (Lamera, 2004).

3.METHODS

3.1 Remote Sensing Methods

3.1.1 Data Acquisition

A combination of remote sensing and fieldwork data have been used for the detection and mapping of the Miocene calderas and hydrothermal alteration areas on Lesvos. The relevant output images were generated from the analysis and processing of:
1. A Landsat–TM satellite image of the Lesvos island, with a spatial resolution of 30x30m pixel size, acquired on 20 August 1999,
2. A SPOT panchromatic (SPOT-Pan) satellite image of the same area, with a spatial resolution of 10x10 m pixel size, acquired on 12 July 1999, and

3. The Digital Elevation Model (DEM) of the study area with a cell size of 30 m.This is a continuous raster layer of the island, in which data values represent elevation, generated from the 30 m contours of the topographic maps.

3.1.2 Digital Data Pre-processing

The satellite data were preprocessed in order to remove the geometric distortions and to bring the remote sensing images into registration with one another. With the use of topographic maps of the study area, the images were geometrically corrected and geo-referenced to the HGRS 87 (Hellenic Geodetic Reference System). The geometric correction had an average RMS error of approximately 0.4 pixels for both images.

With the on-screen interpretation of the Spot-Pan image in the western part of the island we identified a 14 x 10 km elliptical structure open to the northern Aegean (*Sigri caldera*) and two smaller structures south of it (*Agra and Mesotopos calderas*) (Fig. 2). The Stipsi, Skalohorion and Kalloni Gulf calderas were recognized in other parts of the island and the Vatoussa structure (Pe-Piper, 1980b) was identified as a nested caldera within the Sigri (Fig. 2). A slope map with a resolution of 1° was produced from the DEM of the study area. Slope values were averaged within a 30 x 30 m area. The slope map was superimposed on the DEM in order to remove the effect of vegetation, present on the Landsat TM5, obtain a three-dimensional view of the circular features in

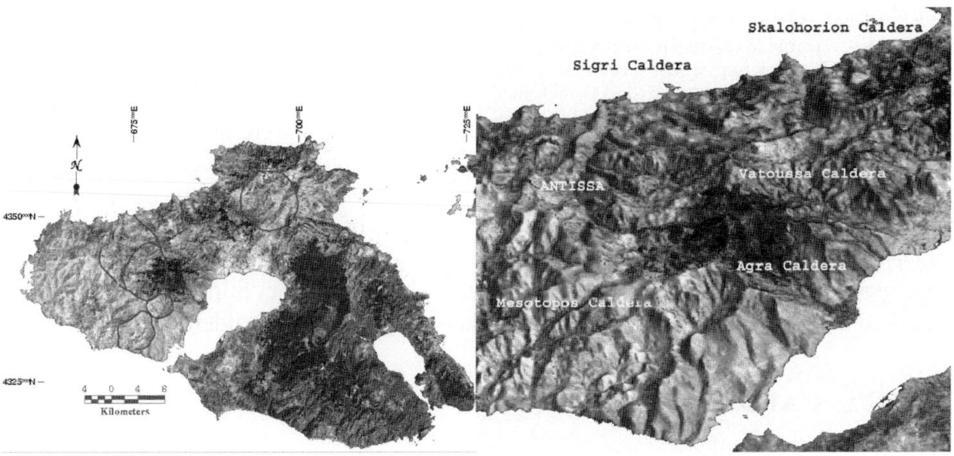

(a) (b)

Fig. 2. (a) False Color Composite 5,7,3 (R,G,B) printed in black and white of the enhanced Landsat-TM satellite image after decorrelation stretching,. The six structures that were interpreted as caldera structures are delineated with red solid lines. The Polychnitos Ignimbrite Formation (PF) is shown in the east, while the Sigri Pyroclastic Formation in the west, exhibits a dendritic drainage network pattern (SPF), (b) A 3-dimensional view of the volcanic structures rotated westwards: overlay of enhanced resolution Landsat-TM FCC 5,7,3 (R,G,B), printed on black and white, on the DEM of western Lesvos. Five caldera structures are delineated with red dashed lines. Internal and external to the calderas, lava domes and flows are clearly revealed. Within this major caldera structure the resurgent lava domes are apparent extruding lavas to the north. The nested Vatoussa caldera displays only a few, small lava domes and a disrupted caldera floor. Altered lava domes of the Mesotopos caldera are emplaced circularly in an intra-caldera ring fault.

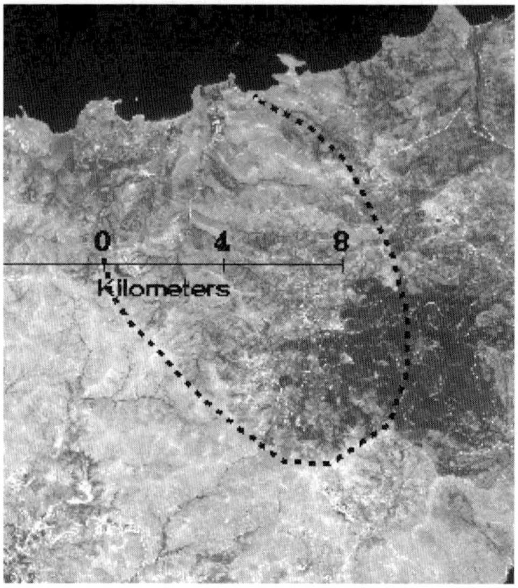

Fig. 3. Spot-Pan image with 10 m resolution of Sigri and Agra calderas. The black dashed line delineates the elliptical shape of the Sigri structure, which was initially identified on this image. To the south the Sigri structure is tangential to one much smaller and obviously circular structure, which encompasses the village of Agra (Agra caldera). The Sigri Pyroclastic Formation, (light gray) occupies almost the whole area west of the Sigri caldera.

combination with the slope information and depict the borders of the volcanic structures with greater accuracy (Fig. 2). We also used the band ratios TM5/TM7 and TM3/TM5 to locate the zones of advanced argillic alteration zone and silicification. In these images the areas of alteration appeared white (Fig. 4).

Finally, the results of the digital image analysis were checked in the field in order to verify in-situ the quality of the interpretation of the geological features.

3.2 Analytical methods

X-ray Diffraction methods were used for the identification of the alteration assemblages. Argillic alteration minerals were further analyzed by Raman spectroscopy to distinguish between polymorphs. In mineralized zones ore mineral assemblages were analyzed by electron microprobe. Bulk specimens from these zones were analyzed for trace elements by INAA. To estimate the temperatures and salinities of the hydrothermal fluid, as well as, its provenance, fluid inclusion and oxygen isotope studies were performed on specimens from alteration zones of Megala Therma in the Stipsi caldera and on silicified dome specimens from the Pterounda dome of the Sigri caldera (Fig. 1).

4. STIPSI CALDERA

The northern part of the island includes a major caldera structure, namely the Stipsi caldera which is intersected by two major NNE and NNW trending fault zones (Fig. 4).

The presence in this area of a similar structure with very different geometry was first reported by Pe-Piper (1998). However, its present extent, extensive hydrothermal alteration along ring and other major faults (see above) and probable association with a high sulphidation epithermal gold system was first reported by Kouli and Seymour (2003) and Vamvoukakis et al. (2001). Intense hydrothermal alteration systems occurs south of the Stipsi village, in the domes of Kapi rhyolites, in the external northern part of the caldera near Megala Therma hot springs (Fig. 4). The alteration zone near Megala

Therma is located on the extension of the NNW trend fault.

4.1 Pyroclastic Formations Associated with the Stipsi caldera

An extensive pyroclastic formation namely the Polychnitos Ignimbrite is located in the southeastern external part of the Stipsi caldera. It consists of nine ignimbrite sheets (Lamera and Seymour, 2003) of shoshonitic composition. Lamera and Seymour (2003) and Kouli and Seymour (2003) using field and digital image processing data respectively, provide evidence that the three most extensive intermediate members of the Polychnitos Ignimbrite, namely MGF I, II and III have originated from the eastern part of the Stipsi caldera. However, one of the lower and most extensive (PU) and the upper (ZV) Polychnitos ignimbrite members have originated from the subsided under water Kalloni Gulf and Pergama caldera respectively (Lamera and Seymour, 2003).

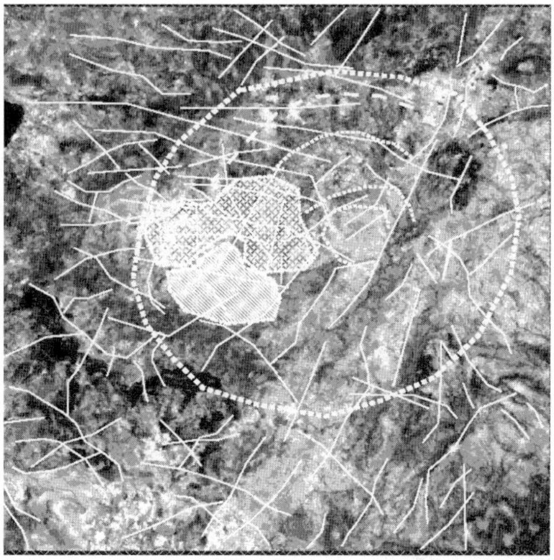

Fig. 4. Tectonic and volcanic structure map in a Lansdat TM 5 image. The thick line represents the Stipsi caldera, the dashed line area represents the silicification and the dotted area the advanced argillic alteration. The white lines represents the faults.

4.2 Epithermal High Sulphidation System Associated with the Stipsi Caldera.

In the extension of a NNW trending fault and outside the rim of the Stipsi caldera (Figs. 1, 4) mineralized quartz veins represent one of the issues of a high sulphidation hydrothermal system. The veins consist of three generations of hydrothermal quartz which has invaded along NNW planes. The early quartz is in the form of milky quartz which is replaced by clear quartz. The third generation of quartz is coarse-grained (spar). Each vein is enclosed in a hydrothermal alteration envelope which displays characteristic zoning.When mineralized, the veins have an ore assemblage of pyrite, chalcopyrite, chalcocite, sphalerite and galena. Electron microprobe analyses of the ore minerals (Table 1) indicate that chalcopyrite and pyrite are the precious metal-bearing ore minerals. As we can see from Table 1, Au concentrations are relatively higher in chalcoryrite and pyrite. Only one measurement in galena shows gold concentrations. Galena, chalcopyrite, and pyrite show Ag values, however, galena is the most important Ag-bearing mineral. Sphalerite grains do not show any gold or silver values.

Fluid inclusion microthermometry on mineralized vein quartz gave a spectrum of temperatures of homogenization (T_h) from 234^0C to 275^0C with a maximum frequency between 240^0C and 244^0C. The salinity of the hydrothermal fluid varies between 0.33-

Table 1. Electron microprobe analyses in sulphides From Megala Therma Quartz veins.

	Fe	As	S	Pb	Ag	Cu	Au	Sn	Co	Zn	Sb	Te	Ni
Chalcopyrite	0	0.12	33.658	0	0.313	68.325	0.035	0	0.006	0	0	0.019	0
	0	0.088	33.695	0	0.306	68.639	0.035	0	0.005	0.095	0.005	0.008	0.013
	0.028	0.138	34.144	0	0.166	69.179	0.046	0.008	0.009	0	0.013	0.058	0
	0	0	33.467	0	0.147	67.915	0.15	0	0.003	0	0	0	0
	0	0.056	33.785	0	0.131	69.097	0.046	0	0	0	0	0	0
Galena	0	0.039	12.777	85.184	0	0	0.028	0.014	0.016	0.008	0.029	0.051	0.001
	0.066	0.054	13.171	86.834	0.021	0	0	0	0	0.039	0.014	0.055	0
	0	0.047	13.331	87.007	0.053	0.547	0	0	0	0	0.374	0.045	0.003
	0	0.049	13.236	87.457	0	0.031	0	0	0	0.031	0	0.058	0.007
Sphalerite	1.6	0.039	32.882	0	0	1.743	0	0	0	63.98	0	0	0
	1.223	0	32.612	0	0	0.925	0	0.023	0.006	65.662	0	0	0.028
	0.721	0	32.204	0	0	0.478	0	0.008	0.006	65.528	0.031	0.005	0.019
Pyrite	46.236	0.032	52.907	0	0.002	0	0.002	0	0.065	0.02	0	0	0
	47.138	0.013	52.91	0	0.015	0.009	0	0	0	0	0	0	0.006
	46.352	0.049	52.686	0	0.005	0.049	0	0	0.007	0	0.005	0.014	0.041
	42.606	0.011	47.143	0	0	0.011	0.005	0.037	0.097	0.081	0.038	0.025	0

4.93 with an average value of 2.00 wt% eq.NaCl. These salinities are similar to those reported from other high sulphidation ore deposits (White and Hedenquist, 1990).

The local geothermal system is still active as evidenced by the hot springs of Megala Therma which also gave the name to the area. The provenance of these hydrothermal waters as it is shown from ^{18}O stable isotope values in vein quartz from this area (Table 2).

Another exit of the geothermal system is near the western-northwestern part of the caldera and the homonymous village of Stipsi built on the caldera wall. In this area there is an exceptionally high frequency of tectonic elements (faults) which intersect with the ring faults of the caldera (Fig. 4). Such passageways have permitted magma ascent in the form of numerous lava domes and then provided a plumbing system for the circulation of the fluids. The domes are penetratively altered. Alteration is of high sulphidation type (Vamvoukakis et al., 2001) and is very extensively developed with widespread silicification and argillic zones with alunite (Fig. 4). Quartz veins when mineralized show chalcopyrite and pyrite (Vamvoukakis et al., 2001).

5. SIGRI CALDERA

In the western part of the island of Lesvos a large elliptical structure (maximum axis 14 Km directed N335^{o} and a small axis 10 Km directed N 65°), open towards the Northern Aegean Sea was first recognized on a SPOT pan satellite image by Kouli and

Table 2. $\delta\ ^{18}O$ analyses from Megala Therma Quartz veins.

Sample	$\delta\ ^{18}O$ ($^{0}/_{00}$ SMOW)
LV 56	+5.4
LV 56 a	+5.0
LV 45 A (milky quartz)	+4.9
LV 45 A (clear quartz)	+4.7
LV 57 A	+5.8
LV 65 D	+ 4.7
LV 56 G	+4.4
LV 56 B	+6.3
LV 60 E	+4.6
LV 56 K	+4.6
LV 56 H	+6.3
LV 61 A	+5.1

Seymour (2003) who have established that this edifice is an eroded caldera and depicted its original geometry.

Presently the caldera displays a low, 700 m (max), discontinuous rim, that is better preserved at its western part near the village of Antissa, which is amphitheatrically built on the inner caldera wall facing east (Fig. 1). The Eastern rim of the caldera is breached by domes and associated couleés (Figs. 2, 3).

The southern border of the Sigri caldera is defined by a chain of lava domes (Figs. 2, 3) and a tangential circular feature, which occupies 40% of the surface extent of the Sigri structure, the nested Vatoussa caldera (Kouli and Seymour, 2003). The Vatoussa caldera was first reported by Piper (1980b) based on extensive alteration and was established as a caldera by Kouli and Seymour (2003). Vamvoukakis and Seymour (2003) argue the reasons for the more intense alteration in this part of the Sigri caldera.

The upper part of the western inner caldera wall displays well exposed radial faults almost to the height of a rather well preserved rim (Figs. 2, 3). The Sigri caldera floor is disrupted by ring and rectilinear faults (Kouli and Seymour, 2003). Such caldera floors are reported by Lipman (1997) as the result of multiple nested or overlapping collapses during successive eruptions.

5.1 Pyroclastic Formations Associated with the Sigri Caldera.

The area externally to the western part of the Sigri caldera exhibits a characteristic texture on satellite images. This area corresponds to the Sigri Pyroclastic Formation (SPF), host to the renowned Sigri lithified forest (Velitzelos et al., 1998; Pe-Piper 1998). The large size of the Sigri caldera justifies the large volume of pyroclastic material comprised in the extensive SPF. The orientation of lithified tree trunks within the park area which are enclosed in the SPF as they are measured in our field work (Seymour et al., 2002) suggests that the source of this portion of SPF is the Sigri caldera, which agrees with a source from the east (Velitzelos et al., 1998).

Fig. 5. (a) A topographic profile (AB) across the DEM of the Mesotopos caldera displays well the flat caldera floor and the regularly emplaced post collapse lava domes, (b) A 3-dimensional view, 5,7,3 (R,G,B) superimposed to the DEM, of the Mesotopos caldera rotated westwards. The caldera walls breached by ash-flow deposits, the coherent flat floor (CF), the radial faults (solid gray lines) and the circularly situated lava domes characterize the Valles-type calderas, (c) Field photo of the mega breccia formation associated with the Mesotopos caldera in a roadcut across its external west wall.

The extensively exposed coarse lithic proximal breccias in the proximity of the western wall of the Sigri caldera (Fig. 5) provide additional proof that this edifice was the source of the Sigri Pyroclastic Formation.

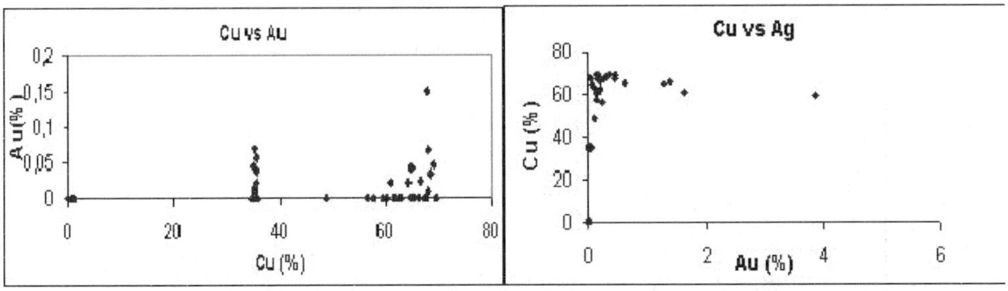

(a) (b)

Fig. 6. Diagram Cu vs As derived from probe analyses in chalcopyrite from Pterounta silicified rocks.

5.2 Epithermal High-Sulphidation System Associated with the Sigri Caldera

In the area of Pterounda in the southeastern part of the Sigri caldera (Figs. 1, 3) where axial faults intercept the Sigri and Vatoussa coalescing boundary ring faults, there is a proliferation of intensely altered lava domes. X-ray diffraction results, from Pterounda dome altered samples indicate assemblages representative of silicification, advanced argillic and argillic zones. Propylitic alteration assemblages are absent (Vamvoukakis and Seymour, 2003). The domes display intense successive zones of alteration with supergene silicification and hematitization occupying the upper parts of the domes. The silicification is in the form of amorphous silica but no quartz veins have been observed. This prohibited the application of fluid inclusion microthermometry. Disseminated grains of pyrite and chalcopyrite pervasively mineralize silicified domes. Pterounda dome specimens have been analysed by INAA for Au, Ag and other trace elements. There is a positive correlation between Au and the "pathfinder" elements As (Fig. 6) and Ba (Fig. 7) within certain concentration limits of Ba. Analyses of bulk specimens for stable isotopes (^{18}O) vary between +7.1 and +10.1 per mil and suggest components for the hydrothermal water of meteoric and Sea-H_2O provenance with small contributions of juvenile water (Table 3).

6. THE MESOTOPOS CALDERA

This is a smaller structure that has been identified south (Figs. 2, 5) of the Sigri caldera (Kouli and Seymour, 2003). The caldera has a flat floor and is open towards the Aegean Sea. Radial and ring faults dictate its drainage pattern and its well-defined intrusions of domes arranged in a circular fashion (Kouli and Seymour, 2003). The lava domes are intensely altered and the argillic alteration assemblages have been studied by Tsoli-Kataga (1981). The Mesotopos epithermal system is presently studied by our group.

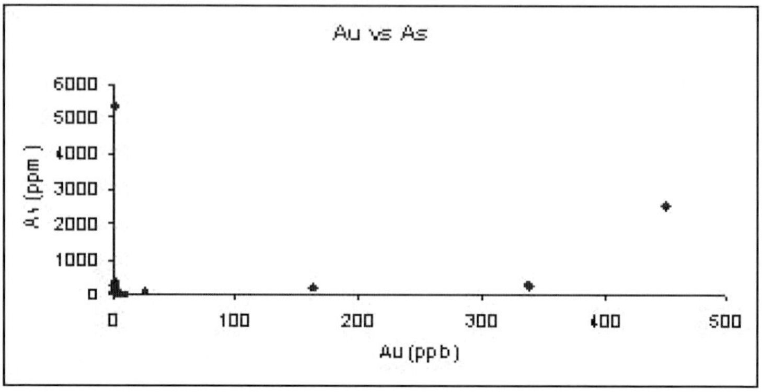

(a)

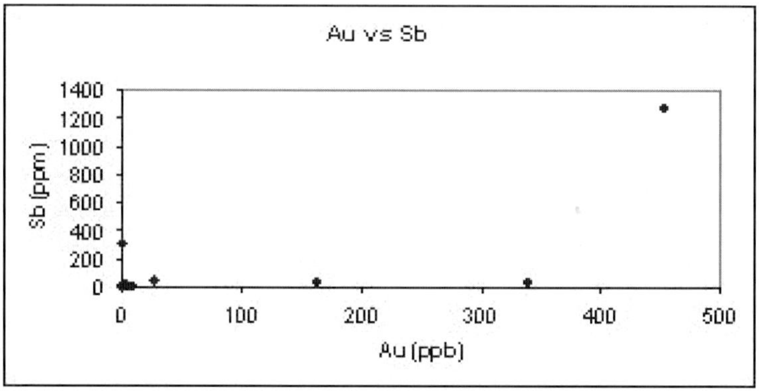

(b)

Fig. 7. Diagram Cu vs Ba derived from probe analyses in chalcopyrite from Pterounta silicified rocks.

7. EVIDENCE FOT THE CALDERA STRUCTURES ON LESVOS ISLAND

The geometry of the Stipsi caldera has been defined with the use of absolute elevations derived from the analysis of the DEM of the area. Radial caldera faults are characteristically displayed in the northern part of the structure (Fig. 4). Chains of lava domes are intruded parallel to the main tectonic directions e.g. lava domes of the Kapi rhyolites which are intruded parallel to a NNE trending graben. Other dome arrangements indicate the center of the Stipsi caldera.

The Sigri caldera is a structure complicated by the multiple intrusion of lava domes and the enclosure of another caldera (nested caldera structure) the Vatoussa caldera. The structure was defined by analysis of the absolute attitudes defined from DEM. Radial caldera faults are particularly well displayed on the remnant western caldera wall and near the village of Antissa. An analysis of the direction and dip of the hydrographic network was also performed in order to define the lowest point of the caldera. The

Table 3: δ ^{18}O analyses from Pterounda Domes.

Sample	δ ^{18}O ($^0/_{00}$ SMOW)
LV 34	+ 9.2
LV 37	+ 10.1
LV 81 B	+ 7.2
LV 94	+ 9.7
LV 95	+ 9.6
LV 96 B	+7.1

arrangement of the domes and ring faults together with the presence of megabreccias on sections of the western caldera wall. Lava domes thickly covered with pine trees separate the Sigri caldera to the south with the much smaller Agra caldera.

The smaller structures of Agra, Mesotopos and Skalohori were identified and measured by the superimposition of the DEM on Landsat TM and SPOT satellite images. In the Mesotopos and Agra calderas the drainage network and tectonic features were analyzed digitally for the correct placement of borders and floors. The Mesotopos caldera is much better preserved and identifiable displaying flat caldera floor and intensely altered lava domes arrange in circle due to intrusion in ring faults. The Mesotopos caldera displays also well developed megabreccias in the external proximity of the caldera wall.

Finally the near land caldera of the Kalloni Gulf is a large circular structure with characteristic radial faults. The voluminous PU unit of the Polychnitos Ignimbrite has its source in this center (Lamera, 2004).

8. CONCLUSIONS

The conclusions of this study are:

(a) Seven calderas structures have been identified on the island namely the Sigri (and Vatoussa), Stipsi, Mesotopos, Kalloni Gulf calderas and Agra and Skalochorion small calderas or vents (Fig. 2). The first three (Sigri, Stipsi and Mesotopos) are being studied by our group for epithermal gold research.

(b) We have first recognized the issue of hot waters and intense alterations in the Stipsi caldera on false color composites with the use of thermal infrared band on a Landsat-TM image. Within the Stipsi caldera intense alteration is concentrated in an area near the village of Stispi where high density tectonic elements intercept caldera ring faults (Fig. 4).

(c) The Sigri structure is a large (~15km) caldera associated with proximal coarse breccias and encloses the smaller nested Vatoussa caldera (Fig. 5). The Vatoussa caldera has been previously recognized and reported in the literature (Pe-Piper, 1980b) on the basis of its intense alteration. We suggest here that the source of the Sigri Pyroclastic Formation which is the host to the renowned Lithified Forest is the Sigri caldera.

(d) In the east of the island members of the Polychnitos Ignimbrite seem to have been derived from more than one sources (Lamera and Seymour, 2003). Application of remote sensing techniques indicated the presence of flow lines in the Polychnitos ignimbrite that originated from the eastern part of the Stipsi caldera. Field investigations, including measurements of maximum lithics, verified the conclusion that the stratigraphically intermediate ignimbrite members (MGF I, II, III) have indeed originated from the Stipsi caldera. Study of remote sensing data indicates that the proximal to the land part of the Gulf of Kalloni is a caldera structure situated on a major graben zone (Kouli and Seymour, 2003). This conclusion was verified by field data (maximum lithics) which indicated that the source to the PU member of the Polychnitos ignimbrite (Lamera and Seymour, 2003) is actually the Kalloni Gulf (Fig. 2).

REFERENCES

Anguita, F., Verma, S.P., Marquez, A., Vasconcelos F.M., Lopez, I. and Laurrieta, A., 2001. Circular features in the Trans-Mexican Volcanic Belt, Journal of Volcanology and Geothermal Research, Vol. 107, 265-274.

Aydar, E., 1998. Early Miocene to Quaternary evolution of volcanism and the basin formations in western Anatolia: a review Journal of Volcanology and Geothermal Research, 85: 69-82.

Hecht, J., 1972-1975. IGME, Geological map of Greece, 1:50.000. Plomari-Mytilene, Ayia Paraskevi, Mithimna, Polychnitos and Eressos sheets.

Katsikatsos, G., 1992. Geology of Greece, Organization for Publishing Didactical Books (O.E.D.B), Athens.

Kelepertzis, A., 1996. Gold and Silver Mineralization Associated with Quartz Veins and Breccia in the Volcanic Rocks of North Lesvos Island. Mineralogy, Hydrothermal Alterations, Geochemistry Bulletin of research information of national and Kapodistrian University of Athens.

Koukouvelas, I.K. and Aydin, A., 2002. Fault structure and related basins of the North Aegean Sea and its surroundings. Tectonics, Vol. 21, 0,10.1029/2001TC901037.

Kouli, M., 2004. Application of Modern Technology to the Solution of Volcanological Problems: I. Use of NDIC to Compare Magma Mixing Plg Textures from the Volcanoes of Teide and Lesvos, II. Identification of Volcanic Structures with the Use of Remote Sensing Technique. Unpublished PhD.

Kouli, M. and Seymour K.St., 2003. Contribution of remote sensing techniques to the identification and characterization of Miocene calderas, Lesvos island, Aegean Sea, Hellas. Conditionally accepted in Canadian Journal of Remote Sensing.

Lamera, S., 2004. The Polychnitos Ignimbrite of Lesvos Island, unpublished PhD.

Lamera, S. and Seymour, K.St., 2003. The volcanic stratigraphy and vent-sources of the ignimbrite members of the Miocene Polychnitos Formation, Lesvos Island, Aegean Sea, Hellas. Submitted to The Bulletin of Volcanology.

Lipman, P.W., 1997. Subsidence of ash-flow calderas: relation to caldera size and magma-chamber geometry, Bulletin of Volcanology, Vol. 59: 198-218.

Pavlides, S., Mountrakis, D., Killias, A. and Tranos, M., 1990. The role of strike-slip movements in the extensional area of northern Aegean (Greece). A case of transtensional tectonics, Ann. Tesctonicae IV V4 (2), pp. 196-211.

Pe-Piper, G., 1980a. Geochemistry of Miocene shoshonites, Lesbos, Greece, Contributions to Mineralogy and Petrology, Vol, 72, pp. 387-396, Berlin.

Pe-Piper, G., 1980b. The Cenozoic volcanic sequence of Lesbos, Greece, Z.Dtsch.Geol.Ges, Vol. 131, pp. 889-901.

Pe-Piper, G., 1992. Geochemical variation with time in the Cenozoic high-K volcanic rocks of the island of Lesbos, Greece; Significance for shoshonitic petrogenesis, Journal of Volcanology and Geothermal Research, Vol. 53, pp. 371-387.

Pe-Piper, G. and Piper D.J.W., 1993. Revised stratigraphy of the Miocene volcanic rocks of Lesbos, Greece, N. Jb. Geol. Paläont. Mh, H2, pp. 97-110.

Pe-Piper, G., 1998. The Petrified Forest of Lesbos, Greece, Newsletter of the Volcanology and Igneous Petrology Division, Geological Association of Canada, November.

Rokos, D., Argialas, D., Mavratza, R., St.-Seymour, K., Vamvoukakis, C., Kouli, M., Lamera, S., Paraskevas, H., Karfakis, I., and Denes, G., 2000. Structural Analysis for Gold Mineralization Using Remote Sensing and Geochemical Techniques in a GIS Environment: Island of Lesvos, Hellas, Natural Resources Research, Vol. 9, pp. 277-293.

Sabins, F.F, 1997. Remote Sensing: Principles and Interpretation", W. H. Freeman & Company, New York, U.S.A., p. 366 – 371.

Smist, W.L., 1974. Remote Sensing Applications for Mineral Exploration, Dowden, Hutchinson and Ross, Inc, Stroudsburg, Pennsylvania.

St. Seymour, K., Kouli, M., Vamvoukakis, C. and Lamera, S., 2002. Detection of Miocene volcanic structures by remote sensing techniques for epithermal gold exploration. Proccedings of Saskatoon 2002.

Tsoli-Kataga, P., 1981. Contribution to the study of the Lesbos Kaoline, I. The Mesotopos Kaoline, Mineral Wealth, Vol. 12, pp. 39-46.

Vamvoukakis, C., 2004. Epithermal mineralogy of Au-Ag on Lesvos Island, unpublished PhD.

Vamvoukakis, C., St-Seymour, K., Williams-Jones, A. and Rokos, D., 2001. The Miocene Caldera of Stipsi on Lesbos Volcano: An Example of "High Sulfidation" hydrothermal system explored for gold, Abstract in Volume, Geological Association of Canada – Mineralogical Association of Canada (GAC-MAC), Annual Meeting May 2001, St. John's, New Foundland, Canada.

Vamvoukakis, C., and St. Seymour, K., 2003. High-Sulphidation epithermal systems associated with caldera structures on Lesvos Island. Submitted to Mineralium Deposita.

Velitzelos, E., Seriades, E and Zouros, N., 1998. The petrified forest of Lesbos – new results on the Paleochloris emphasize the outstanding universal value of the monument", 3rd International Symposium. Protected Areas Management and Natural Monuments, sponsored by UNESCO, Molyvos, Lesbos, 13-15 July 1998.

White, N.C., Hedenquist, J.W., 1990. Epithermal environments and styles of mineralization: variations and their causes, and guidelines for exploration. Journal of Geochemical Exploration, Vol. 36, pp. 445 – 474.

Tertiary high-Mg volcanic rocks from Western Anatolia and their geodynamic significance for the evolution of the Aegean area

S. Agostini[1,3], C. Doglioni[2], F. Innocenti[1,3], P. Manetti[3], M.Y. Savascin[4], S.Tonarini[3]

[1]Dipartimento di Scienze della Terra, Università di Pisa, Italy.

[2]Dipartimento di Scienze della Terra, Università La Sapienza, Roma, Italy.

[3]Istituto di Geoscienze e Georisorse, CNR, Pisa Italy.

[4]Geothermal Energy Research and Application Center, Dokuz Eylul University, Izmir, Turkey.

ABSTRACT

Scattered Late Miocene high-Mg basaltic andesites to dacites can be found in Western Anatolia. These rocks display Mg#>65, high CaO/Al2O3 ratio, low alkalies and TiO2 contents. Trace element distribution shows typical orogenic signature with higher values of Fluid Mobile Elements and lower values of HREE and HFSE with respect to Early Miocene Western Anatolia calc-alkaline rocks. The $^{87}Sr/^{86}Sr$ and $^{143}Nd/^{144}Nd$ ratios virtually overlap the values of the less evolved calc-alkaline rocks. The variations observed in this association have been attributed to an FC process combined with interaction with crustal material. A thermal anomaly affecting a depleted mantle source has been invoked for the genesis of these products. Such an anomaly was produced by the ascent of deep sub-slab mantle, which replaced the underthrust lithosphere, already thinned and stretched by extensional process.

1. INTRODUCTION

The Anatolian-Aegean region is a key area for understanding the complex geodynamic history of the eastern Mediterranean. It is characterized by a protracted subduction process, marked by multiple continental collisions, by a southwest post-Cretaceous migration of the subduction hinge, and by a back-arc extension that led to the formation of the Aegean area (e.g. Doglioni et al., 2002). This intricate geotectonic history is matched by a widespread Tertiary-Quaternary magmatic activity whose space-time development and petrogenetic changes can be used to constrain the evolutionary model of the area.

Overall, three broad types of mantle-derived magmas have been identified in the volcanism of western Anatolia: a) an orogenic magma that generated a calc-alkaline to shoshonite association, active from the Upper Eocene to Middle Miocene, which is part of a magmatic belt extending from the Rhodopean-Thracian zone to Central Anatolia; b) a magma of lithospheric origin forming a Middle Miocene, volumetrically and spatially restricted lamproitic association, and c) a Quaternary Na-alkaline magma, heralded by scattered Late Miocene alkali basalts whose geochemical and petrological features point to an interaction process between OIB-type melts and subduction-related products of the previous activity.

Investigation of the volcanism of Western Anatolia (WA) has led to the discovery of some unusually high-Mg basaltic andesites and andesites that were erupted within a short period of time (Late Miocene). In the present paper we describe this magmatism and discuss its relationships with near-coeval high-Mg products detected in the central Aegean area (Pe-Piper and Piper, 1994; Pe-Piper, 1991), within the framework of the Tertiary and Quaternary geodynamic evolution of the Anatolian-Aegean area.

2. GEOLOGICAL AND VOLCANOLOGICAL OUTLINES

The architecture of the Aegean-West Anatolian system is the result of the NE subduction of Africa beneath Eurasia, which has continued since at least the Cretaceous. The subduction migrated southwards, producing orogens that were subsequently affected by extensional processes that reached their maximum intensity in the Aegean area. The study region (Fig. 1) lies between the Menderes Massif to the south and the Ankara-Izmir zone to the north, representing an ophiolitic suture of Paleocene-Eocene age (Sengör and Ylmaz, 1981). The extensional tectonics affecting the area since Early Miocene led to the formation of several E-W and NW-SE basins. These basins are filled by continental sediments, characterized by the occurrence of volcanic and volcanoclastic intercalations (Seyitoglu and Scott, 1991).

The area was affected by a Tertiary extensive volcanism with an orogenic signature. The oldest products are of Upper Eocene-Oligocene age and are exposed in limited volumes in the northern part of WA and Thrace (Yilmaz et al., 2001); the volcanic activity reached a climax in the Lower Miocene and was exhausted by the Middle Miocene. The Lower-Middle Miocene volcanic products spread into the Central-Northern Aegean area, forming an orogenic- magmatic arc extending eastwards as far as the NS Afyon-Isparta junction, where the volcanic activity persisted up to the Late Pliocene (Francalanci et al., 2000). The Tertiary orogenic association, which also includes co-genetic plutonic rocks (Altunkaynak and Yilmaz, 1999; Genç, 1998), ranges in composition from high-K andesite to rhyolite. The volcanic products are dominated by lava domes and lava flows with associated minor pyroclastics and breccias. Felsic ignimbritic covers outcrop widely throughout WA and the north-eastern Aegean Sea (e.g. the islands of Lesbos and Limnos (Innocenti et al., 1994). The final activity of this

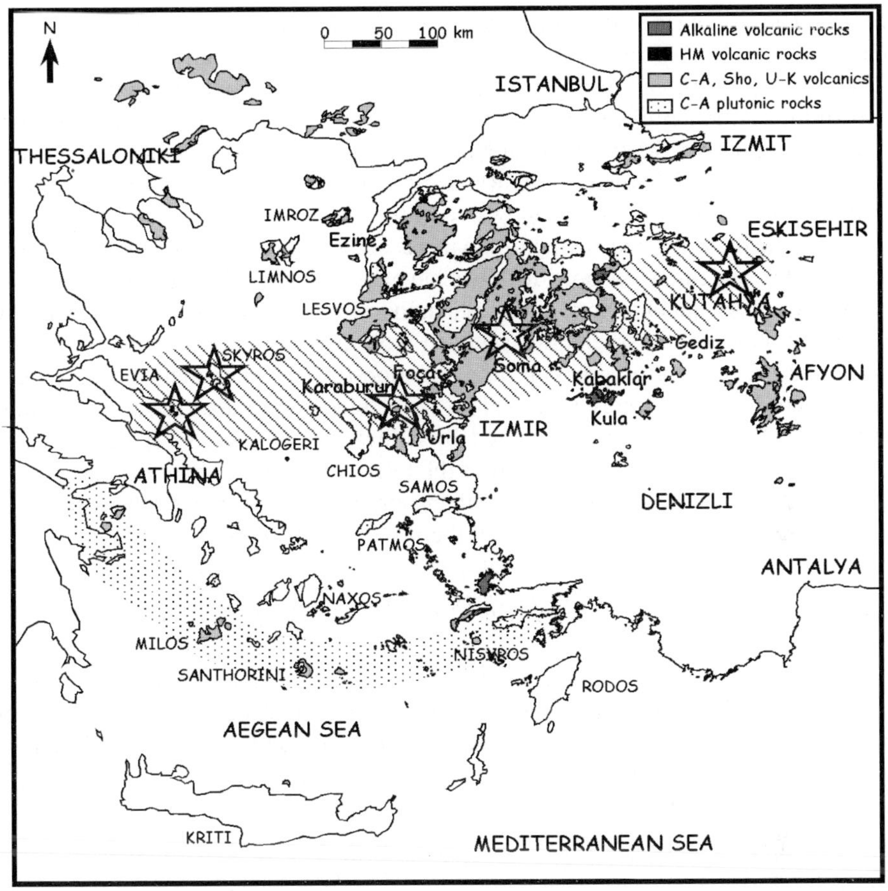

Fig. 1. Map of western Anatolia and Aegean area showing the distribution of Tertiary and Quaternary igneous products and the location of HM rocks (stars). Oblique lines highlight the distribution area of HM rocks. Dotted area: South Aegean Volcanic Arc.

orogenic cycle is characterized by the presence of K-rich shoshonites and latites, sometimes with ultrapotassic character. Minor lamproites, such as dykes and lava flows, are also found in a restricted region of central western Anatolia (Gediz area), at about 14-15 Ma (Innocenti et al., 2004). Typically, these rocks are primitive with high Mg# (~80) and compatible-element contents, and are enriched in K, Rb, Cs and Ba and markedly depleted in Nb and Ta; the mineralogical assemblage is characterized by the occurrence of Mg-rich olivine, poikilitic sanidine, flogopite, clinopyroxene and high-Cr# spinel.

The volcanic activity ceased after the Middle Miocene. The area was subsequently affected, during the Late Miocene, by an alkali basaltic activity that produced scattered and volumetrically limited eruptions (e.g. Ezine, Kabaklar, Foca, Urla, and some central

Aegean islands such as Patmos and Kaloyeri (Francalanci et al., 1991); these lavas generally exhibit an OIB geochemical imprinting, although a slight subduction-related geochemical signature can in some cases be identified. The volcanic activity ended in WA with the formation of the widespread Quaternary volcanic field of Kula, made up of relatively primitive (Mg# up to 75) Na-alkaline lavas (from basanites and hawaiites to phonotephrites; (Richardson-Bunbury, 1996), displaying a clear OIB geochemical character.

Overall, the different geochemical signatures documented in the Tertiary-Quaternary WA volcanics and their isotopic features have been interpreted as the result of the activation of three main mantle sources (Innocenti et al., 2004). The high-K calc-alkaline and shoshonite associations seem to have derived from a wedge mantle that was enriched to varying degrees by a subduction-related component; these orogenic magmas were also moderately modified by their interaction with crustal material. The primitive lamproitic rocks, characterized by a high Sr and low Nd isotope ratio, are regarded as lithospheric magmas, whereas the OIB-type magma show geochemical and isotopic characteristics indicative of an asthenospheric (sub-slab) mantle source.

This picture has been complicated by the discovery of high-Mg (HM) intermediate lavas with peculiar geochemical and isotopic features that were erupted after the end of the main calc-alkaline and shoshonitic associations. Their stratigraphic position is thus interposed between the main orogenic cycle and the first appearance of OIB-type products. These HM rocks will be described in the following sections.

3. HIGH-MGO VOLCANIC ROCKS FROM WESTERN ANATOLIA

3.1 Analytical methods

Major elements and trace elements were performed at the Dipartimento di Scienze della Terra, Università di Pisa. Major elements were determined by XRF on fused disks and FeO by titration. Loss-on-ignition was measured by gravimetry at 1000°C after pre-heating at 110°C. Trace elements were determined by ICP-MS (VG PQII Plus STE). Analytical precisions is generally between 2 and 5 % RSD, except for Gd (6%), Tm (7%), Pb and Sc (8%). For further analytical details see D'Orazio (1995).

Mineral analyses were performed by WDS-electron mocroprobe (JEOL JXA-8600) at Istituto Geoscienze e Georisorse - CNR (Firenze). Calibrations were carried out with natural silicate standards; matrix effects were corrected according to Bence and Albee (1968).

Sr and Nd isotope analyses were carried out at Istituto Geoscienze e Georisorse - CNR (Pisa). The powders were leached with warm 6.2 N HCl for 45 minutes then rinsed throughly in pure sub-boiled water. Measurements were made using a Finnigan MAT 262V multi-collector thermal ionization mass-spectrometer. Measured $^{87}Sr/^{86}Sr$ ratios have been normalized to $^{86}Sr/^{88}Sr = 0.1194$; $^{143}Nd/^{144}Nd$ ratios to $^{146}Nd/^{144}Nd = 0.7219$. The quoted error is the standard deviation of the mean (2sm) and refers to the last digit. Replicate analyses, during the period of the measurement of the samples NIST

987 and La Jolla international reference standards gave average values of $^{87}Sr/^{86}Sr$ = 0,71024x±0,000004 (standard deviation of the mean, n= 10) and $^{143}Nd/^{144}Nd$ = 0,511848±0,000002 (standard deviation of the mean, n=10), respectively. No age corrections were made on $^{87}Sr/^{86}Sr$ and $^{143}Nd/^{144}Nd$.

3.2 Classification

An unusual magmatism characterized by relatively evolved MgO-rich products has been identified in several convergent zones where the oceanic lithosphere is subducted under oceanic or continental plates (Arndt, 2003). These rocks, typified by their andesitic compositions, have been given different names in different places; for example, they are called bononites (Bonin islands), bajaites (Baja California), and sanukitoids (SSW Japan), but are generally grouped together as high-Mg andesites (HMA) (Tatsumi and Hanyu, 2003), a name that we have adopted in this paper. According to Crawford et al. (1989), the HMA group from WA includes the rocks with SiO2 of >53%, MgO of >8% and Mg-value (Mg#) of >60. These criteria match the parameters accepted by the IUGS Subcommission on the Systematics of Igneous Rocks for high-Mg volcanic rocks (Le Maitre, 2002), except for TiO2, which ranges between 0.5 and 1%. The WA-HMA also include some relatively evolved lavas (SiO2 of 60-65%) with MgO of <8%, but with Mg# of >60 and a distinctly higher MgO content than the other orogenic rocks with comparable silica content (Table 1).

3.3 Distribution

The location of the HM rocks from Western Anatolia is shown in Fig. 1. These products are scattered over a relatively wide area in central WA, between Bergama, Kutahiya and the Karaburun Peninsula. They are made up of low-volume lava flows with minor breccias, generally resting on Middle Miocene continental deposits (e.g. Ilica, south of Kutahiya and Kücükbahçe in the Karaburun Peninsula). The area with the most extensive exposure of HMA is located in the NE-SW trending Soma basin (Fig. 1), where feeding dykes and relatively thin lava flows occur. The lavas are interbedded in a lignite-bearing continental sequence of middle-late Miocene age, as suggested by paleobotanical and palynological evidence (Inci, 1998). On the basis of the stratigraphic relationships and the occurrence of local contact metamorphism, the Soma volcanics have been ascribed to the Late Miocene. HM-evolved lavas (dacites) have been found in Mersinli, about 20 km west of Kula, where they rest on a Miocene continental sequence. Other dacitic products akin to Kula magnesian dacites have been reported in the Gördes area by Ercan et al. (1983).

In the Aegean area, Miocene magnesian andesites and dacites have been described in Evia (Pe-Piper and Piper, 1994) and Skyros (Pe-Piper, 1991). On the basis of age and geochemical features, these rocks can be considered equivalent to the western Anatolia HM rocks and are therefore regarded as part of the same HM association in the following sections.

Table 1. Chemical analyses of HMA from Western Anatolia.

Place	Soma			Ilica (Kütahia)		Kkaraburun	Kula	
Label	IZ 193	IZ 194	IZ 195	IZ 183	IZ 184	K83	K 99/16	K 99/20
SiO_2	50.40	50.98	50.66	53.92	54.16	53.53	65.55	63.46
TiO_2	0.63	0.63	0.65	0.78	0.80	0.92	0.59	0.69
Al_2O_3	12.07	12.18	12.14	13.53	13.45	14.06	15.44	15.64
Fe_2O_3	2.18	4.27	3.80	2.66	2.54	3.74	3.45	3.75
FeO	5.55	3.81	4.27	4.30	4.52	3.41	0.94	1.23
MnO	0.14	0.13	0.14	0.13	0.11	0.13	0.07	0.10
MgO	9.07	11.87	12.10	9.03	10.33	8.96	3.40	3.34
CaO	8.76	8.98	9.08	8.73	7.40	8.12	4.18	4.74
Na_2O	1.73	1.65	1.75	2.28	2.47	2.60	2.75	2.74
K_2O	1.55	1.38	1.42	2.05	2.01	2.52	2.26	2.58
P_2O_5	0.14	0.14	0.14	0.19	0.20	0.39	0.15	0.19
LOI	8.27	3.58	4.19	2.94	1.49	1.17	1.22	1.24
Tot	100.49	99.60	100.34	100.53	99.48	99.55	100.00	99.70
Mg#*	70.95	75.84	76.10	73.20	75.43	72.79	62.97	59.47
Trace elements (ppm)								
Li		39.66						
Be		1.54	1.55	2.28	2.37	2.57		3.42
Sc		31	30	26	25	26		15
V		184	177	160	165	170		76
Cr		778	562	527	619	552		172
Co		39	39	37	35	33		13
Ni		239	229	185	158	188		9
Cu			47.0	27.7	24.2	52.5		
Ga		13.2	11.2	12.6	14.8	13.9		
Rb		38.6	38.3	74.9	65.7	90.6		87.0
Sr		894	999	904	848	765		511
Y		16.3	16.2	21.1	21.7	20.1		20.7
Zr		67	67	115	115	165		95
Nb		4.05	3.91	8.33	8.47	10.91		12.19
Cs		5.72	3.43	28.72	53.70	4.75		3.76
Ba		839	901	1088	1092	1659		614
La		19.2	19.0	36.4	37.6	37.0		39.3
Ce		39	36	68	70	72		74
Pr		4.3	4.3	7.8	7.9	8.4		8.6
Nd		16.7	16.5	29.0	29.9	32.1		31.9
Sm		3.1	3.5	5.8	5.7	6.0		6.0
Eu		0.58	0.29	1.08	1.13	0.97		1.22
Gd		2.8	3.1	4.6	4.8	4.9		4.5
Tb		0.46	0.46	0.71	0.68	0.68		0.68
Dy		2.72	2.53	3.66	3.89	3.59		3.61
Ho		0.58	0.59	0.73	0.75	0.71		0.71
Er		1.55	1.50	1.95	2.02	1.87		1.84
Tm		0.22	0.23	0.29	0.30	0.29		0.28

Table 1. Chemical analyses of HMA from Western Anatolia (continued).

Place Label	Soma IZ 193	IZ 194	IZ 195	Ilica (Kütahia) IZ 183	IZ 184	Kkaraburun K83	Kula K 99/16	K 99/20
Yb		1.53	1.56	1.81	1.82	1.79		1.62
Lu		0.22	0.20	0.24	0.27	0.25		0.24
Hf		1.83	1.98	3.01	3.00	4.53		2.78
Ta		0.28	0.27	0.71	0.72	0.72		0.96
Tl		0.37	0.23	2.72	0.52	0.47		0.48
Pb		21	22	52	45	25		53
Th		7.6	7.4	14.5	14.9	14.3		14.6
U		3.71	3.64	5.80	5.96	3.10		3.86
Sr/Sr		0.707244	0.707208	0.708555	0.708488	0.708438		0.710349
±2s		0.000013	0.000010	0.000011	0.000010	0.000010		0.000017
Nd/Nd		0.512460	0.512462	0.512392	0.512387	0.512297		0.512274
±2s		0.000008	0.000009	0.000009	0.000008	0.000008		0.000009

Iz 193, IZ 194, IZ195: Soma, lava flows cutting lower coal unit (Inci, 1998)
IZ 183, IZ184: Ilica, about 20 km N of Kutahia: lava dome intruding lacustrine
K 83: Küçük Bahçe, 10 km SW of Karaburun; lava flow resting on Carbonuferous carbonate rocks
K99/16, K99/20: Mersinli: 2o km W of Kula; la flows in Upper Miocene continental clastic sediments

3.4 Petrography and chemistry

The WA HMAs display porphyritic textures with a phenocryst assemblage made up of olivine, clinopyroxene and plagioclase. In addition to the same phases found as phenocryst, the matrix includes Mg-orthopyroxene, opaques and scattered interstitial alkali feldspar. The glass is very scarce or absent; when present, it is generally transformed into smectites, which may also occur as alteration products of femic minerals.

Olivine is Mg-rich and generally slightly zoned; the highest magnesian compositions are found in Soma rocks (Fo 87-90) where inclusions of Cr-rich spinel (Cr# ~73) are observed; the clinopyroxene has an endiopside-augite composition (Wo42-45 En42-51 Fs6-13) and the plagioclase has a bytownite-labradorite composition (Appendix 1).

In the Ilica and Karaburun lavas the olivine is more Fe-rich than in Soma (Fo 75-83) and includes Cr-spinel with variable composition (Cr# 66-76); the clinopyroxene is augitic (Wo36-46 En43-50 Fs9-14) and plagioclase exhibits labradoritic composition (An54-65). Ti-magnetite and orthopyroxene are ubiquitous in the groundmasses (En ~77 in Soma; En ~70 in Ilica and Karaburun). Ilmenite also occurs in the Ilica and Karaburun rocks (Appendix 1).

The HMA of the study areas are characterized by SiO_2 contents ranging between 53 and 56% (water-free), except for the Kula lavas, which have SiO_2 of ~ 66% (Table 1); these rocks also exhibit an elevated MgO content, far higher than the values observed in the rocks of the calc-alkaline and shoshonite associations (Fig. 2), whereas Al_2O_3, TiO_2 and K_2O show relatively low abundances. CaO contents are restricted to the range 7.5 - 9% and the CaO/Al_2O_3 ratio remains relatively high (0.55-0.74).

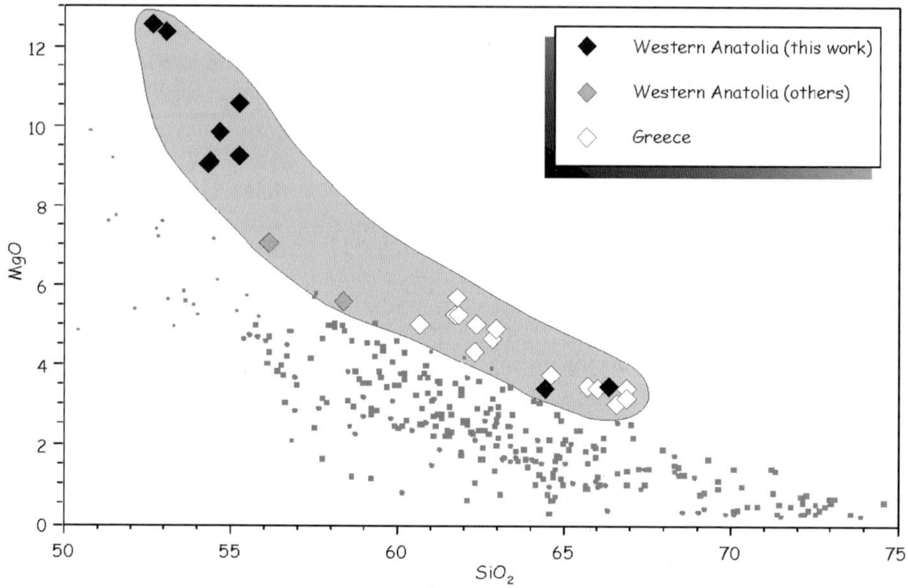

Fig. 2. MgO vs SiO2 diagram for HM rocks from WA and Greece. The calc-alkaline and shoshonitic rocks from WA are reported for comparison purposes (data base of published and unpublished analyses, available upon request). Gray squares and circles are WA calc-alkaline and shoshonitic rocks, respectively.

The K_2O versus SiO_2 diagram of Fig. 3 highlights the low K_2O contents of the HM rocks with respect to other orogenic rocks. These samples, in fact, fall astride the boundary line dividing calc-alkaline and high-K calc-alkaline fields.

The HMA exhibit high Ni (~240-100 ppm) and Cr (780-250 ppm) contents. The compatible trace elements (Ni, Cr, Co, V and Sc) correlate positively with MgO or Mg# and negatively with SiO_2 (Fig. 4).

The incompatible trace elements show a roughly positively correlation with silica, except Sr, which decreases significantly with increasing SiO_2. Taking into account the HM dacites from Kula and the central Aegean area, we can observe a general inversion of the correlation from positive to negative, as illustrated in the Zr vs. SiO_2 diagram of Fig. 4.

The REE distribution (Fig. 5) is characterized by prominent LREE ($La_n/Sm_n \approx 3.4$-4.1) and moderate HREE fractionations ($Tb_n/Yb_n \approx 1.3$-1.8). The patterns are similar in shape to those observed in the WA calc-alkaline rocks (basaltic andesites), with a slightly higher HREE fractionation. It is worth noting the occurrence of a significant Eu negative anomaly in most primitive rocks, which decrease with increasing degree of evolution (Eu/Eu* from 0.27 to 0.64 for a silica range of 52.7 - 55.3 wt. %).

Multi-element patterns are displayed in Fig. 5. The HMA lavas show a typical orogenic signature with a high LILE/HFSE ratio (e.g. Th/Ta≈27-20) and negative spikes of Ta, Nb, P and Ti. Compared with calc-alkaline basaltic andesites, they exhibit lower

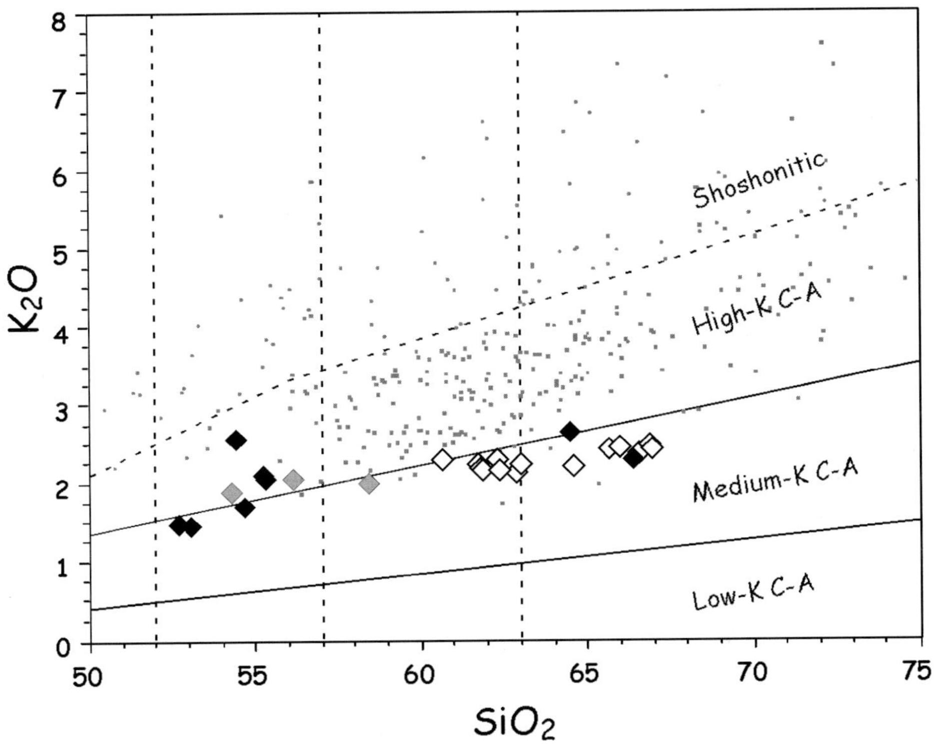

Fig. 3. K_2O vs SiO_2 classification diagram for HM rocks from WA and Greece. Symbols as in Fig. 2.

HFSE and HREE abundances; as a consequence the ratios between mobile and immobile elements (e.g., Ba/Nb and Rb/Zr) are higher than observed in calc-alkaline and shoshonitic rocks.

The Sr and Nd isotope ratios are reported in Table 1; they vary between 0.70721 and 0.71035 and 0.51246 and 0.51227, respectively. In the classic isotope diagram Sr-Nd (Fig. 6) they fall in the enriched quadrant and overlap the field defined by the WA Tertiary rocks. In Fig. 7 the isotope compositions are plotted against silica contents and compared with data for calc-alkaline rocks. As a whole, the HM rocks display higher $^{87}Sr/^{86}Sr$ and lower $^{143}Nd/^{144}Nd$ values with respect to the calc-alkaline samples, along with good correlations with silica content. Furthermore, the most primitive lavas from Soma exhibit isotope compositions that overlap the less evolved rocks of the entire orogenic association (Innocenti et al., 2004).

4. DISCUSSION

In WA the HM volcanism developed during the Middle-Late Miocene, with a time gap between HM activity and the final products of the main orogenic cycle (mainly

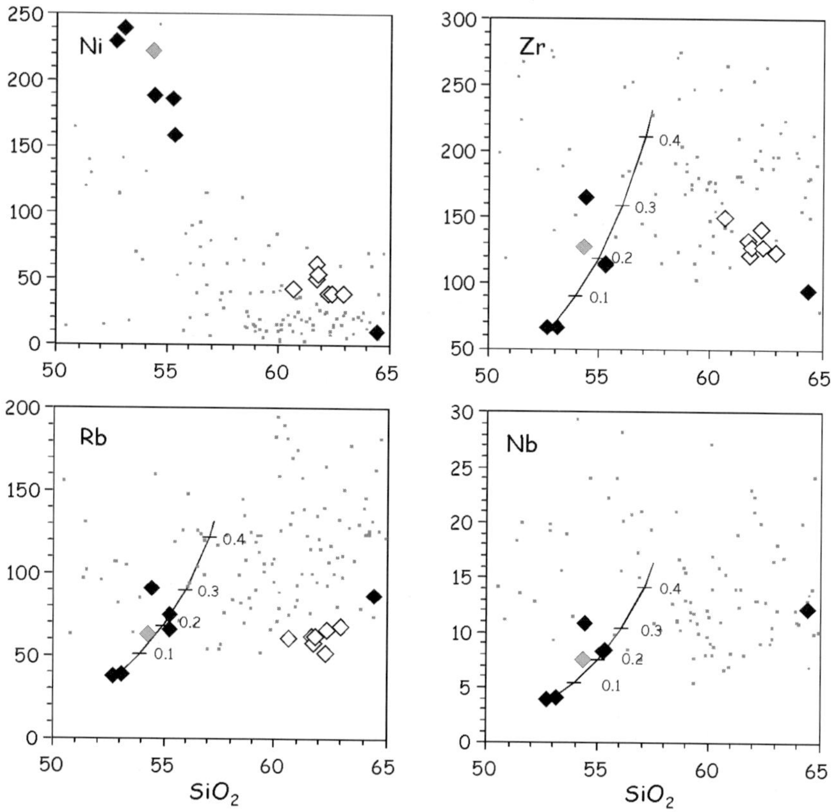

Fig. 4. Ni, Zr, Rb and Nb vs SiO2 diagrams for HM rocks from WA and Greece. Small dots are calk-alkaline and shoshonitic samples (Symbols as in Fig. 2). AFC curves are drawn for Zr, Rb and Nb, assuming a fractionaction of melagabbroic assemblage (olivine 34%, clinopyroxene 34%, plagioclase 25% and Ti-magnetite 7%), assimilation of Bulk Continental Crust (Plank and Langmuir, 1998), and r= 0.5; numbers on the curves are the amount of fractionated material. Bulk distribution coefficients for trace elements were calculated according to distribution coefficients taken from http://earthref.org.

shoshonites), which affected the region as a consequence of the subduction of the African plate under the Aegean-Anatolian blocks. The HM rocks show distinctive geochemical features with respect to calk-alkaline and shoshonitic associations. These products have, in fact, higher contents of MgO and compatible elements such as Ni, Cr and Sc, with silica mainly spanning the compositional range of basaltic andesites to andesites. Even though the mantle-normalized profiles of the incompatible trace elements show a strong similarity with calc-alkaline rocks, the HM volcanics exhibit distinctly lower abundances of REE and HFSE, whereas some LILE, mainly fluid-mobile elements (Rb, Ba, Th, U) virtually overlap the calc-alkaline values (Fig. 5).

In the following sections we discuss the processes responsible for the compositional variability of the HM rocks, and the significance of the geochemical features of the

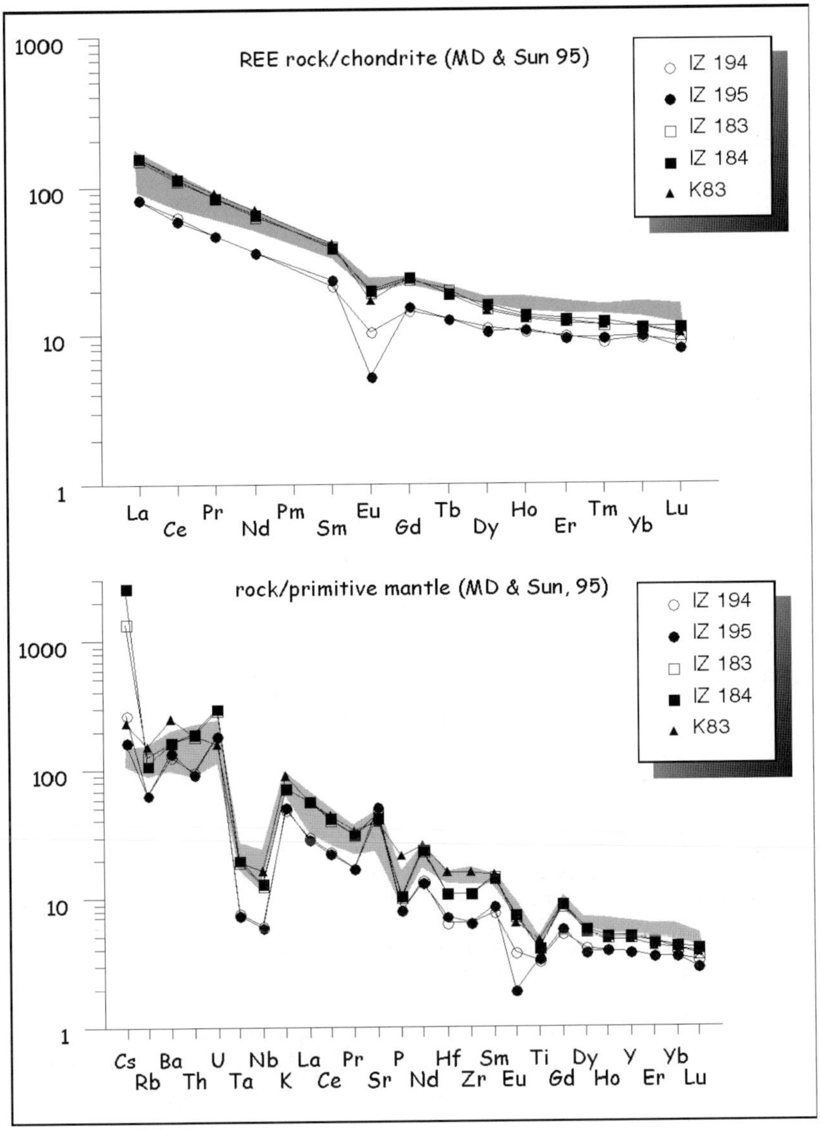

Fig. 5. Multielement patterns for REE and incompatible elements normalized to CI chondrites and Primordial Mantle, respectively (McDonough and Sun, 1995).

most primitive rocks for the characterization of the geochemical nature of the mantle source. Finally, a geodynamic model will be proposed to explain their occurrence within the framework of the evolution of Tertiary and Quaternary volcanism in the Aegean-Western Anatolia area.

4.1 Geochemical variability of the HMA

Taking the entire Aegean-Anatolian HM association into consideration, we noted that the variation diagrams of the major elements reveal a good overall correlation between degree of evolution (i.e. SiO_2) and element abundances. MgO, FeO*, and CaO in particular, and, to a lesser extent TiO_2, exhibit negative correlations with silica whereas K_2O is positively correlated. Al_2O_3 and Na_2O show a more complex trend; they are positively correlated up to about 60% SiO2 and then show a decrease in the more evolved rocks. On the whole, these variations are compatible with a fractionation process involving a solid assemblage mainly made up of mafic phases, olivine and clinopyroxene, plus plagioclase, which dominates in the last segment of the trend. Mass balance calculations suggest that the transition from the most primitive (IZ 195) to more evolved (IZ 183) HM rock requires the fractionation of about 24% of a solid of melagabbroic composition, made up of Mg-rich olivine (~34%), clinopyroxene (~34%), bytownite (~25%) and Ti-magnetite (~7%) (SSR, sum of the squared residuals between calculated and observed values, equals 0.63). The transition from HMA to dacites (K99/20) is more difficult to explain in terms of fractionation process in that it would require an unreasonable amount of removed solid (~ 92% with SSR ~3).

Trace element distribution in the HMA, although qualitatively coherent with the model involving melagabbro fractionation, shows an enrichment factor of incompatible elements (EF= element abundance in a sample relative to the abundance of the most primitive sample IZ 194) that does not match the amount of removed solid indicated by

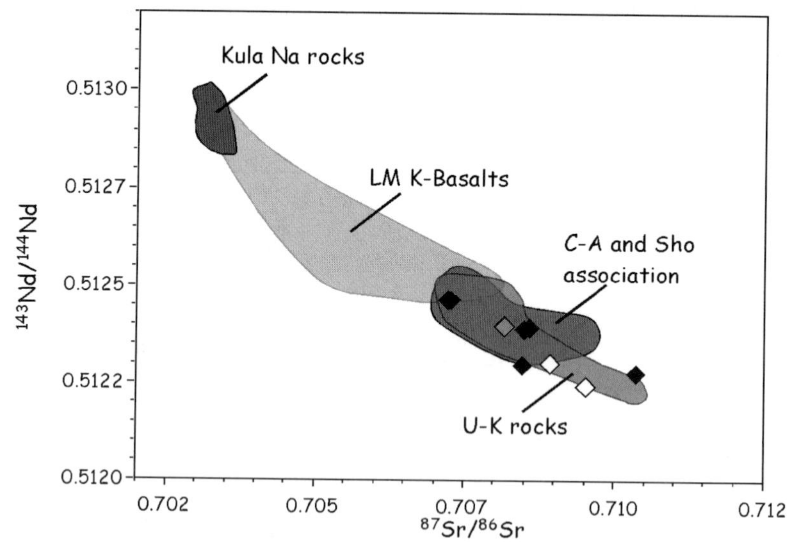

Fig. 6. 143Nd/144Nd versus 87Sr/86Sr diagram for HM rocks from WA and Greece (data from Pe-Piper and Piper, 2001). The variation field of the Kula Na-rocks, Late Miocene potassic basalts (LM K-basalts), calc-alkaline (CA), shoshonitic (Sho) and ultra-potassic (U-K) associations: after (Innocenti et al., 2004). Symbols as in Fig. 2.

major element modeling. The most incompatible elements, such as Rb, La, Th, Pb, Nb and Ta, display EF ranging from 1.9 to 2.5, which is far higher than the values suggested by the fractionation model (~1.3). Furthermore, the relatively wide variations of isotope composition argue against a crystal fractionation process in a closed system. The Sr and Nd isotopes are well correlated with silica (correlation coefficient r^2 0.75 and 0.64, respectively), suggesting that assimilation of crustal material, together with fractional crystallization (AFC), has been a dominant process in the evolution of HM rocks (Fig. 7).

In order to better constrain the AFC process, model curves (DePaolo, 1981) were calculated (Fig. 4) using the most primitive lava of our data set (IZ 195) and the composition of the bulk continental crust (Plank and Langmuir, 1998) as end members. Fig. 4 reveals that the geochemical and isotopic variations inside the HMA can be satisfactorily reproduced by a 26% melagabbroic fractionation combined with assimilation of a crustal component, assuming a ratio between assimilation rate and fractional crystallization rate of 0.5. However, the subsequent evolution from andesites to dacites is not so well constrained quantitatively by this model probably as a consequence of the lack of precise compositional information on the assimilated material involved in the process. A mixing process between Mg-rich andesitic magma and felsic melts has, moreover, been documented in the Aegean HM rocks (Pi-Piper and Piper, 1994).

On the whole, the HM rocks appear to have experienced a complex petrogenetic evolution in which assimilation of crustal material and fractional crystallization of gabbroic assemblage played a major role. In the more evolved dacitic rocks this process was further complicated by the concurrent occurrence of interaction (mixing) with felsic melts.

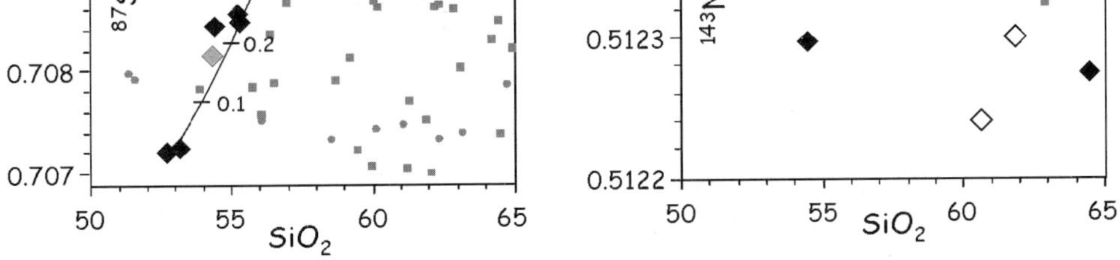

Fig. 7. Sr and Nd isotope composition vs silica diagram for HM rocks. Symbols as in Fig. 2. The curves represent the AFC model calculated by means of the parameters reported in the caption to Fig. 4; numbers on the curves are the amount of fractionated material. Isotope composition of bulk crust: 87Sr/86Sr=0.7210; 143Nd/144Nd=0.51215.

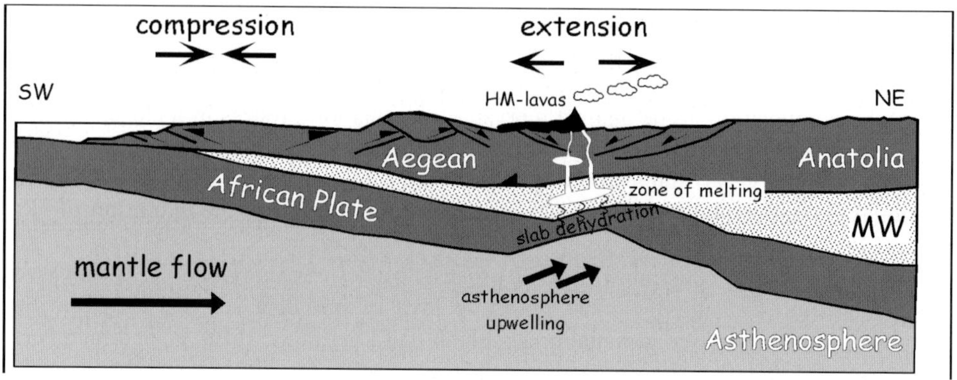

Fig. 8. Schematic cross-section of the Western Anatolian region during the Late Miocene: extensional dynamics in the back-arc region led to the deformation and stretching of the Lithosphere, Mantle Wedge (MW) and subducting African slab, triggering an upwelling of the asthenosphere. HM magmas (white area) are generated in the MW and reach the surface directly or after interacting with crustal material.

4.2 Constraints on source region composition of HM magmas

Information on the geochemical features of the mantle source can be inferred from the HM basaltic andesites of Soma, whose characteristics (olivine F90, Cr-rich spinel, Mg# 76, high Ni and Cr abundances) indicate that they were in equilibrium with mantle parageneses. The low alumina abundance, high CaO/Al_2O_3 ratio, and low content of HFS elements strongly suggests a depletion of the mantle source. The relatively high LILE content does, however, contrast with the refractory nature of the peridotite mantle. The Sr and Nd isotope data ($^{87}Sr/^{86}Sr$= 0.70721 and $^{143}Nd/^{144}Nd$ 0.51246) are very close to the lowest Sr and highest Nd isotope values of WA calc-alkaline and shoshonitic associations (Fig. 6), which suggests that the mantle source of the HM magmas was modified by subduction-related components as in the case of the source of the calc-alkaline magmas. The Eu-negative anomaly that also occurs in the most primitive HM rocks lends support to the hypothesis of a dominant sedimentary origin for the slab-derived component, as proposed for the preceding WA orogenic magmatism (Innocenti et al., 2004). The strong enrichment of fluid mobile elements with respect to incompatible but fluid-immobile elements, as typified by the Ba/Nb (>200), Th/Ta (>25) and Rb/Zr (>0.55) ratios, suggests that a hydrous fluid phase was involved, rather than a slab melt, as stressed also by some experimental data (see Arndt, 2003 and references therein).

Overall, these geochemical and isotopic indicators depict a scenario in which the mantle source of the HM magmas appears to be a residual mantle wedge that underwent earlier melt extraction to generate Lower-Middle Miocene calc-alkaline and shoshonite volcanism. The relatively high LILE enrichment of these magmas with respect to the calc-alkaline association indicates a late contribution of hydrous fluids, which favored the partial melting of a refractory source.

4.3 Implications for geodynamic evolution of Anatolia-Aegean region

The convergence between Africa and Anatolia and Greece is a phenomenon that is considered to be under way since at least Upper Cretaceous-Eocene; assuming a conservative convergence rate of around 3 cm/y, this process has therefore produced a subducted slab of at least 1000 km in length. The seismological data indicate that the initial dip of the slab, up to about 200 km, is relatively low (≈30°, Papazachos et al., 2000) becoming more horizontal later on (Christova and Nikolova, 1993). Doglioni et al. (2002) reviewed the geodetic data in order to define the relative motion between Europe, Africa, Greece and Anatolia with greater precision. Considering Africa to be fixed, they came to the conclusion as the Aegean microplate is moving faster than Anatolia over the African plate. The differential overriding velocity is considered to be responsible for the Aegean extension and for the formation of a wide area of internal deformation between the two microplates. Thus, the hanging lithosphere was stretched. This process was matched at depth by the stretching and folding of the underlying slab which facilitated the rise of sub-slab mantle and compensated for the lithosphere thinning. The extensional process at depth eventually led to a break in the subducted slab, with a consequent formation of an horizontal slab window; the relatively deep mantle found a passage through the latter, permitting the generation of Plio-Quaternary OIB-type magma of Kula region and neighboring areas.

Within the framework of this model, the generation of HM magmas can be considered the effect of a local thermal anomaly located where the extensional process produced the greatest thinning of the underthrusting lithosphere, with a consequent ascent of the deep mantle. The data collected indicate that the source of mantle-derived HM rocks showed isotopic and trace element patterns in multielement diagrams that are very close to that of the mantle that produced the older calc-alkaline magmas. Furthermore, this mantle source region was probably depleted by earlier melt extraction as suggested by the low HREE and HFSE abundances. We can thus conclude that mantle source of the HM magmas is represented by the mantle wedge sited above the leading edge of subducting deformed slab (Fig. 8).

The peculiar geochemical character of HM rocks was produced by a process of partial melting in a mantle region chemically modified by a fluid component transferred from subducted slab. On the base of the decoupling of the geochemical behavior of incompatible fluid-mobile (enriched) elements with respect to fluid-immobile (depleted) elements, we can argue that this component was essentially represented by a hydrous fluid. The enrichment event affecting wedge mantle could be linked to the thermal anomaly produced by the rise of the asthenosphere, which facilitated the release of residual volatiles stored in the subducted slab; alternatively, it could be related to the preceding metasomatic episode that favored the production of calc-alkaline magmas. This hypothesis, illustrated in Fig. 8, is supported by two further considerations:

(i) The age of HM rocks: although age constraints are relatively poor, they were erupted between the time of calc-alkaline-shoshonitic volcanism and the OIB-type magmatism (Fytikas et al., 1984); this period is related to the activation of an asthenospheric source whose partial melting was facilitated by the opening of an horizontal slab-window (Innocenti et al., 2004);

(ii) The spatial distribution of HM rocks (Fig. 1) which are located along a narrow belt, close to or overlapping the region where OIB-type rocks were erupted.

5. CONCLUSIONS

The study of the WA Tertiary and Quaternary volcanism and its relation to the subduction of Africa under Aegean and Anatolian microplates has revealed the occurrence of unusual high-magnesium magmatism. The limited volume of erupted products, as well as their space-time distribution shows that the conditions that facilitated their formation took place within a short time span and within a specific area. We propose a scenario in which the source of HM magmas can be attributed to a subduction-modified mantle wedge that was previously depleted by calc-alkaline melt extraction. The partial melting process was caused by a local thermal anomaly originated by upwelling of asthenosphere, a process that was, in its turn, triggered by extensional dynamics that affected Aegean-Anatolia area since at least the Lower Miocene.

Acknowledgements

The authors are grateful to M. Helen Dickson for checking the English; M. Bertoli and M. Tamponi (Dip. Di Scienze della Terra, University of Pisa) are thanked for major element analyses. We are indebted also to M. D'orazio for his assistance during ICP-MS analyses. An anonymous referee is acknowledged for the constructive review of the manuscript. The research has been financially supported by MIUR (Cofin 40%) and CNR.

REFERENCES

Altunkaynak, S. and Yilmaz, Y., 1999. The Kozak Pluton and its emplacement. Geol. J., 34: 257-274.

Arndt, N.T., 2003. Komatiites, kimberlites and boninites. J. Geophys. Res., 108(B6): 2293, doi:10.1029/2002JB002157.

Bence, A.E. and Albee, A.L., 1968. Empirical correction factors for the electron micro-analyses of silicates and oxides. J. Geol., 76: 382-483.

Christova, C. and Nikolova, S.B., 1993. The Aegean region: deep structures and seismological properties. Geophys. J. Int., 115: 635-653.

Crawford, A.J., Falloon, T.J. and Green, D.H., 1989. Classification, petrogenesis and tectonic setting of boninites. In: A.J. Crawford (Editor), Boninites. Unwin Hyman, London, pp. 1-49.

D'Orazio, M., 1995. Trace element determination in igneous rocks by ICP-MS: results on ten international reference samples. Per. Mineral., 64: 315-328.

DePaolo, D.J., 1981. Trace element and isotopic effects of combined wallrock assimilation and fractional crystallisation. Earth Planet. Sci. Lett., 53: 189-202.

Doglioni, C., Agostini, S., Crespi, M., Innocenti, F., Manetti, P., Riguzzi, F., and Savascin, Y., 2002. On the extension in western Anatolia and the Aegean sea. J. Virtual Expl., 8: 169-184.

Ercan, T., 1983. Gördes volkanitlerinin (Manisa) petrolojisi ve koekensel yorumu. (Petrology of volcanic rocks of Gordes and their genetic implications). Türkiye Jeol. Kurumu Bülteni (Bull. Geol. Soc. Turkey), 26: 41-48.

Francalanci, L., Civetta, L., Innocenti, F. and Manetti, P., 1991. Tertiary-Quaternary alkaline magmatism of the Aegean-Western Anatolian area: a poetrological study in the light of new geochemical and isotopic data. In: M.Y. Savasçin, &, Eronat, A.H. (Editor), IESCA 1990, pp. 385-396.

Francalanci, L., Innocenti, F., Manetti, P. and Savasçin, M.Y., 2000. Neogene alkaline volcanism of the Afyon-Isparta area, Turkey: petrogenesis and geodynamic implications. Mineral. Petrol., 70: 285-312.

Fytikas, M., 1984. Tertiary to Quaternary evolution of volcanism in the Aegean region. In: J.E. Dixon and A.H.F. Robertson (Editors), The Geological evolution of the Eastern Mediterranean. Geol. Soc. London, Spec. Publ., London, pp. 687-699.

Genç, S.C., 1998. Evolution of Bayramiç magmatic complex, north western Anatolia. J. Volcanol. Geotherm. Res., 85: 233-249.

Inci, U., 1998. Lignite and carbonate deposition in Middle Lignite succession of the Soma Formation, Soma coalfield, western Turkey. Int. J. Coal Geol., 37: 287-313.

Innocenti, F., Agostini, S., Di Vincenzo, G., Doglioni, C., Manetti, P., Savascin, M.Y., and Tonarini, S., 2004. Neogene and Quaternary volcanism in Western Anatolia: magma sources and geodynamic evolution. Mar. Geol., in press.

Innocenti, F., Manetti, P., Mazzuoli, R., Pertusati, P., Fytikas, M. and Kolios, N., 1994. The Geology and geodynamic Significance of the Island of Limnos, north Aegean Sea, Greece. N. Jb. Palaeont. Geol.Mh., H.11: 661-691.

Le Maitre, R.W., 2002. Igneous rocks - A classification and glossary of terms. Cambridge Univ. Press, cambridge, 236 pp. pp.

McDonough, W.F. and Sun, S.-S., 1995. The composition of the Earth. Chem. Geol., 120: 223-253.

Papazachos, B.C., Karakostas, V.G., Papazachos, C.B. and Scordilis, E.M., 2000. The geometry of the Wadati-Benioff zone and lithospheric kinematics in the Hellenic arc. Tectonophysics, 319: 275– 300.

Pe-Piper, G., 1991. Magnesian andesites from the island of Skyros, Greece: geochemistry and regional significance. Geol. Mag., 128: 585-593.

Pe-Piper, G. and Piper, D.J.W., 1994. Miocene magnesian andesites and dacites, Evia, Greece: adakites associated with subducting slab detachment and extension. Lithos, 31: 125-140.

Pe-Piper, G. and Piper, D.J.W., 2001. Late Cenozoic, post collisional Aegean igneous rocks: Nd, Pb and Sr isotopic constraints on petrogenetic and tectonic models. Geol. Mag., 138: 653-668.

Plank, T. and Langmuir, C.H., 1998. The geochemical composition of subducting sediment and its consequences for the crust and mantle. Chem. Geol., 145: 325-394.

Richardson-Bunbury, J.M., 1996. The Kula volcanic field, western Turkey: the development of a Holocene alkali basalt province and the adjacent normal-faulting graben. Geol. Mag., 133: 275-283.

Sengör, A.M.C. and Ylmaz, Y., 1981. Tethyan evolution of Turkey: A plate tectonic approach. Tectonophysics, 75: 181-214.

Seyitoglu, G. and Scott, B.C., 1991. Late Cenozoic crustal extension and basin formation in west Turkey. Geol. Mag., 128: 155-166.

Tatsumi, Y. and Hanyu, T., 2003. Geochemical modeling of dehydration and partial melting of subducting lithosphere: Toward a comprehensive understanding of high-Mg andesite formation in the Setouchi volcanic belt, SW Japan. Geochem. Geophys. Geosyst., 4(9): 1081, doi:10.1029/2003GC000530.

Yilmaz, Y., Genç, S.C., Karacik, Z. and Altunkaynak, S., 2001. Two contrasting magmatic association of NW Anatolia and their tectonic significance. J. Geodynamics, 31: 243-271.

A mathematical model for the morphological evolution of a volcano on an island

G.Aim. Skianis*, D. Vaiopoulos and V. Tsarbos

Remote Sensing Laboratory, Department of Geology and Geoenvironment, University of Athens

ABSTRACT

The present paper attempts a mathematical description, in two dimensions, of the morphological evolution of a volcano, as a result of erosional processes and volcanic activity. The whole morphological evolution is represented by a partial differential equation, in which erosion and volcanic activity are represented by the erosion coefficient K and a (mass) transfer function *Trsf*, respectively. The transfer function expresses the deposition rate of lava and pyroclastic material at the slopes of the volcano. The boundary conditions of the differential equation express a volcano which crowns an island area with length L and a time constant sea level. The solutions of the differential equation represent the morphological evolution of the volcano through time, under different initial states and geological conditions. It is concluded that the altitude of a dead volcano tends to zero with time. On the other hand, an active volcano with a time constant transfer function tends to a steady state of dynamic equilibrium. The dimensions of the profile at the steady state depend on the mass transfer rate and the erosion coefficient. The time at which the volcano comes to the steady state is proportional to the square length L^2 and inversely proportional to the erosion coefficient. The results and conclusions of this paper may be useful in understanding, in quantitative terms, how the relief of a volcano may evolve in time and which factors control the whole process.

Keywords: erosion coefficient, transfer function, steady state, volcanic island, differential equation, boundary condition.

1. INTRODUCTION

As early as the beginning of the sixties, it was suggested that landform evolution could be modelled according to the principle of the conservation of mass, which may be

* Corresponding author: e-mail: skianis@geol.uoa.gr

expressed as the continuity equation, combined with a slope-dependent transport law (Culling, 1960; Scheidegger, 1961). Since then, this idea has been mainly applied in modelling the morphological evolution of hillslopes in two dimensions (Culling, 1963, 1965; Kirkby, 1971; Hirano, 1975; Trofimov and Moskovkin, 1984; Ahnert, 1976, 1987; Fernandes and Dietrich, 1997), three dimensional hills or mountains (Culling, 1963; Hirano, 1976; Vaiopoulos et al., 2001), drainage basins (Armstrong, 1976; Willgoose et al., 1990; Howard, 1994), as well as larger landscapes at regional and continental scales (Howard et al., 1994; Kooi and Beaumont, 1996; Kirkby, 1999).

We have not found references about the mathematical modelling of the geomorphological development of volcanoes. Recently, Vaiopoulos et al. (2002), proposed a mathematical model of the evolution of a dead volcanic cone, based on the diffusion equation. In that model, erosional processes were mainly taken into account. In this paper is presented a two-dimensional model of a volcanic island. Two factors are taken into account in order to model the evolution of the volcano: erosion and volcanic activity. First, the differential equation of the whole process is formulated, with the proper boundary conditions. Then, the differential equation is solved and the profile of the volcano is found, for various time instants and geological conditions. Characteristic curves are calculated to demonstrate graphically how erosion and volcanic activity influence the whole geomorphological process.

The results and conclusions of this paper may be useful in understanding how a volcano at an island area may evolve in time and which factors control the whole process. If the numerical values of certain parameters are known, this model may help in describing in quantitative terms and predicting the future development of the South Aegean Volcanic Arc or of other volcanoes in different environments.

2. THE MATHEMATICAL MODEL

Mathematical models of landform development are generally derived by the following differential equation (Lawrence, 1996):

$$\frac{\partial y}{\partial t} = -(\nabla \cdot q)\mathbf{m} + F \qquad (1)$$

y is the elevation of a point at the surface of the landform, q is the flux of the eroded material, **m** is a unity vector describing the surface gradient and F expresses an external factor which supplies mass and/or energy to the landform.

In the two dimensional model of a volcano, F is the Transfer Function $Trsf(x, t)$, which expresses the magma transfer from the magma chamber to the surface of the volcano and the deposition of lava and pyroclastic material on the slopes and around the volcano. In this model, the only independent variable is location x of a point at horizontal axis, therefore equation (1) becomes:

$$\frac{\partial y(x,t)}{\partial t} = -\frac{\partial q(x,t)}{\partial x} + Trsf(x,t) \qquad (2)$$

q is often considered to be proportional to the slope (Culling, 1963; Hirano, 1976; Fernandes and Dietrich, 1997), which means that:

$$q = -K\frac{\partial y}{\partial x} \qquad (3)$$

K is the erosion coefficient, which expresses how fast is the eroded material transported from high to low elevations. This coefficient is often called "diffusion coefficient" (Culling, 1960, Fernandes and Dietrich, 1997), but we think that the term "erosion coefficient" is more appropriate, since it expresses directly the physical process which takes place.

Combining (2) and (3) gives:

$$\frac{\partial y}{\partial t} = K\frac{\partial^2 y}{\partial x^2} + Trsf(x,t) \qquad (4)$$

This is the differential equation which describes the temporal evolution of the profile of the volcano, as a result of erosional processes and volcanic activity, which are expressed with the first and second term of the right part of relation (4), respectively. If the volcano crowns an island area with length L and the sea level is constant with time, the following boundary condition has to be introduced (Culling, 1963):

$$y(x=0,t) = y(x=L,t) = 0 \qquad (5)$$

The physical meaning of this boundary condition is that at the coast, the deposited pyroclastic or eroded material is removed by the wind or the sea waves and the horizontal dimensions of the island do not change considerably with time. The differential equation (4), combined with the boundary condition (5), describes the morphological evolution of a volcanic island. The initial state of the volcano, at time t equal to zero, is represented in Fig. 1. The maximum elevation of the volcano is y_0 and the base of the volcanic cone is considered to be equal to one tenth of the length of the island. Horizontal distance x is measured from the left edge of the island area L.

The solution of the differential equation is well known from the theory of heat transport (e.g. Trachanas, 2001) and, making certain necessary modifications, it can be written in the form:

$$y(x,t) = \sum_n f_n \exp(-\lambda_n t)\sin\frac{n\pi x}{L} + \sum_n \exp(-\lambda_n t)a_n(t)\sin\frac{n\pi x}{L} \qquad (6)$$

where:

$$\lambda_n = Kn^2\pi^2 / L^2 \qquad (7)$$

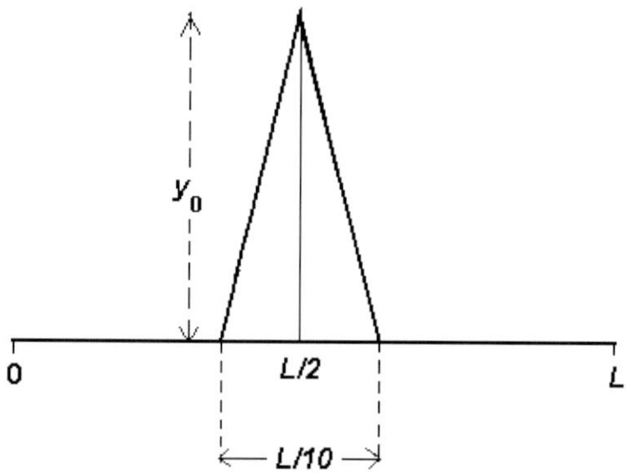

Fig. 1. The model of the volcanic island.

$$a_n(t) = \int_0^t \exp(\lambda_n t') Trsf_n(t') dt' \tag{8}$$

$$Trsf_n(t) = \frac{2}{L}\int_0^L Trsf(x,t)\sin\frac{n\pi x}{L}dx \tag{9}$$

$$f_n = \frac{2}{L}\int_0^L f(x)\sin\frac{n\pi x}{L}dx \tag{10}$$

$f(x)$ is the profile of the volcano in it's initial state (see Fig. 1), therefore it is defined by:

$$\begin{aligned} &f(x) = y_0 - (20/L)y_0\left|x - L/2\right| \quad for \quad \left|x\right| \le L/2 \\ &f(x) = 0 \quad for \quad \left|x\right| > L/2 \end{aligned} \tag{11}$$

Combining relations (10) and (11) and performing the necessary integrations, it can be found that coefficients f_n are given by:

$$f_n = \frac{80 y_0}{n^2\pi^2}\sin\frac{n\pi}{2}(1-\cos\frac{n\pi}{20}) \tag{12}$$

Based on relation (6), it is possible to describe the development of the volcano, for various initial states, mass transfer rates and erosion coefficients.

3. THE CASE OF NO MASS TRANSFER (DEAD VOLCANO)

If the volcano is dead, no mass transfer takes place, therefore:

$$Trsf(x,t) \equiv 0 \tag{13}$$

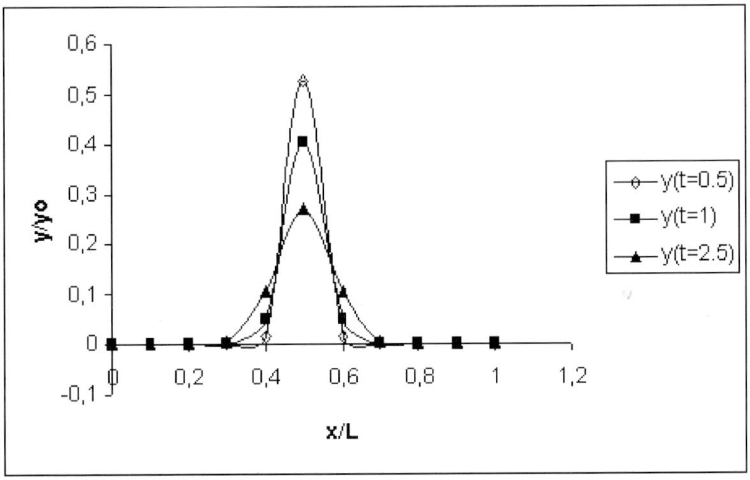

Fig. 2. The temporal evolution of the profile of the dead volcano.

Combining relations (6), (12) and (13) gives:

$$y(x,t) = \sum_n f_n \exp(-\lambda_n t) \sin\frac{n\pi x}{L} \qquad (14)$$

which describes the evolution of the profile of the volcano. In Fig. 2 are presented profiles at various time instants. In order to calculate the y values, the first hundred terms of the series were summed. It was realized that 100 terms were quite enough to approximate satisfactorily the function $f(x)$ of relation (11), putting t=0 in relation (14). For t values greater than zero, the contribution of high order values f_n on the formation of $y(x,t)$ diminishes, since they are multiplied by the factor exp($-\lambda_n t$), which is less than unity. Therefore, a hundred summation terms are sufficient to describe the evolution of the volcano, according to the relation (14).

In Fig. 2, it can be observed that the maximum elevation of the volcano reduces with time and the slope gets concave. The concavity is due to the initial state, at which the basis of the volcano was quite smaller than the length L of the island area (see Fig. 1). If the basis of the volcano were equal to L, the slope would become convex as time passes (Vaiopoulos et al., 2002). In Fig. 3 is presented the elevation of the peak of the volcano (maximum elevation) against time. The maximum elevation can be calculated by relation (14), for x equal to $L/2$. It can be observed that the maximum elevation tends to zero with time, because of the erosional process which transports material from higher to lower elevations and flattens the whole landform. The decrease of the maximum elevation is faster, as long as the erosion coefficient K increases.

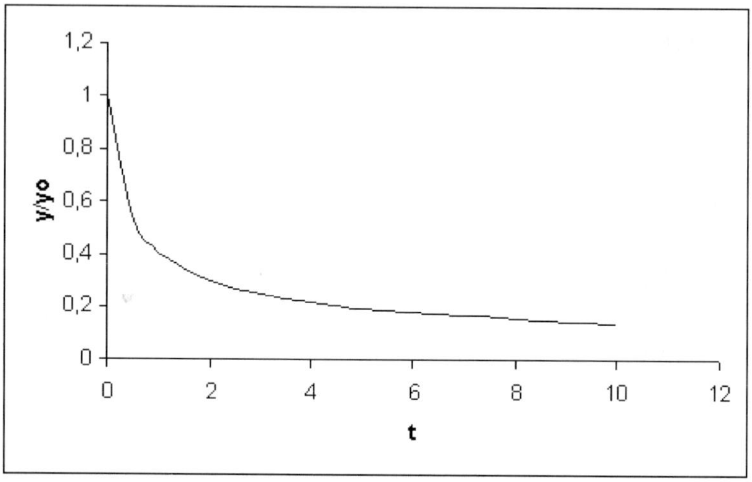

Fig. 3. Height variation against time.

4. THE CASE OF AN ACTIVE VOLCANO WITH A TIME CONSTANT MASS TRANSFER RATE

If the volcano is active, lava and pyroclastic material are deposited at the slopes and around the volcanic cone. It is reasonable to assume that the rate of deposition of material tends to decay as the horizontal distance from the crater increases. This assumption is supported by the fact that the thickness of the pyroclastic material which is deposited at the slope of the volcano decreases downslope (Kyriakopoulos, 2003). If the mass transfer rate is constant with time, the expression for the transfer function *Trsf*(*x*) may be the following:

$$Trsf\left(x\right) = T.\exp\left(-d\left|x - L/2\right|\right) \tag{15}$$

T is the deposition rate of magmatic material immediately around the crater. *d* is a positive constant which controls the maximum distance at which lava or pyroclastic material is deposited. As *d* increases, the maximum distance decreases. The exponential function expresses the decrease of deposition rate with distance *x*. If *d* is several times (for example ten times) larger than *L*, the deposition rate of the pyroclastic material at the edges of the island (the coast) is practically zero.

Combining relations (6), (8) and (15), and making the necessary integrations, it can be derived the following expression for the development of the volcanic profile with time:

$$y(x,t) = \sum_n \left(f_n \exp(-\lambda_n t) + \frac{2T}{\lambda_n} \exp(-dL/2) \frac{[1-\exp(-\lambda_n t)]n\pi[1-\cos(n\pi)]}{d^2L^2 + n^2\pi^2} \right) \sin\frac{n\pi x}{L} \tag{16}$$

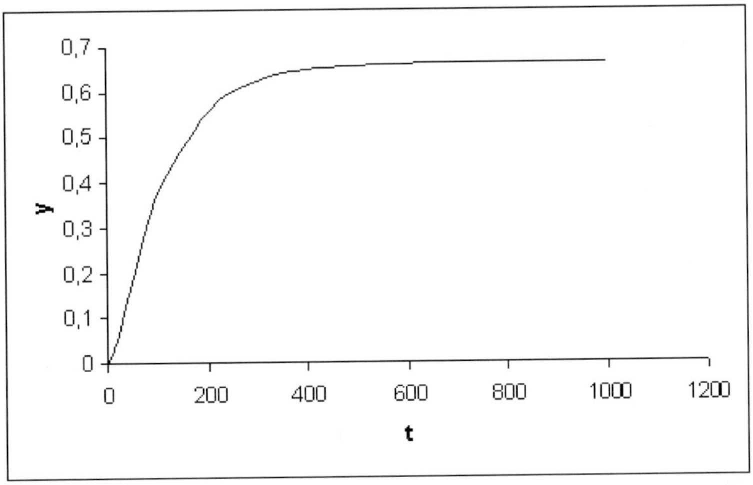

Fig. 4. Height variation against time, for a constant mass transfer rate. c=0, T=10, K=0.001, y0=0, L=1.

If it is assumed that the height y_0 of the volcano is zero at it's initial state ($t = 0$), the f_n coefficients are zero and the expression for the elevation $y(x, t)$ becomes:

$$y(x,t) = 2T.\exp(-dL/2)\sum_n \frac{[1-\exp(-\lambda_n t)]n\pi[1-\cos(n\pi)]}{\lambda_n(d^2L^2+n^2\pi^2)}\cdot\sin\frac{n\pi x}{L} \tag{17}$$

An initial state with a null elevation represents the moment at which the formation of the volcano begins. In Fig. 4 is presented the maximum elevation variation against time, according to relation (17). It can be observed that the maximum elevation tends to a constant value. For a non zero initial elevation y_0, the evolution of the volcano is described by relation (16). The curve may then be different from that of Fig. 4 in the beginning, but it will have the same asymptotic behavior as t increases. The physical meaning of this behavior is that if the mass transfer rate is time constant, mass transfer from the magma chamber to the crater is balanced by mass transport due to erosion, after a certain time. Therefore, a steady state of dynamic equilibrium is achieved and the profile of the volcano does not change with time. This steady profile may be defined by relation (16) or (17), putting a very high value for t. Since the factor $f_n\exp(-\lambda_n t)$ has a decay tendency with time and practically diminishes for high values of t, both relations are expected to give the same results, if t is very high.

In Fig. 5 is presented the profile of the volcano at the steady state. It can be observed that the profile is flat at the middle and the slopes are very steep and convex. The elevation at locations x=0 and x=L remains equal to zero, because of the boundary condition (5). This characteristic of the profile may be questionable for a critical reader, who might argue that erosion and transportation proceeds out of the interval [0, L] and the elevation at these locations may change with time. We think that the boundary condition has a physical meaning, as long as the horizontal dimensions of the island do

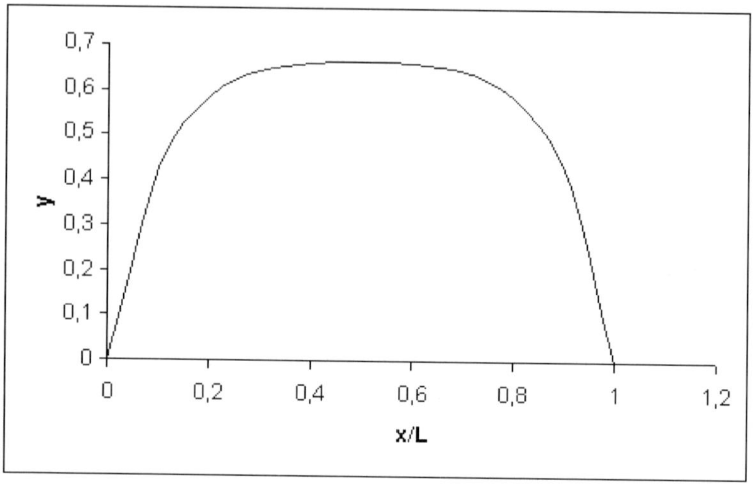

Fig. 5. The profile of the volcano at it's steady state. c=0, d=10, T=10, K=0.001, y0=0, L=1, t=500.

not change considerably.

The maximum elevation at the steady state, y_{eq}, is the limit of relation (17) when t tends to infinity and x is equal to $L/2$. It can be proved that y_{eq} is given by:

$$y_{eq} = \frac{2T}{K} \exp(-dL/2) \sum_n \frac{L^2[1-\cos(n\pi)]}{n\pi(d^2L^2 + n^2\pi^2)} \sin\frac{n\pi}{2} \tag{18}$$

According to relation (18), for a steady L, there is a linear relation between the logarithm of the maximum elevation at the steady state and the logarithm of T/K. As long as the quantity dL decreases, y_{eq} increases.

Putting L=10 km, T=10 cm/yr and K=1 m^2/yr in relation (18), we obtain $y_{eq} \cong 660$ m. The value of T was chosen in such a way that the mean annual thickness of the deposited pyroclastic material at a radius equal to 0.5 km round the crater be equal to 1 mm. The value of K, according to available references, varies for several orders of magnitude. We chose a value which is suggested by Flemings and Grotinzger, 1996.

According to the relation (6), the evolution of the profile is controlled by the factor $\exp(-\lambda_1 t)$. This factor practically gets equal to zero when the product $\lambda_1 t$ is equal to or greater than 5. When this factor is zero, the terms $f_n\exp(-\lambda_n t)$, which contain information about the initial state of the volcano practically vanish and the profile of the landform remains steady. Therefore, the time τ at which the steady state is reached or the landform becomes flat and time invariant, can be practically estimated by the following expression:

$$\lambda_1 \tau = 5 \tag{19}$$

According to relation (7), and for n equal to 1, we have:

$$\lambda_1 = K\pi^2 / L \tag{20}$$

According to relations (19) and (20), and taking into account that $\pi^2 \cong 10$, time τ is given by:

$$\tau = 0.5L^2 / K \tag{21}$$

From this relation, it is possible to calculate the time at which the volcano comes at it's steady state, if the length L and the erosion coefficient K are known. For example, according to relation (21), for L=10 km and K = 10 m^2/yr, we obtain τ= 5 million years. τ is about 500,000 years for $K \sim 10^2$ m^2/yr. If L is one order of magnitude smaller, equilibrium is obtained at time τ equal to 2 orders of magnitude smaller. For example, for $K \sim 100$ m^2/yr and $L \sim 1$ km, τ is about 5000 years. The values for K are taken by Flemings and Grotzinger, 1996. Other authors suggest K values which are several orders of magnitude smaller and correspond to very high τ values. In this case, a steady state is practically never reached.

5. A "BURST-LIKE" APPROACH OF THE EVOLUTION OF THE VOLCANO

In real volcanoes, mass supply does not occur continuously, but in discrete time moments, when eruptions take place. In such a case, the transfer function may be defined by a series of δ functions in the following way:

$$Trsf(x,t) = \sum_i T_{p,i} \exp(-d\,|\,x - L/2\,|)\delta(t - t_i) \tag{22}$$

$T_{p,i}$ is the deposition rate of magmatic material immediately around the crater, at time t_i at which a volcanic eruption takes place. i takes values 1,2,3,.... t_i is the time moment of the first eruption after t=0.

For t=0 the profile of the volcano is given by the relation (11). For $t \in (0, t_1)$, which expresses the time interval between the initial state and the first eruption, only erosional processes take place. According to the relation (14), the profile $y(x,t)$ is given by:

$$y(x,t) = y_0(x,t) = \sum_n f_n \exp(-\lambda_n t)\sin\frac{n\pi x}{L} \tag{23}$$

For t=t_1, an eruption takes place and pyroclastic material is deposited at the volcano. The profile of the volcano at time t_1 becomes:

$$y(x,t) = f^{(1)}(x) = y_0(x,t_1) + T_{p,1} \exp(-d\,|\,x - L/2\,|) \tag{24}$$

For $t \in (t_1, t_2)$ the profile of the volcano evolves as a result of erosional processes and it is described by the relation:

$$y(x,t) = y_1(x,t) = \sum_n f_n^{(1)} \exp[-\lambda_n (t - t_1)] \sin \frac{n\pi x}{L} \tag{25}$$

The coefficients $f_n^{(1)}$ are given by:

$$f_n^{(1)} = \frac{2}{L} \int_0^L f^{(1)}(x) \sin \frac{n\pi x}{L} dx \tag{26}$$

Generally, at time t equal to t_i, the profile of the volcano is given by:

$$y(x,t) = f^{(i)}(x) = y_{i-1}(x,t_i) + T_{p,i} \exp(-d \,|\, x - L/2 \,|) \tag{27}$$

And at the time interval (t_i, t_{i+1}) the profile of the volcano is:

$$y(x,t) = y_i(x,t) = \sum_n f_n^{(i)} \exp[-\lambda_n (t - t_i)] \sin \frac{n\pi x}{L} \tag{28}$$

where $f_n^{(i)}$ is given by:

$$f_n^{(i)} = \frac{2}{L} \int_0^L f^{(i)}(x) \sin \frac{n\pi x}{L} dx \tag{29}$$

Relations (27), (28) and (29) provide a recursive function to describe the morphological evolution of the volcano as a result of sudden eruptions (burst-like activity) and continuous erosional processes. The factor $\exp[-\lambda_n(t-t_i)]$ in relation (28) expresses a decay tendency of the altitude, between two successive eruptions. The factor $T_{p,i}.\exp(-d|x-L/2|)$ expresses an elevation increase, during the eruption. If the eruptions take place at more or less equal time intervals and the value of $T_{p,i}$ does not change considerably, the evolution of the profile of the volcano is expected to be similar with that described by relations (16) and (17). The maximum elevation variation against time will basically have the behavior of Fig. 4, but the curve will be more complicated. It will present rapid increases at time moments t_i and decay tendencies at the time intervals (t_i, t_{i+1}). As long as the middle deposition rate, between two successive eruptions, is larger than the mean erosion rate at the same time, the general tendency is expected to be a time increase of the maximum elevation.

A more systematic study of the behavior of the recursive function $y(x, t)$, according to the relations (27), (28) and (29), would exceed the intentions of the present paper. It is reasonable to assume however, that a continuous mass supply, which is described by the relation (15), gives qualitatively similar results with the "burst-like" transfer function of relation (19), if T and $T_{p,i}$ fulfill the following relation:

$$T \cong T_{p,i} / (t_{i+1} - t_i) \tag{30}$$

The "burst-like" approach should be preferable when the time scale at which the temporal evolution of the volcano is studied is not much bigger than the time interval between two successive eruptions. If the time scale is several times bigger than this time interval, we think that the continuous mass supply approach is not less reliable than the "burst-like" model.

6. THE CASE OF AN ACTIVE VOLCANO WITH A TIME DECAYING MASS TRANSFER RATE

For a continuous time decaying mass transfer rate, the transfer function *Trsf(x, t)* may have the form:

$$Trsf(x,t) = T\exp\left(-d\left|x - L/2\right|\right)\exp(-ct) \tag{31}$$

c is a positive constant. The factor exp(-*ct*) expresses the decay tendency of the volcanic activity with time. When *c* increases, the decay tendency is stronger. Combining relations (6), (8), (9) and (31) gives:

$$y(x,t) = \sum_n \left(f_n \exp(-\lambda_n t) + 2T\exp(-dL/2)\frac{\left[\exp(-ct) - \exp(-\lambda_n t)n\pi\left[1-\cos(n\pi)\right]\right]}{(\lambda_n - c)\left(d^2L^2 + n^2\pi^2\right)} \right) \sin\frac{n\pi x}{L} \tag{32}$$

which is the expression for the profile of the volcano at various time instants.

If the initial maximum elevation y_0 is equal to zero, the expression for *y(x, t)* becomes:

$$y(x,t) = 2T\exp(-dL/2)\sum_n \frac{[\exp(-ct) - \exp(-\lambda_n t)]n\pi[1-\cos(n\pi)]}{(\lambda_n - c)(d^2L^2 + n^2\pi^2)}\sin\frac{n\pi x}{L} \tag{33}$$

In Fig. 6 is presented the time variation of the maximum elevation *y*(*x* = *L*/2) against time, according to relation (21). It can be observed that in the beginning the maximum elevation increases, because the mass transfer rate is higher than the erosion rate. At the peak of the graph, these two rates are equal. At higher time values, the erosion rate is higher than the mass transfer rate, therefore the maximum elevation decreases and tends to zero.

In Fig. 7 are presented profiles of the volcano at various time instants. In it's initial state, the volcano is assumed to have a maximum elevation y_0 equal to one. The maximum elevation decreases rapidly in the beginning and then increases and takes it's maximum value at about one hundred time units. Later on, the erosional processes prevail and the elevation tends to zero. For *t* equal to 1 and to 5 time units, the respective profiles present local maxima at the edges of the island area. These maxima express accumulation of material.

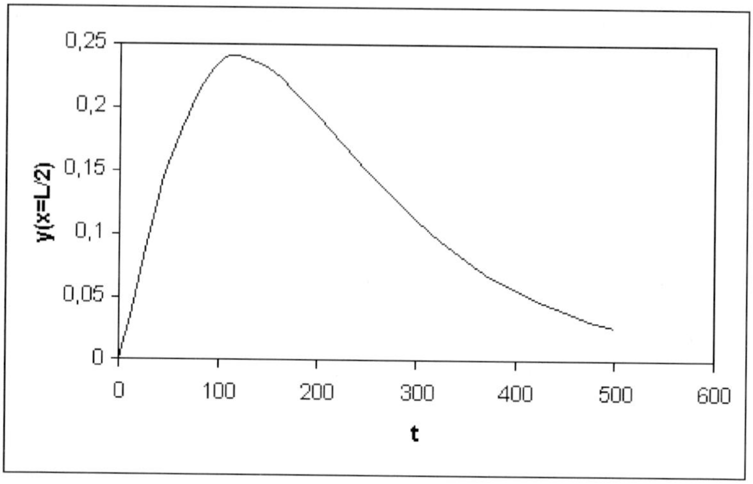

Fig. 6. Height variation against time, for a time decaying mass transfer rate. y0=0, c=0.01, d=10, T=10, L=1, K=0.001.

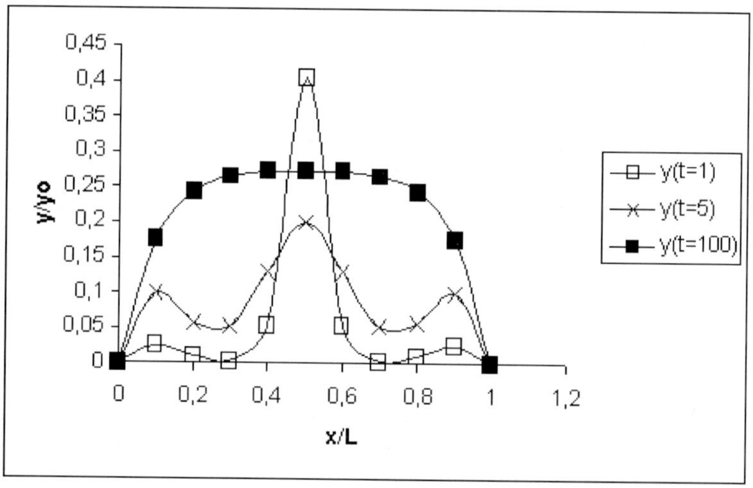

Fig. 7. Profiles of the volcano at various time instants. y0=1, c=0.01, d=10, T=10, L=1, K=0.001.

7. CONCLUSIONS

According to the mathematical analysis which was carried out, the following conclusions may be stated:

The maximum elevation of a dead volcano decays with time and the whole landform is converted to a flat surface with a null elevation at all points.

An active volcano with a time constant mass transfer rate of the form of relation

(15), tends to a steady state of dynamic equilibrium. The maximum altitude at the steady state is proportional to the ratio *T/K*.

If the mass transfer rate decreases with time, the maximum elevation of the volcano tends to zero.

The time at which a steady state is reached, is proportional to the square length L^2 of the platform which supports the volcano and inversely proportional to the erosion coefficient *K*.

The boundary condition of the differential equation of the landform development, expresses a volcano which crowns an island area with a time constant sea level. The parameters of the evolution of the volcano are expected to vary with the tectonic environment and the magma composition. It is important to know the exact values of the parameters of the specific landform development in it's specific geological environment. In the international literature, there are some estimates of the erosion coefficient in various geomorphological systems (for example Nash, 1980; McKean et al., 1993; Flemings and Grotzinger, 1996), which differ several orders of magnitude. On the other hand, we have not in our disposal some data about the parameters of the transfer function. Because of these reasons, we have only made some rough numerical estimates about the evolution of the landform. Laboratory and/or field data for the parameters involved in the geomorphological process would be very useful in making reliable numerical calculations of the profile of the volcano at different time instants.

The results and conclusions of the mathematical analysis which was carried out in this paper may contribute in understanding, in quantitative terms, how the profile of a volcano may evolve in time and which factors control it's development. Reliable estimates of the parameters of the landform development may help in making reliable predictions about the evolution of the South Aegean Volcanic Arc, as well as other volcanoes in different environments.

REFERENCES

Ahnert, F., 1976. Brief description of a comprehensive three-dimensional model of landform development. Zeitschrift für Geomorphologie, NF Supplement Band, 25: 29-49.

Ahnert, F., 1987. Approaches to dynamic equilibrium in theoretical simulations of slope development. Earth Processes and Landforms, 12: 3-15.

Armstrong, A., 1976. A three-dimensional model of slope forms. Zeitschrift für Geomorphologie, NF Supplement Band, 25: 20-28.

Culling, W.E.H., 1960. Analytical theory of erosion. Journal of Geology, 68: 336-344.

Culling, W.E.H., 1963. Soil creep and the development of hill-side slopes. Journal of Geology, 71: 127-161.

Culling, W.E.H., 1965. Theory of erosion on soil-covered slopes. Journal of Geology, 73: 230-254.

Fernandes, N.F. and Dietrich, W.E., 1997. Hillslope evolution by diffusive processes: The timescale from equilibrium adjustments. Water Resources Research, 33

(6): 1307-1318.
Flemings, P.B. and Grotzinger, J.P., 1996. STRATA: Freeware for analyzing classic stratigraphic problems. GSA TODAY, 6 (12): 1-7.
Hirano, M., 1975. Simulation of development processes of interfluvial slopes with reference to graded form. Journal of Geology, 83: 113-123.
Hirano, M., 1976. Mathematical model and the concept of equilibrium in connection with slope shear ratio. Zeitschrift für Geomorphologie, NF Supplement Band, 25: 50-71.
Howard, A.D., 1994. A detachment-limited model of drainage basin evolution. Water Resources Research, 30: 2261-2285.
Howard, A.D., Dietrich, W.E. and Seidl, M.A., 1994. Modeling fluvial erosion on regional to continental scales. Journal of Geophysical Research, 99 (B7): 13791-13986.
Kirkby, M.J., 1971. Hillslope process-response models based on the continuity equation. In: D. Brunsden (Editor), Slopes: Form and Process. Institute Institute of British Geographers, Special Publication, 3, pp 15-30.
Kirkby, M.J., 1999. Landscape Modelling at Regional to Continental Scales. In: S. Hergarten and H. Neugebauer (Editors), Process Modelling and Landform Evolution. Springer, pp 189-203.
Kooi, H. and Beaumont, C., 1996. Large-scale geomorphology: Classical concepts reconciled and integrated with temporary ideas via a surfaced process model. Journal of Geophysical Research, 101 (B2): 3361-3386.
Kyriakopoulos, K., 2003. Volcanology, Athens.
Lawrence, D.S.L., 1996. Physically Based Modelling and the Analysis of Landscape Development. In the Scientific Nature of Geomorphology: Proceedings of the 27th Binghamton Symposium in Geomorphology, 27-29 September 1996, edited by Bruce L. Rhoads and Colin E. Thorn. John Wiley & Sons Ltd. pp 273-288.
McKean, J.A., Dietrich, W.E., Finkel, R.C., Southon, J.R. and Caffee, M.W., 1993. Quantification of soil production and downslope creep rates from cosmogenic ^{10}Be accumulations on a hillslope profile. Geology, 21: 343-346.
Nash, D., 1980. Morphologic dating of degraded normal fault scarps. Journal of Geology, 88: 353-360.
Scheidegger, A.E., 1961. Mathematical models of slope development. Bulletin of the Geological Society of America, 72: 37-59.
Trachanas, S., 2001. Partial Differential Equations. University Publications of Crete (in Greek).
Trofimov, A.M. and Moskovkin, V.M., 1984. Diffusion models of slope development. Earth Surface Processes and Landforms, 9: 435-453.
Vaiopoulos, D.A., Skianis, G. Aim., and Tsarbos, V., 2001. A 3-d model of the morphological evolution of a mountain, as a result of fluvial processes. Proceedings of the 9th International Congress of the Geological Society of Greece, Athens, September 2001, Vol. 1, pp 363-370 (in Greek, with an abstract in English).
Vaiopoulos, D.A., Skianis, G. Aim., and Tsarbos, V., 2002. A mathematical study of the

morphological evolution of a volcanic cone, as a result of erosional processes. Proceedings of the 6th Panhellenic Geographical Conference of the Hellenic Geographical Society, Thessaloniki, 3-6 October 2002, Vol. II, pp 79-86 (in Greek, with an abstract in English).

Willgoose, G., Bras, R.L. and Rodriguez-Iturbe, I., 1990. A model of river basin evolution. EOS, Transactions of the American Geophysical Union, 71: 1806-1807.

Subject index

Authors' index